Springer Monographs in Mathematics

More information about this series at http://www.springer.com/series/3733

Mikhail S. Agranovich

Sobolev Spaces, Their Generalizations, and Elliptic Problems in Smooth and Lipschitz Domains

Springer

Mikhail S. Agranovich
Moscow
Russia

Russian original version published as Sobolevskie prostranstva, ikh obobshcheniya i ellipticheskie zadachi v oblastyakh s gladkoj i lipshitsevoj granitsej, Moscow Center for Continuous Mathematical Education 2013

ISSN 1439-7382   ISSN 2196-9922   (electronic)
Springer Monographs in Mathematics
ISBN 978-3-319-35183-4   ISBN 978-3-319-14648-5   (eBook)
DOI 10.1007/978-3-319-14648-5

Mathematics Subject Classification (2010): 31-xx, 35-xx, 45-xx, 46-xx, 47-xx

Springer Cham Heidelberg New York Dordrecht London
© Springer International Publishing Switzerland 2015
Softcover reprint of the hardcover 1st edition 2015

Printed on acid-free paper

Springer International Publishing AG Switzerland is part of Springer Science+Business Media
(www.springer.com)

Contents

Preface

This book consists of four chapters and two additional sections.

The first two chapters contain introductory courses. Chapter 1 presents the theory of Sobolev-type spaces H^s ($s \in \mathbb{R}$) on $\mathbb{R}^n$, on a smooth closed manifold, and on a smooth bounded domain. Chapter 2 outlines the foundations of the theory of general linear elliptic equations on smooth closed manifolds and the theory of general elliptic boundary value problems in smooth bounded domains in the spaces H^s.

Chapter 3 presents the theory of basic boundary value problems for second-order strongly elliptic systems in the same spaces in a bounded Lipschitz domain; its boundary can generally be very nonsmooth.

Chapter 4 is of review character. Here we briefly consider the more general Bessel potential spaces H_p^s and the Besov spaces $B_p^s = B_{p,p}^s$, $1 < p < \infty$, and describe their applications to the same problems.

The Sobolev spaces W_p^s are the spaces H_p^s for integer $s \geq 0$. The spaces B_p^s coincide with the Slobodetskii spaces W_p^s for noninteger $s > 0$. For $p = 2$, the spaces H_2^s and B_2^s coincide with H^s.

This is a great honor for the author that the Springer publishing company has decided to include this book in the series "Springer Monographs in Mathematics." But I must mention that the book is based on lectures delivered by the author at the Independent University of Moscow in 2005–2006, substantially extended and finalized. As well as [18], this book was originally mainly intended for beginning mathematicians and is accessible in Russia to four-year students (familiar, in particular, with the basic notions of functional analysis, such as Lebesgue measure, Hilbert and Banach spaces, compact operators and their spectra, etc.).

References to the required material of undergraduate mathematical courses are usually not given.

First of all, this book may be useful to students and postgraduate students specializing in partial differential equations and functional analysis. But it may also interest those specializing in other areas of mathematics, including geometry and applied mathematics, and in physics.

We now describe the contents of the book in more detail.

In the first chapter we consider, in particular, the spaces H^s (the simplest Bessel potential spaces) of negative order, which are not Sobolev–Slobodetskii spaces; they are used in Chapter 3. Even in the case of the half-space $\mathbb{R}^n_+$, some statements of Sections 3 and 4 require fairly delicate proofs. The construction of a bounded universal extension operator for functions defined on a domain to the entire space, proposed by Rychkov in [320], is postponed to Section 10, because this operator is constructed for Lipschitz domains.

In the second chapter, we proceed to the general theory of elliptic equations and elliptic boundary value problems. It is developed in detail in Sections 6 and 7 with minimal generality, but the main generalizations are stated or, at least, mentioned and explained. Note that the theory of elliptic equations and problems constructed on the basis of the theory of Sobolev-type spaces, in turn, substantially influenced the latter by providing, first, important questions and, second, some convenient results.

Our point of view on what topics form the foundations of the theory of elliptic equations is, possibly, not quite canonical. As essential elements of this theory we regard the theory of equations elliptic with parameter and the "more classical" theory of strongly elliptic equations, in which the key role is played by uniquely solvable (rather than Fredholm) equations and problems. The classical variational Dirichlet and Neumann problems for strongly elliptic equations are first considered in smooth domains in Section 8. In our view, strongly elliptic equations presently constitute a rich of content section of the theory of elliptic equations rather than a past stage of its development.

The third chapter begins with Section 9, in which we first discuss the specifics of Lipschitz domains and surfaces. In particular, we show, following Stepanov's paper [363], that any Lipschitz function is differentiable almost everywhere; we also show that any convex function is Lipschitz. Then we explain the basic facts of the theory of H^s spaces on Lipschitz domains and Lipschitz surfaces. In Section 10 we, following [320], discuss discrete norms and a discrete representation of functions, which are popular in the current literature (preliminaries on these norms and representation are given in Section 1.14) and construct a universal extension operator (but only for the H^s spaces, with some simplifications in comparison with [320]).

Sections 11 and 12 present, against the background of the already described theory of general elliptic problems, the theory of basic boundary value problems in Lipschitz domains in the same simplest spaces for second-order strongly elliptic systems. It is possibly a central place in the book. These sections develop and supplement the material of the excellently written book [258]. Much progress in this theory has been made during the past 15–20 years. The author's personal achievements are mentioned in Section 19. In our exposition of this theory, its technical details were methodologically purged. This theory can hardly be considered quite complete. Sections 11 and 12 and their continuation in Sections 16 and 17 may be interesting for specialists in the theory of partial differential equations, functional analysis, and probably other fields of mathematics or physics.

Some decades ago the author had an occasion to participate in the development of the general theory of (pseudodifferential) elliptic boundary value problems and the corresponding spectral problems, and in recent years, of the theory of boundary value problems in Lipschitz domains. The latter theory impelled him to reevaluate, to some extent, classical things, which was one of the stimuli for writing this book. Hopefully, it will also be interesting to the reader to compare these theories.

The priority for the author was to present the exposition accessible. For this reason, statements and proofs are not always given in full generality.

A beginning mathematician deserves not only a self-contained exposition of the simplest version of a theory but also explanations of what lies further, as far as possible without unnecessary details. For this reason, our book contains the survey fourth chapter devoted to generalizations of the material of the first three chapters to the Bessel potential spaces H_p^s and the Besov spaces B_p^s. This is an important but not easy material, and its complete exposition with all proofs would require too much space. Certainly, the exposition is accompanied by literature references, in particular, to many monographs. This chapter begins with an essay on interpolation theory in Section 13, which is written from the standpoint of a "consumer" of this remarkable theory. Although this section is of review character, it contains proofs of several very useful theorems. In Sections 16 and 17 important theorems supplementing the material of Chapter 3 are proved by using elements of interpolation theory.

We deviate from the tradition of proving everything, which is usually followed by the authors of mathematical books, but try to comment all definitions and facts presented in the book. The character of the exposition is closer to that of a lecture course than of a monograph. For a lecturer, the informal clarity of exposition is more important than its formal completeness.

This book cannot be used as a handbook on Sobolev-type spaces and elliptic problems.

We use notions of distribution theory; a brief introduction to this theory is contained in the author's preceding book [18], but it has not been translated into English. The reader may use other books, e.g., classical books by L. Schwartz [335] and Gel'fand and Shilov [165]. On the other hand, the author is planning to write a continuation of the present book, the third book [26]. There, in particular, the calculus of pseudodifferential elliptic operators will be touched on, while here we restrict ourselves to the classical method of freezing coefficients, which has not lost its significance, is transparent, and deserves to be mastered. In the author's opinion, for a beginning analyst, it is more useful to learn this method before studying the calculus of pseudodifferential operators, not to mention the more complicated calculus of pseudodifferential elliptic boundary value problems. The method of freezing coefficients is used not only to study "smooth" elliptic problems but also to obtain the Gårding inequality in Lipschitz domains.

However, pseudodifferential operators are repeatedly mentioned in the book; therefore, the main definitions and a few basic facts of their theory are given in Section 18.

To orient the reader, we touch on spectral problems at several places, but we do not give detailed proofs of theorems on which the study of these problems is based. We plan to present these theorems in [26]. In particular, asymptotic formulas for eigenvalues are given without proof.

Section 18 contains reference information from the general theory of linear operators. First, statements concerning Fredholm operators and the Lax–Milgram theo-

rem are presented with proofs. Then some basic facts of the spectral theory of operators (Section 18.3) and results related to pseudodifferential operators (Section 18.4)
are given without proof.

Section 19 is devoted to comments on the preceding sections. Here references
to the literature are detailed and some directions in the theory of elliptic equations
and related problems are mentioned, which border on the material of the preceding
sections but are not touched upon there.

Sections 18 and 19 are not included in Chapter 4.

The author again thanks his listeners at the Independent University of Moscow,
especially Polina Vytnova, Nikolai Gorev, Vasilii Novikov, and Mikhail Surnachev;
discussions with them have been of great help in selecting the material and searching
for an accessible form of exposition.

Special thanks are due to Prof. V. I. Ovchinnikov for reading and criticizing the
initial version of Section 13 and to Prof. S.E. Mikhailov for useful remarks to the
initial version of Sections 11.1 and 11.2.

It was the greatest luck for the author that Prof. T. A. Suslina agreed to take on
the scientific editing of the book. She was the first reader of the yet raw Russian text.
Thanks to her highest mathematical level, exceptional scrupulousness, and interest
in the topic, many refinements and substantial improvements were made.

In the English version, Chapters 3 and 4 are somewhat reorganized. The number
of sections has increased by one. The text of the book has underwent new refinements and improvements; in particular, Sections 17.4–17.6 are added. Again, of
invaluable help were Suslina's comments and remarks.

The author's work was supported by grants of the Russian Foundation for Basic
Research and of the National Research University Higher School of Economics.

The author would be grateful for any remarks and comments; the readers are
kindly asked to send them to the address magran@orc.ru

Preliminaries

The integer part of a positive number s is denoted by $[s]$.

A functional $\phi(x)$ on a linear space X is said to be *antilinear*, or *conjugate-linear*, if

$$\phi(\alpha x + \beta y) = \bar{\alpha}\phi(x) + \bar{\beta}\phi(y)$$

for any $x, y \in X$ and any complex numbers α and β. A functional $\Phi(x,y)$ is *sesquilinear* if it is linear in the first argument and antilinear in the second.

All subspaces (in Banach or Hilbert spaces) are assumed to be closed.

A function $f(x) = f(x_1, \ldots, x_m)$ of several real or complex variables is said to be positive homogeneous of degree h ($\in \mathbb{R}$) if $f(tx) = t^h f(x)$ for $t > 0$.

Two norms on a Banach space are *equivalent* if their ratio is between two positive constants.

Constants C_k are usually numbered within a coherent argument; when the topic changes, they are numbered anew.

When considering the space of vector-valued functions of dimension m with elements in X, some authors write X^m. We do not specify dimension, it is always clear from the context. The norm of a vector-valued function $f = (f_1, \ldots, f_m)$ can be defined as

$$\|f\| = [\|f_1\|^p + \ldots + \|f_m\|^p]^{1/p}$$

for any $p \geq 1$, such norms with different p are equivalent.

All integrals written without specifying the domain of integration are over $\mathbb{R}^n$.

Derivatives are denoted by $D_j = -i\partial/\partial x_j = -i\partial_j$.

Let $F(u)$ and $G(u)$ be two numerical functions of an arbitrary function u in some space, and let $G(u)$ be nonnegative. If $|F(u)| \leq CG(u)$, where the constant C does not depend on u, then we say that $G(u)$ *dominates* $F(u)$, or $F(u)$ is *dominated* by $G(u)$.

Chapter 1
The Spaces H^s

1 The Spaces $H^s(\mathbb{R}^n)$

1.1 Definition and Simplest Properties

Our initial definition of the *Bessel potential space* $H^s(\mathbb{R}^n) = H^s(\mathbb{R}^n_x)$ is as follows. Let s be a real number. First, we introduce the space $\widehat{H}^s(\mathbb{R}^n_\xi)$ of Lebesgue measurable complex-valued functions $\widehat{u}(\xi)$ for which

$$\int (1+|\xi|^2)^s |\widehat{u}(\xi)|^2 \, d\xi \qquad (1.1.1)$$

is finite. This is a linear manifold in the Schwartz space $S'(\mathbb{R}^n_\xi)$ of tempered distributions. The space $H^s(\mathbb{R}^n_x)$ consists of those distributions $u \in \hat{S}' = S'(\mathbb{R}^n_x)$ for which the Fourier transform

$$\widehat{u}(\xi) = Fu(\xi) = \int e^{-i\xi \cdot x} u(x) \, dx \qquad (1.1.2)$$

in the sense of distributions (see [18], [335], or [165]) belongs to $\widehat{H}^s(\mathbb{R}^n_\xi)$. Quantity (1.1.1) is taken for the squared norm on $H^s(\mathbb{R}^n_x)$ and, simultaneously, on $\widehat{H}^s(\mathbb{R}^n_\xi)$:

$$\|u\|^2_{H^s(\mathbb{R}^n_x)} = \|\widehat{u}\|^2_{\widehat{H}^s(\mathbb{R}^n_\xi)} = \int (1+|\xi|^2)^s |\widehat{u}(\xi)|^2 \, d\xi. \qquad (1.1.3)$$

Obviously, $H^s(\mathbb{R}^n)$ is a normed linear space. Its Fourier image $\widehat{H}^s(\mathbb{R}^n)$ is a weighted space of L_2 type. As is known, such a space is complete and, therefore, Banach (see, e.g., [137, Vol. I, Chap. III, Sec. 6]). Hence $H^s(\mathbb{R}^n)$ is a Banach space, too. Moreover, this space, as well as $\widehat{H}^s(\mathbb{R}^n)$, is a *Hilbert space* with inner product

$$(u,v)_{s,\mathbb{R}^n_x} = (\widehat{u},\widehat{v})_{s,\mathbb{R}^n_\xi} = \int (1+|\xi|^2)^s \widehat{u}(\xi)\overline{\widehat{v}(\xi)} \, d\xi. \qquad (1.1.4)$$

© Springer International Publishing Switzerland 2015
M.S. Agranovich, *Sobolev Spaces, Their Generalizations and Elliptic Problems in Smooth and Lipschitz Domains*, Springer Monographs in Mathematics, DOI 10.1007/978-3-319-14648-5_1

In what follows, we omit the index $s = 0$.

Obviously, $\|u\|_{H^s(\mathbb{R}^n)} \leq \|u\|_{H^\sigma(\mathbb{R}^n)}$ if $s < \sigma$, so that the space with larger index is continuously embedded in the space with smaller index. The space $H^0(\mathbb{R}^n)$ coincides with $L_2(\mathbb{R}^n)$ in view of Parseval's identity[1]

$$\int |Fu|^2 \, d\xi = (2\pi)^n \int |u|^2 \, dx \tag{1.1.5}$$

(and the norms on these spaces coincide up to an inessential numerical factor). Therefore, *all spaces* $H^s(\mathbb{R}^n)$ *with nonnegative s consist of usual square integrable* (i.e., having integrable square of absolute value) *complex-valued functions*. The delta-function $\delta(x)$ belongs to $H^s(\mathbb{R}^n)$ with $s < -n/2$, because its Fourier transform is 1.

Next, the first derivatives (in the sense of distributions) of functions in $H^1(\mathbb{R}^n)$ belong to $L_2(\mathbb{R}^n)$. Indeed, $D_j u(x) = -i\partial u(x)/\partial x_j$ is the inverse Fourier transform of the square integrable function $\xi_j \widehat{u}(\xi)$. Thus, we can give the second definition of the space $H^1(\mathbb{R}^n)$: it consists of those functions $u(x) \in L_2(\mathbb{R}^n)$ for which the first derivatives $D_j u(x)$ in the sense of distributions belong to $L_2(\mathbb{R}^n)$ as well; the norm is defined by

$$\|u\|^2_{H^1(\mathbb{R}^n)} = \int \left[|u(x)|^2 + \sum_1^n |D_j u(x)|^2 \right] dx. \tag{1.1.6}$$

The coincidence (up to a numerical factor) of this expression with expression (1.1.3) for $s = 1$ follows from Parseval's identity.

The following more general assertion is easy to verify.

Theorem 1.1.1. *For any positive integer* m, *the space* $H^m(\mathbb{R}^n)$ *consists of all square integrable functions whose derivatives in the sense of distributions up to order* m *are square integrable. The norm on* $H^m(\mathbb{R}^n)$ *can be defined by*

$$\|u\|'^2_{H^m(\mathbb{R}^n)} = \int \sum_{|\alpha|\leq m} |D^\alpha u(x)|^2 \, dx. \tag{1.1.7}$$

The corresponding inner product has the form

$$(u,v)'_{m,\mathbb{R}^n} = \int \sum_{|\alpha|\leq m} D^\alpha u(x) \cdot \overline{D^\alpha v(x)} \, dx. \tag{1.1.8}$$

As usual,

$$\alpha = (\alpha_1,\ldots,\alpha_n) \in \mathbb{Z}_+^n, \quad |\alpha| = \alpha_1 + \ldots + \alpha_n, \quad \text{and} \quad D^\alpha = D_1^{\alpha_1} \ldots D_n^{\alpha_n}.$$

The norms $\|u\|'_m$ *and* $\|u\|_m$ *are equivalent*; this follows from the obvious inequality

---

[1] We could remove the multiplier $(2\pi)^n$ from (1.1.5) (and make the Fourier transform unitary) by writing $(2\pi)^{-n/2}$ in front of the integral sign in (1.1.2), but we follow the notation of [165] and [18].

$$0 < C_1 \le \frac{\sum\limits_{|\alpha| \le m} |\xi^\alpha|^2}{(1 + |\xi|^2)^m} \le C_2, \tag{1.1.9}$$

where $\xi^\alpha = \xi_1^{\alpha_1} \dots \xi_n^{\alpha_n}$ and the constants C_1 and C_2 do not depend on ξ.

This definition essentially coincides with Sobolev's (Sobolev assumed m to be a nonnegative integer, but he also considered spaces W_p^m with nonnegative integer m and $p > 1$, which we shall discuss in Section 14).

We also mention that it is possible to retain only the terms with $\alpha = (0, \dots, 0)$, $(m, 0, \dots, 0), \dots, (0, \dots, 0, m)$ in the sums on the right-hand sides of (1.1.7) and (1.1.8). Such a definition is equivalent to the initial one as well.

Remarks. 1. It also follows from (1.1.9) that *any partial differential operator with constant coefficients of order m is bounded as an operator from $H^s(\mathbb{R}^n)$ to $H^{s-m}(\mathbb{R}^n)$ for any $s \in \mathbb{R}$.* A little further, in Section 1.9, we generalize this assertion to operators with "sufficiently good" variable coefficients.

2. As is seen from the definition, for positive noninteger s, that is, for $s = m + \theta$, where $m \in \mathbb{Z}_+$ and $0 < \theta < 1$, the space $H^s(\mathbb{R}^n)$ consists of all functions in $H^m(\mathbb{R}^n)$ whose derivatives of order m belong to $H^\theta(\mathbb{R}^n)$.

Problem. Show that the norms on $H^s(\mathbb{R}^n)$ are translation invariant, that is, for $u_h(x) = u(x + h)$,

$$\|u_h\|_{H^s(\mathbb{R}^n)} = \|u\|_{H^s(\mathbb{R}^n)} \quad \text{and} \quad \|u_h\|'_{H^m(\mathbb{R}^n)} = \|u\|'_{H^m(\mathbb{R}^n)}. \tag{1.1.10}$$

We also set

$$H^\infty(\mathbb{R}^n) = \bigcap H^s(\mathbb{R}^n) \quad \text{and} \quad H^{-\infty}(\mathbb{R}^n) = \bigcup H^s(\mathbb{R}^n). \tag{1.1.11}$$

1.2 Embedding Theorems

It is seen from the above considerations that the regularity of elements in $H^s(\mathbb{R}^n)$ increases with s. This is even more evident from *embedding theorems* proved by Sobolev for positive integers s. The simplest embedding theorem is as follows.

Theorem 1.2.1. *Let s be a real number, and let $s > n/2$. Then any function $u \in H^s(\mathbb{R}^n)$, corrected if necessary on a set of Lebesgue measure zero, is uniformly continuous and bounded, and*

$$\sup |u(x)| \le C_1 \|u\|_{H^s(\mathbb{R}^n)}, \tag{1.2.1}$$

where C_1 is a constant independent of u. Moreover, for

$$0 < \vartheta < s - n/2, \quad \vartheta < 1, \tag{1.2.2}$$

$u(x)$ satisfies the uniform Hölder condition of order ϑ, and

$$\sup_{x \neq y} \frac{|u(x) - u(y)|}{|x - y|^\vartheta} \le C_2 \|u\|_{H^s(\mathbb{R}^n)}, \tag{1.2.3}$$

where C_2 is a constant independent of u.

Proof. Under the above assumptions, the Fourier transform $\widehat{u}(\xi)$ of $u(x)$ is an integrable function, because, by virtue of the Schwarz inequality [2], we have

$$\int |\widehat{u}(\xi)| \, d\xi \le \left(\int (1 + |\xi|^2)^s |\widehat{u}(\xi)|^2 \, d\xi \right)^{1/2} \left(\int (1 + |\xi|^2)^{-s} \, d\xi \right)^{1/2} = C_3 \|u\|_{H^s(\mathbb{R}^n)}, \tag{1.2.4}$$

since the last integral converges. Therefore,

$$u(x) = \frac{1}{(2\pi)^n} \int e^{ix \cdot \xi} \widehat{u}(\xi) \, d\xi \tag{1.2.5}$$

almost everywhere. Indeed, $u(x)$ treated as a distribution is reconstructed from its Fourier transform by the formula

$$\langle u, \varphi \rangle = \langle \widehat{u}, F^{-1}[\varphi] \rangle = \int \widehat{u}(\xi) \frac{1}{(2\pi)^n} \int e^{ix \cdot \xi} \varphi(x) \, dx \, d\xi.$$

Here φ is any function in the Schwartz space $S = S(\mathbb{R}^n)$ of infinitely differentiable functions rapidly decreasing together with all derivatives (see [18], [335], or [165]). The integrals on the right-hand side can be interchanged by Fubini's theorem:

$$\langle u, \varphi \rangle = \int \varphi(x) \frac{1}{(2\pi)^n} \int e^{ix \cdot \xi} \widehat{u}(\xi) \, d\xi \, dx.$$

Thus, we see that the function $u(x)$ must coincide with the right-hand side of (1.2.5) almost everywhere. Now we can use the fact that the (inverse) Fourier transform of an integrable function is a bounded continuous function; this is well known from Fourier analysis and can be easily verified. Moreover, (1.2.5) and (1.2.4) imply (1.2.1).

The uniform continuity of $u(x)$ can be verified by using the estimate

$$|u(x) - u(y)| \le \frac{1}{(2\pi)^n} \int |e^{i(x-y) \cdot \xi} - 1| |\widehat{u}(\xi)| \, d\xi.$$

We decompose the integral on the right-hand side into integrals over the ball

$$O_R(0) = \{\xi : |\xi| \le R\}$$

and over its complement. Given $\varepsilon > 0$, the integral over the complement is less than $\varepsilon/2$ provided that R is large enough (see (1.2.4)). For fixed R, the integral over the ball is less than $\varepsilon/2$ provided that $|x - y|$ is small enough.

---

[2] For brevity, we use this term for the classical Cauchy–Bunyakovskii–Schwarz inequality, omitting the names of Bunyakovskii and Cauchy.

Of course, uniform continuity follows also from inequality (1.2.3). To prove this inequality, we note that

$$
\begin{aligned}
|e^{iz\cdot\xi}-1|^2 &= [\cos(z\cdot\xi)-1]^2+\sin^2(z\cdot\xi) \\
&= 2-2\cos(z\cdot\xi) = 4\sin^2[(z\cdot\xi)/2].
\end{aligned}
\tag{1.2.6}
$$

Therefore,

$$
\frac{|u(x)-u(y)|}{|x-y|^\vartheta} \le (2\pi)^{-n}\int \frac{2|\sin[(x-y)\cdot\xi/2]|}{|x-y|^\vartheta|\xi|^\vartheta}|\xi|^\vartheta|\widehat{u}(\xi)|\,d\xi.
\tag{1.2.7}
$$

It is easy to check that the fraction in the integrand is bounded by a constant:

$$
\frac{2|\sin[(x-y)\cdot\xi/2]|}{|x-y|^\vartheta|\xi|^\vartheta} \le C_4.
\tag{1.2.8}
$$

Indeed, it suffices to use the inequalities $|\sin t| \le |t|$ for $|t| \le \pi/2$ and $|\sin t| \le 1$ for other $|t|$ and note that the Schwarz inequality implies

$$
\int (1+|\xi|^2)^{\vartheta/2}|\widehat{u}(\xi)|\,d\xi \le \left(\int (1+|\xi|^2)^s|\widehat{u}(\xi)|^2\,d\xi\right)^{1/2}\left(\int (1+|\xi|^2)^{-s+\vartheta}\,d\xi\right)^{1/2},
$$

where, by assumption, $-s+\vartheta < -n/2$.   □

Problem 1. Using an approximation of $\widehat{u}(\xi)$ in $\widehat{H}^s(\mathbb{R}^n)$ by compactly supported smooth functions, show that the assumptions of the theorem imply also that $u(x) \to 0$ as $x \to \infty$.

Example. We verify the sharpness of the condition $s > n/2$ as follows. Suppose that $n = 1$ and $s = 1/2$. Consider the family of functions $u_t(x)$ with Fourier transforms $\widehat{u}_t(\xi) = (1+|\xi|)^{-t}$, $t > 1$. All these functions belong to $H^{1/2}(\mathbb{R})$. For this family of functions, inequality (1.2.1) with $s = 1/2$ does not hold. Indeed, since the functions $\widehat{u}_t(\xi)$ are absolutely integrable, it follows that

$$
u_t(0) = (2\pi)^{-1}\int (1+|\xi|)^{-t}\,d\xi.
$$

This quantity equals $(t-1)^{-1}$ up to a constant factor. On the other hand, the squared norm of $u_t(x)$ in $H^{1/2}(\mathbb{R})$ is of the same order for $t \to 1$: the corresponding integral

$$
\int (1+|\xi|^2)^{1/2}(1+|\xi|)^{-2t}\,d\xi
$$

is easy to evaluate after replacing $1+|\xi|^2$ by $(1+|\xi|)^2$, and it has order $(t-1)^{-1}$. It remains to note that the inequality

$$
(t-1)^{-1} \le C(t-1)^{-1/2}
$$

with any constant C is certainly false for small $t-1$.

Let us introduce *spaces* $C_b^s(\mathbb{R}^n)$, $s \geq 0$. The space $C_b^m(\mathbb{R}^n)$ with nonnegative integer $s = m$ consists of all functions having bounded continuous derivatives up to order m and is endowed with the norm

$$\|u\|_{C_b^m(\mathbb{R}^n)} = \sup|D^\alpha u(x)|, \tag{1.2.9}$$

where the upper bound is taken over all $x \in \mathbb{R}^n$ and all α with $|\alpha| \leq m$. The space $C_b^{m+\vartheta}(\mathbb{R}^n)$, $0 < \vartheta < 1$, which is called the *Hölder space*, consists of all functions in $C_b^m(\mathbb{R}^n)$ whose higher derivatives satisfy the uniform Hölder condition of order ϑ; the norm on $C_b^{m+\vartheta}(\mathbb{R}^n)$ is defined by

$$\|u\|_{C_b^{m+\vartheta}(\mathbb{R}^n)} = \|u\|_{C_b^m(\mathbb{R}^n)} + \sup \frac{|D^\alpha u(x) - D^\alpha u(y)|}{|x-y|^\vartheta}, \tag{1.2.10}$$

where the upper bound is over all x and y such that $x \neq y$ and all α with $|\alpha| = m$. For $s = 0$, we write $C_b(\mathbb{R}^n)$. The subscript b indicates uniform boundedness. If it is omitted, then the space consists of all functions of specified smoothness in $\mathbb{R}^n$.

Remark. The spaces $C_b^{m+\vartheta}(\mathbb{R}^n)$ are also denoted by $C_b^{m,\vartheta}(\mathbb{R}^n)$. Such a space is also defined for $\vartheta = 1$ as the space of functions $u(x)$ belonging to $C_b^m(\mathbb{R}^n)$ whose derivatives of order m satisfy the uniform *Lipschitz condition*, that is, the Hölder condition of order 1. The norm on this space is defined by (1.2.10) with $\vartheta = 1$. According to a well-known theorem, any function satisfying a Lipschitz condition is differentiable almost everywhere and has bounded first derivatives. (We prove this theorem in Section 9.) Therefore, all functions in $C_b^{m,1}(\mathbb{R}^n)$ have $(m+1)$th derivatives almost everywhere, and these derivatives are bounded.

Theorem 1.2.1 can be stated as follows: *For $s \geq n/2 + \theta$, where $0 < \theta < 1$, the space $H^s(\mathbb{R}^n)$ is continuously embedded in $C_b^\vartheta(\mathbb{R}^n)$ provided that $0 < \vartheta < \theta$.* It is easy to obtain the following generalization.

Theorem 1.2.2. *If $s \geq n/2 + m + \theta$, where m is a nonnegative integer and $0 < \theta < 1$, then the space $H^s(\mathbb{R}^n)$ is continuously embedded in $C_b^{m+\vartheta}(\mathbb{R}^n)$ for $0 < \vartheta < \theta$.*

Here, as well as at other places in similar assertions, a possible correction of functions on a set of measure zero is assumed.

Problem 2. Verify the assertion of Theorem 1.2.2.

Remark. The results mentioned above can also be stated in the following form: *For $s > n/2 + t$, where $t > 0$, the space $H^s(\mathbb{R}^n)$ is continuously embedded in $C_b^t(\mathbb{R}^n)$.*

Moreover, this assertion can be strengthened as follows: *If t is noninteger, then continuous embedding holds also for $s = n/2 + t$*; see [376, Sec. 2.8.1]. The proof employs additional technical tools, and we do not reproduce it here.

Corollary 1.2.3. *Any function $u \in H^\infty(\mathbb{R}^n)$, possibly being corrected on a set of measure zero, is infinitely differentiable, and any of its derivatives is bounded and square integrable.*

Sobolev also proved the following theorem for positive integers s.

Theorem 1.2.4. *If $0 < s < n/2$ and*

$$s \geq \frac{n}{2} - \frac{n}{p}, \quad 2 < p < \infty, \tag{1.2.11}$$

then the space $H^s(\mathbb{R}^n)$ is continuously embedded in $L_p(\mathbb{R}^n)$.

This means that

$$\|u\|_{L_p(\mathbb{R}^n)} = \left(\int |u(x)|^p \, dx \right)^{1/p} \leq C \|u\|_{H^s(\mathbb{R}^n)}. \tag{1.2.12}$$

We outline the proof of a somewhat weakened version of this theorem (with the strict inequality for s in (1.2.11)) in Section 1.15 with a reference to the Young inequality, which we obtain in Section 13.2 by using interpolation arguments. Elementary facts about the L_p spaces are collected in Section 13.1. Here and in Section 1.15, it suffices to know that $L_p(\mathbb{R}^n)$ with $1 \leq p < \infty$ is a Banach space with norm defined by the middle expression in (1.2.12) and that the set of compactly supported smooth functions is dense in this space.

1.3 The Spaces $H^s(\mathbb{R}^n)$ of Negative Order

Theorem 1.3.1. *For a positive integer m, the elements of the space $H^{-m}(\mathbb{R}^n)$ are finite sums of derivatives (in the sense of distributions) of order up to m of functions from $L_2(\mathbb{R}^n)$.*

Proof. As mentioned at the end of Section 1.1, differentiation of order m maps functions in $L_2(\mathbb{R}^n)$ to functions in $H^{-m}(\mathbb{R}^n)$. Now take $u \in H^{-m}(\mathbb{R}^n)$ and $\widehat{u}(\xi) = (Fu)(\xi)$. We have

$$w(\xi) = (1 + |\xi_1|^m + \ldots + |\xi_n|^m)^{-1} \widehat{u}(\xi) \in L_2(\mathbb{R}^n),$$

because

$$0 < C_1 \leq \frac{1 + |\xi_1|^m + \ldots + |\xi_n|^m}{(1 + |\xi|^2)^{m/2}} \leq C_2,$$

where the constants do not depend on ξ. Therefore,

$$\widehat{u}(\xi) = w(\xi) + \xi_1^m w_1(\xi) + \ldots + \xi_n^m w_n(\xi)$$

almost everywhere with $w_j(\xi) = (|\xi_j|/\xi_j)^m w(\xi)$ if $\xi_j \neq 0$ and $w_j(\xi) = 0$ otherwise. The functions w_j belong to $L_2(\mathbb{R}^n)$. Hence

$$u = u_0 + D_1^m u_1 + \ldots + D_n^m u_n$$

in the sense of distributions, where all $u_j(x)$ belong to $L_2(\mathbb{R}^n)$. $\qquad\square$

Corollary 1.3.2. *The space $H^{-\infty}(\mathbb{R}^n)$ contains all compactly supported distributions.*

Indeed, any compactly supported continuous function belongs to $L_2(\mathbb{R}^n)$. Hence it suffices to apply a theorem about the structure of a compactly supported distribution, which asserts that any such distribution is a finite sum of derivatives (in the sense of distributions) of some orders of compactly supported continuous functions.

□

1.4 Isometric Isomorphisms Λ^t

The following proposition is a model for some theorems of the theory of elliptic operators presented in Section 6. By I we denote the identity operator and by Δ the Laplace operator.

Theorem 1.4.1. *The operator $I - \Delta$ defines an isometric isomorphism $H^s(\mathbb{R}^n) \to H^{s-2}(\mathbb{R}^n)$ for any s.*

More precisely, we have the isometry if our initial definition of the spaces $H^s(\mathbb{R}^n)$ is used. Indeed, $I - \Delta$ acts on the Fourier images as multiplication by $1 + |\xi|^2$.

The spaces $H^s(\mathbb{R}^n)$ are very convenient for studying this operator. They are also convenient for constructing a general theory of elliptic pseudodifferential operators.

An immediate and obvious generalization is as follows.

Theorem 1.4.2. *Let s and t be any real numbers. Then the operator*

$$\Lambda^t = F^{-1}(1 + |\xi|^2)^{t/2} F \tag{1.4.1}$$

defines an isometric isomorphism $H^s(\mathbb{R}^n) \to H^{s-t}(\mathbb{R}^n)$. Its inverse is the operator Λ^{-t}.

As can be seen from our survey in Section 18.4, (1.4.1) is an example of an elliptic pseudodifferential operator of order t. Actually, this is the $(t/2)$th power of $I - \Delta$. In particular, $\Lambda^2 = I - \Delta$.

The function $\Lambda^t g$, where $g \in L_2(\mathbb{R}^n)$, is sometimes referred to as a *Bessel potential of order t*, at least for negative t. The space $H^s(\mathbb{R}^n)$ can be defined as the image $\Lambda^{-s} L_2(\mathbb{R}^n)$ of the space $L_2(\mathbb{R}^n)$ under the action of the operator Λ^{-s}. Apparently, it is for this reason that the spaces $H^s(\mathbb{R}^n)$ are called Bessel potential spaces. The more general Bessel potential spaces H_p^s will be considered in Section 14, where the alternative names of these spaces will also be mentioned.

1.5 Dense Subsets

The space $S = S(\mathbb{R}^n)$ is obviously contained in all spaces $H^s(\mathbb{R}^n)$.

Theorem 1.5.1. *For any s, the space $H^s(\mathbb{R}^n)$ is the completion of S with respect to the norm $\|\cdot\|_{H^s(\mathbb{R}^n)}$.*

Proof. Let $u \in H^s(\mathbb{R}^n)$. Then the function $(1 + |\xi|^2)^{s/2}\widehat{u}(\xi)$ belongs to $L_2(\mathbb{R}^n_\xi)$ and can be approximated there by compactly supported C^∞ functions $v_k(\xi)$. Let us write v_k in the form

$$v_k(\xi) = (1 + |\xi|^2)^{s/2} w_k(\xi), \quad \text{where } w_k(\xi) = \frac{v_k(\xi)}{(1 + |\xi|^2)^{s/2}}.$$

The functions $w_k(\xi)$ are compactly supported and C^∞, and they converge to $\widehat{u}(\xi)$ in the space $\widehat{H}^s(\mathbb{R}^n)$. Their Fourier preimages belong to S and converge to u in $H^s(\mathbb{R}^n)$.   □

Corollary 1.5.2. *For $\sigma > s$, the space $H^\sigma(\mathbb{R}^n)$ is dense in $H^s(\mathbb{R}^n)$.*

Now we strengthen Theorem 1.5.1.

Theorem 1.5.3. *For any s, the space $H^s(\mathbb{R}^n)$ is the completion of the linear manifold $C_0^\infty(\mathbb{R}^n) = \mathcal{D} = \mathcal{D}(\mathbb{R}^n)$ with respect to the norm of $H^s(\mathbb{R}^n)$.*

Proof. It suffices to show that any function $u \in S$ can be approximated by compactly supported infinitely differentiable functions in the norm of $H^s(\mathbb{R}^n)$. Moreover, it suffices to show this for positive integer s, which we do below. Let $\varphi(x)$ be a function in $\mathcal{D}$ equal to 1 at $|x| \le 1$, and let

$$u_\varepsilon(x) = u(x)\varphi(\varepsilon x).$$

Then $u_\varepsilon \to u$ in $L_2(\mathbb{R}^n)$ as $\varepsilon \to 0$, and for any α with $0 < |\alpha| \le s$ we have

$$D^\alpha[u_\varepsilon(x) - u(x)] = D^\alpha u(x) \cdot [\varphi(\varepsilon x) - 1] + \sum c_{\beta\gamma} D^\beta u(x) \cdot D^\gamma \varphi(\varepsilon x), \quad (1.5.1)$$

where $\alpha = \beta + \gamma$ and $\gamma \ne 0$. Clearly, since

$$D^\gamma \varphi(\varepsilon x) = \varepsilon^{|\gamma|}(D^\gamma \varphi)(\varepsilon x),$$

it follows that the right-hand side in (1.5.1) tends to 0 in $L^2(\mathbb{R}^n)$ as $\varepsilon \to 0$. Therefore, $u_\varepsilon \to u$ in $H^s(\mathbb{R}^n)$.   □

This theorem will be supplemented in Section 1.11.

Problem. Verify that if $s > 0$, $u \in H^s(\mathbb{R}^n)$, $u_k \in C_0^\infty(\mathbb{R}^n)$, and $u_k \to u$ in $H^s(\mathbb{R}^n)$, then $D^\alpha u_k \to D^\alpha u$ in $H^{s-|\alpha|}(\mathbb{R}^n)$ for $|\alpha| \le s$. (See Remark 1 in Section 1.1.)

1.6 Continuous Linear Functionals on $H^s(\mathbb{R}^n)$

Consider the problem of describing, or realizing, the space dual to $H^s(\mathbb{R}^n)$, i.e., the space of continuous linear functionals on $H^s(\mathbb{R}^n)$. This space has two realizations.

The first is obvious: since $H^s(\mathbb{R}^n)$ is a Hilbert space with inner product (1.1.4), *the general form of a continuous linear functional $f(u)$ on $H^s(\mathbb{R}^n)$ is*

$$f(u) = (u, v)_{s, \mathbb{R}^n}, \qquad (1.6.1)$$

where v is any fixed element of $H^s(\mathbb{R}^n)$; it is uniquely determined by f.

The second uses the *duality between $H^s(\mathbb{R}^n)$ and $H^{-s}(\mathbb{R}^n)$*. The point is that the form $(u, v)_{\mathbb{R}^n} = (u, v)_{0, \mathbb{R}^n}$ on $H^0(\mathbb{R}^n)$ (we shall omit the subscript 0) *can be extended to the direct product $H^s(\mathbb{R}^n) \times H^{-s}(\mathbb{R}^n)$ as*

$$(u, v)_{\mathbb{R}^n} = \int \widehat{u}(\xi) \cdot \overline{\widehat{v}(\xi)} \, d\xi; \qquad (1.6.2)$$

moreover, the form thus extended satisfies the generalized Schwarz inequality

$$|(u, v)_{\mathbb{R}^n}| \le \|u\|_{H^s(\mathbb{R}^n)} \|v\|_{H^{-s}(\mathbb{R}^n)}, \qquad (1.6.3)$$

which is obtained by applying the usual Schwarz inequality to

$$\int \widehat{u}(\xi) \cdot \overline{\widehat{v}(\xi)} \, d\xi = \int (1 + |\xi|^2)^{s/2} \widehat{u}(\xi) \cdot (1 + |\xi|^2)^{-s/2} \overline{\widehat{v}(\xi)} \, d\xi.$$

Theorem 1.6.1. *For any s, each continuous linear functional $f(u)$ on $H^s(\mathbb{R}^n)$ can be represented in the form*

$$f(u) = (u, w)_{\mathbb{R}^n}, \qquad (1.6.4)$$

where w is an element of $H^{-s}(\mathbb{R}^n)$ uniquely determined by f. Conversely, for any $w \in H^{-s}(\mathbb{R}^n)$, expression (1.6.4) defines a continuous linear functional on $H^s(\mathbb{R}^n)$.

Proof. The second assertion is obvious. To verify the first, let $u_1 = \Lambda^s u$ (see (1.4.1)). Then $u_1 \in H^0(\mathbb{R}^n)$, and $g(u_1) = f(u)$ is a continuous linear functional on $H^0(\mathbb{R}^n)$. Therefore,

$$f(u) = \int \widehat{u}_1(\xi) \overline{\widehat{v}_1(\xi)} \, d\xi = \int (1 + |\xi|^2)^{s/2} \widehat{u}(\xi) \overline{\widehat{v}_1(\xi)} \, d\xi,$$

where $v_1 \in H^0(\mathbb{R}^n)$. It follows that

$$f(u) = (u, w)_{\mathbb{R}^n}, \quad \text{where } w = \Lambda^s v_1 \in H^{-s}(\mathbb{R}^n). \qquad \square$$

Remark 1.6.2. Obviously, these two realizations of the same continuous linear functional on $H^s(\mathbb{R}^n)$ are related as

$$(u, v)_{s, \mathbb{R}^n} = (u, w)_{\mathbb{R}^n}, \quad \text{where } w = \Lambda^{2s} v. \qquad (1.6.5)$$

Remark 1.6.3. If $\{u_k\}$ and $\{v_k\}$ are sequences of functions in $C_0^\infty(\mathbb{R}^n)$ converging to u and v in $H^s(\mathbb{R}^n)$ and $H^{-s}(\mathbb{R}^n)$, respectively, then

$$(u, v)_{\mathbb{R}^n} = \lim_{k \to \infty} (u_k, v_k)_{\mathbb{R}^n}.$$

Of course, the form under the sign of limit can be rewritten as the usual integral of $u_k \overline{v_k}$ over $\mathbb{R}^n$.

Using similar approximations, we can show that if, say, φ is a function in $C_0^\infty(\mathbb{R}^n)$, then

$$(\varphi u, v)_{\mathbb{R}^n} = (u, \overline{\varphi} v)_{\mathbb{R}^n}$$

for $u \in H^s(\mathbb{R}^n)$ and $v \in H^{-s}(\mathbb{R}^n)$. This, in turn, readily implies that if u and v have disjoint supports, then $(u, v)_{\mathbb{R}^n} = 0$.

Here and, as a rule, in what follows, we use inner products, i.e., sesquilinear forms, rather than the bilinear form $\langle \cdot, \cdot \rangle$. We could obviate this inconsistency between expositions of distribution theory and the theory of Sobolev-type spaces, but our approach is more traditional.

1.7 Norms of Positive Fractional Order

As we saw, for positive integer s, there is a norm on $H^s(\mathbb{R}^n)$ which can be written without using the Fourier transform. Below, we indicate such a norm for positive noninteger s.

Theorem 1.7.1. *Let $s = m + \theta$, where m is a nonnegative integer and $0 < \theta < 1$. Then the norm on $H^s(\mathbb{R}^n)$ is equivalent to the norm defined by*

$$\|u\|'^2_{H^s(\mathbb{R}^n)} = \|u\|'^2_{H^m(\mathbb{R}^n)} + \sum_{|\alpha|=m} \iint \frac{|D^\alpha u(x) - D^\alpha u(y)|^2}{|x-y|^{n+2\theta}}\, dx\, dy. \qquad (1.7.1)$$

The spaces $H^s(\mathbb{R}^n) = W_2^s(\mathbb{R}^n)$ with noninteger $s > 0$ as spaces with norms (1.7.1) were introduced by Slobodetskii in [350]. These norms are translation invariant, as well as those introduced above.

Proof of the theorem. Consider the case $m = 0$. Let us evaluate the integral

$$\iint \frac{|u(x) - u(y)|^2}{|x-y|^{n+2\theta}}\, dx\, dy \qquad (1.7.2)$$

via the Fourier transform $\widehat{u}(\xi)$ of the function $u(x)$ under the assumption that $u(x)$ is compactly supported and C^∞. The result to be obtained is given in (1.7.5) below. The integral (1.7.2) absolutely converges; indeed, on the compact set $\operatorname{supp} u(x) \times \operatorname{supp} u(y)$, the singularity at $x = y$ is integrable (it has order $n - 2 + 2\theta$, because $u(x)$ is smooth), and outside this set, $|u(x)|^2|x-y|^{-n-2\theta}$ and $|u(y)|^2|x-y|^{-n-2\theta}$ are integrable. Setting $y = x + z$, we rewrite this integral as the double integral

$$\int \frac{1}{|z|^{n+2\theta}}\, dz \int |u(x) - u(x+z)|^2\, dx. \qquad (1.7.3)$$

By Parseval's identity (1.1.5), the inner integral is equal (up to a constant factor) to

$$\int |1 - e^{iz\cdot\xi}|^2 |\widehat{u}(\xi)|^2 \, d\xi.$$

Here, according to (1.2.6),

$$|1 - e^{iz\cdot\xi}|^2 = 4\sin^2[(z\cdot\xi)/2].$$

Substituting all this into (1.7.3) and changing the order of integration by Fubini's theorem, we obtain the integral

$$4\int |\widehat{u}(\xi)|^2 \, d\xi \int \frac{\sin^2[(z\cdot\xi)/2]}{|z|^{n+2\theta}} \, dz. \qquad (1.7.4)$$

Writing the inner integral in the form

$$\int \sin^2\left(\frac{1}{2}|z||\xi|\cos\varphi\right) \frac{dz}{|z|^{n+2\theta}},$$

where φ is the angle between the vectors z and ξ, and passing to the spherical coordinates $z = r\omega$, $|\omega| = 1$, $dz = r^{n-1} \, dr \, dS$, we obtain the double integral

$$\int_S dS \int_0^\infty \sin^2\left(\frac{1}{2}r|\xi|\cos\varphi\right) \frac{dr}{r^{1+2\theta}},$$

where S denotes the unit sphere. In the inner integral, we make the change $\frac{1}{2}r|\xi|\cos\varphi = \tau$, that is, pass from the variable r to the variable τ, which yields

$$2^{-2\theta}|\xi|^{2\theta} \int_S \cos^{2\theta}\varphi \, dS \int_0^\infty \frac{\sin^2\tau}{\tau^{1+2\theta}} \, d\tau = C_\theta'|\xi|^{2\theta},$$

where the constant C_θ' does not depend on ξ. We have shown that

$$\iint \frac{|u(x) - u(y)|^2}{|x - y|^{n+2\theta}} \, dx \, dy = 4C_\theta \int |\xi|^{2\theta} |\widehat{u}(\xi)|^2 \, d\xi, \qquad (1.7.5)$$

where the constant C_θ does not depend on ξ. The obtained result is extended to any $u \in H^\theta(\mathbb{R}^n)$ by passing to the limit. This essentially proves the theorem for $m = 0$. The extension of the proof to nonzero m is obvious.   □

Problem. Write the inner product corresponding to the norm (1.7.1).

1.8 Estimates of Intermediate Norms

Proposition 1.8.1. *Let $\tau < s < \sigma$. Then, for any $\varepsilon > 0$, there exists a $C_\varepsilon > 0$ such that the norms of functions $u \in H^\sigma(\mathbb{R}^n)$ satisfy the inequality*

$$\|u\|_{H^s(\mathbb{R}^n)} \le \varepsilon\|u\|_{H^\sigma(\mathbb{R}^n)} + C_\varepsilon\|u\|_{H^\tau(\mathbb{R}^n)}. \tag{1.8.1}$$

Proof. An equivalent assertion is that

$$\|u\|^2_{H^s(\mathbb{R}^n)} \le \varepsilon^2\|u\|^2_{H^\sigma(\mathbb{R}^n)} + C'_\varepsilon\|u\|^2_{H^\tau(\mathbb{R}^n)} \tag{1.8.2}$$

for any $\varepsilon > 0$, and inequality (1.8.2) follows from the obvious fact that, given $\varepsilon > 0$, there exists a C'_ε for which

$$(1 + |\xi|^2)^s \le \varepsilon^2(1 + |\xi|^2)^\sigma + C'_\varepsilon(1 + |\xi|^2)^\tau. \qquad \square$$

1.9 Multipliers

Now let us discuss conditions sufficient for the operator of multiplication by a function $a(x)$ to be bounded in the space $H^s(\mathbb{R}^n)$ (i.e., for $a(x)$ to be a multiplier in this space) and estimate the norm of this operator.

It is well known that *the operator of multiplication by a bounded measurable function is bounded in $L_2(\mathbb{R}^n)$, and its norm does not exceed the least upper bound of the absolute value of this function.* Of course, it suffices to assume the function to be only essentially bounded, i.e., bounded outside a set of measure zero. This observation (together with Leibniz' product rule for derivatives) immediately implies the following theorem.

Theorem 1.9.1. *Let m be a positive integer, and let $a(x)$ be a function in $C_b^{m-1,1}(\mathbb{R}^n)$. Then the operator of multiplication by $a(x)$ is bounded in $H^m(\mathbb{R}^n)$. Moreover,*

$$\|au\|_{H^m(\mathbb{R}^n)} \le C_1 \sup|a(x)|\|u\|_{H^m(\mathbb{R}^n)} + C_2\|a\|_{C_b^{m-1,1}(\mathbb{R}^n)}\|u\|_{H^{m-1}(\mathbb{R}^n)}, \tag{1.9.1}$$

where the constants C_1 and C_2 do not depend on u and a.

In particular, for this operator to be bounded, it is sufficient that the function $a(x)$ belong to the space $C_b^m(\mathbb{R}^n)$. Now consider noninteger s.

Theorem 1.9.2. *Let m be a nonnegative integer, and let $s = m + \theta$, where $0 < \theta < 1$. Suppose that a function $a(x)$ belongs to the space $C_b^{m+\vartheta}(\mathbb{R}^n) = C_b^{m,\vartheta}(\mathbb{R}^n)$, where $\vartheta \in (\theta, 1)$. Then the operator of multiplication by $a(x)$ is bounded in $H^s(\mathbb{R}^n)$. Moreover,*

$$\|au\|_{H^{m+\theta}(\mathbb{R}^n)} \le C_1 \sup|a(x)|\|u\|_{H^{m+\theta}(\mathbb{R}^n)} + C_2\|a\|_{C_b^{m+\vartheta}(\mathbb{R}^n)}\|u\|_{H^m(\mathbb{R}^n)}, \tag{1.9.2}$$

where the constants C_1 and C_2 do not depend on u and a.

Proof. We prove the theorem for $m = 0$. Given $u \in H^\theta(\mathbb{R}^n)$, we have to estimate the expression

$$\iint \frac{|a(x)u(x) - a(y)u(y)|^2}{|x-y|^{n+2\theta}} \, dx \, dy \qquad (1.9.3)$$

in terms of $\|u\|^2_{H^\theta(\mathbb{R}^n)}$. Clearly,

$$|a(x)u(x) - a(y)u(y)|^2 \le 2|a(x)[u(x) - u(y)]|^2 + 2|[a(x) - a(y)]u(y)|^2.$$

Therefore, (1.9.3) does not exceed

$$2\sup|a(x)|^2 \|u\|^2_{H^\theta(\mathbb{R}^n)} + 2\int |u(y)|^2 I(y) \, dy,$$

where

$$I(y) = \int \frac{|a(x) - a(y)|^2}{|x-y|^{n+2\theta}} \, dx, \qquad (1.9.4)$$

and the problem reduces to estimating this integral in terms of $\|a\|^2_{C_b^\vartheta(\mathbb{R}^n)}$. Take $C_1 > 0$. The integrand does not exceed $C_2'|x-y|^{-n+2(\vartheta-\theta)}$ if $|x-y| \le C_1$ and $C_3'|x-y|^{-n-2\theta}$ if $|x-y| \ge C_1$, where C_2' and C_3' are proportional to $\|a\|^2_{C_b^\vartheta(\mathbb{R}^n)}$; this gives the desired result. $\qquad \square$

Estimate (1.9.2), together with Proposition 1.8.1, implies that, given any $\varepsilon > 0$, there exists a constant $C_3(\varepsilon)$ for which

$$\|au\|_{H^{m+\theta}(\mathbb{R}^n)} \le (C_1 \sup|a(x)| + \varepsilon)\|u\|_{H^{m+\theta}(\mathbb{R}^n)} + C_3(\varepsilon)\|u\|_{H^0(\mathbb{R}^n)}. \qquad (1.9.5)$$

In the theorems stated above, of interest are, first, the sufficient smoothness of the multiplier and, second, the coefficient of the "principal part" in the estimate of its norm.

Obviously, the condition that the function $a(x)$ and its derivatives up to order m are bounded and continuous is sufficient for the operator of multiplication by $a(x)$ to be bounded in $H^s(\mathbb{R}^n)$ with noninteger $s \in (0, m)$.

For the purposes of the theory of elliptic equations presented in Chapter 2, we need to develop the obtained result.

Lemma 1.9.3. *Suppose that a bounded linear operator T in $H^s(\mathbb{R}^n)$ satisfies the condition*

$$\|Tu\|_{H^s(\mathbb{R}^n)} \le K_1\|u\|_{H^s(\mathbb{R}^n)} + K_2\|u\|_{H^{s-1}(\mathbb{R}^n)} \qquad (1.9.6)$$

for $s_0 \le s \le s_1$, where K_1 and K_2 are constants independent of s. Then, for any $\varepsilon > 0$, the operator T admits the representation

$$T = T_1 + T_2, \qquad (1.9.7)$$

where T_2 is a bounded operator from $H^{s-1}(\mathbb{R}^n)$ to $H^s(\mathbb{R}^n)$ and T_1 satisfies the inequality

$$\|T_1 u\|_{H^s(\mathbb{R}^n)} \le (K_1 + \varepsilon)\|u\|_{H^s(\mathbb{R}^n)} \tag{1.9.8}$$

for the same s.

Proof. Let $\psi(t) = \psi_R(t)$ be an infinitely differentiable function on the half-axis $\mathbb{R}_+$, which takes the value 1 at $t \le R$, 0 at $t \ge R+1$, and values between 0 and 1 at $R < t < R+1$. We set

$$T_2 = TF^{-1}\psi(|\xi|)F \quad \text{and} \quad T_1 = T - T_2. \tag{1.9.9}$$

Since the function $\psi(|\xi|)$ is compactly supported, it follows that T_2 has the required boundedness property. Next, for $u \in H^s(\mathbb{R}^n)$ and $v = (1 - F^{-1}\psi(|\xi|)F)u$, we have

$$\|T_1 u\|_{H^s(\mathbb{R}^n)} = \|Tv\|_{H^s(\mathbb{R}^n)} \le K_1\|v\|_{H^s(\mathbb{R}^n)} + K_2\|v\|_{H^{s-1}(\mathbb{R}^n)}$$

$$\le K_1\|u\|_{H^s(\mathbb{R}^n)} + K_2(1 + R^2)^{-1/2}\|u\|_{H^s(\mathbb{R}^n)}.$$

It remains to take sufficiently large R. □

Corollary 1.9.4. *Let $a(x)$ be a function in $C_b^\infty(\mathbb{R}^n)$. Then, for any $s_1 > 0$ and any $\varepsilon > 0$, the operator T of multiplication by $a(x)$ in $H^s(\mathbb{R}^n)$, where $0 \le s \le s_1$, admits representation (1.9.7), where*

$$\|T_1 u\|_{H^s(\mathbb{R}^n)} \le (C \sup |a(x)| + \varepsilon)\|u\|_{H^s(\mathbb{R}^n)} \quad \text{and} \quad \|T_2 u\|_{H^s(\mathbb{R}^n)} \le C(\varepsilon)\|u\|_{H^{s-1}(\mathbb{R}^n)}, \tag{1.9.10}$$

with constants independent of u and s; C is also independent of a.

Now consider the question of when the operator of multiplication by $a(x)$ is bounded in $H^s(\mathbb{R}^n)$ with *negative* s. Recall that the spaces $H^s(\mathbb{R}^n)$ and $H^{|s|}(\mathbb{R}^n)$ are dual with respect to the extension of the form $(\cdot, \cdot)_{\mathbb{R}^n}$. Obviously, for compactly supported smooth functions u and v, we have

$$(au, v)_{\mathbb{R}^n} = (u, \bar{a}v)_{\mathbb{R}^n}; \tag{1.9.11}$$

therefore, the operator adjoint to the operator of multiplication by a in $H^s(\mathbb{R}^n)$ must be the operator of multiplication by $\bar{a}$ in $H^{|s|}(\mathbb{R}^n)$. An operator in this space is bounded if and only if the adjoint operator is bounded in the dual space; moreover, these operators have the same norm. Obviously, the operators of multiplication by $a(x)$ and by $\overline{a(x)}$ are bounded or unbounded in $H^{|s|}(\mathbb{R}^n)$ simultaneously. We have obtained the following quite convenient result.

Theorem 1.9.5. *Suppose that s is negative and the operator of multiplication by $a(x)$ is bounded in $H^{|s|}(\mathbb{R}^n)$. Then this operator, being defined in $H^s(\mathbb{R}^n)$ by relation (1.9.11), is bounded in this space and has the same norm as in $H^{|s|}(\mathbb{R}^n)$.*

Corollary 1.9.4 can be extended to negative s.

To be more precise, under the passage from an operator T in a space with positive index $|s|$ to an operator in the dual space with negative index s, the decomposition $T = T_1 + T_2$ is preserved; the second term in this decomposition remains a smoothing operator (namely, an operator from $H^s(\mathbb{R}^n)$ to $H^{s+1}(\mathbb{R}^n)$), while the first term retains the norm estimate.

Now consider a linear partial differential operator with variable coefficients of the form

$$a(x, D) = \sum_{|\alpha| \leq m} a_\alpha(x) D^\alpha. \tag{1.9.12}$$

Obviously, if all $a_\alpha(x)$ are bounded measurable functions, then (1.9.12) is a bounded operator from $H^m(\mathbb{R}^n)$ to $H^0(\mathbb{R}^n) = L_2(\mathbb{R}^n)$.

Corollary 1.9.6. *An operator of the form* (1.9.12) *with coefficients having bounded continuous derivatives of any order is a bounded operator from $H^s(\mathbb{R}^n)$ to $H^{s-m}(\mathbb{R}^n)$ for any s.*

Further information about multipliers can be found in [254] and [369].

1.10 Traces on Hyperplanes

Now let us investigate the possibility of considering the trace of a function in $H^s(\mathbb{R}^n)$ on the hyperplane $\mathbb{R}^{n-1}$ determined, for simplicity, by $x_n = 0$. The trace of a usual continuous function on such a hyperplane is merely its restriction to this hyperplane.

We set $x = (x', x_n)$ and denote the Fourier transform of functions of x' into functions of ξ' by F'. First, we prove the following theorem.

Theorem 1.10.1. *Let $s > 1/2$. Then the functions $u \in \mathcal{D}$ satisfy the inequality*

$$\|u(x', 0)\|_{H^{s-1/2}(\mathbb{R}^{n-1})} \leq C\|u(x)\|_{H^s(\mathbb{R}^n)}, \tag{1.10.1}$$

where C is a constant not depending on u.

Proof. We represent $(F'u)(\xi', x_n)\big|_{x_n=0}$ in the form

$$(F'u)(\xi', 0) = (2\pi)^{-1} \int (Fu)(\xi)\, d\xi_n.$$

By the Schwarz inequality,

$$|(F'u)(\xi', 0)|^2 \leq (2\pi)^{-2} \int |(Fu)(\xi)|^2 (1 + |\xi|^2)^s\, d\xi_n \int (1 + |\xi|^2)^{-s}\, d\xi_n.$$

The last integral is evaluated by substituting $\xi_n/(1 + |\xi'|^2)^{1/2} = \tau$, it equals $(1 + |\xi'|^2)^{-s+1/2}$ up to a constant factor. It remains to divide both sides by this expression and integrate them with respect to ξ'. $\qquad\square$

This result allows us to give the following definition of the trace of a function u from $H^s(\mathbb{R}^n)$ on $\mathbb{R}^{n-1}$ for $s > 1/2$.

Definition. Let $\{u_l\}$ be a sequence of functions in $\mathcal{D}$ converging to u in $H^s(\mathbb{R}^n)$, $s > 1/2$. Then this sequence is fundamental, and (1.10.1) implies that the traces on $\mathbb{R}^{n-1}$ of the functions u_l form a fundamental sequence in $H^{s-1/2}(\mathbb{R}^{n-1})$. This sequence has a limit in the space $H^{s-1/2}(\mathbb{R}^{n-1})$, because this space is complete. Moreover, it follows from (1.10.1) that this limit does not depend on the choice of the approximating sequence, because any two such sequences approach each other arbitrarily closely in $H^s(\mathbb{R}^n)$. It is this limit which is called the *trace* of the function u on $\mathbb{R}^{n-1}$; we denote it by $u(x', 0)$ or $(\gamma u)(x')$.

Inequality (1.10.1) is extended to functions from $H^s(\mathbb{R}^n)$ by passing to the limit.

Obviously, for usual continuous functions in $H^s(\mathbb{R}^n)$, the trace thus defined coincides with the usual one. By virtue of (1.10.1), *for $s > 1/2$, the trace operator acts boundedly from $H^s(\mathbb{R}^n)$ to $H^{s-1/2}(\mathbb{R}^{n-1})$.*

A little later we shall introduce spaces H^s on the half-space $\mathbb{R}^n_+$ and on a bounded domain, and there it will be natural to talk about boundary values of functions instead of their traces. We shall also show there that the trace operator has a bounded right inverse.

Theorem 1.10.1 was obtained by Slobodetskii in [350]. The same paper also contains an example showing that inequality (1.10.1) becomes false at $s = 1/2$. We briefly recall this example.

Example. Let $n = 2$, and let $u(x, y) = \varphi(x)\varphi(y)$, where

$$\varphi(y) = \frac{1}{2\pi} \int \frac{e^{i\eta y} \, d\eta}{(1 + \eta^2)^{1/2}[\ln(2 + \eta^2)]^q}, \tag{1.10.2}$$

$1/2 < q < 1$. The Fourier image of φ is

$$\frac{1}{(1 + \eta^2)^{1/2}[\ln(2 + \eta^2)]^q},$$

and it is easy to verify that it belongs to $\widehat{H}^{1/2}(\mathbb{R})$. This follows from the convergence of the integral

$$\int\limits_1^\infty \frac{dt}{(1 + t)[\ln(1 + t)]^{2q}}$$

checked by substituting $\ln(1 + t) = \tau$. Therefore, φ belongs to $H^{1/2}(\mathbb{R})$ and $u(x, y)$ belongs to $H^{1/2}(\mathbb{R}^2)$. On the other hand, the integral (1.10.2) converges conditionally at $y \neq 0$ and diverges at $y = 0$. Essentially for this reason (cf. [350]), we have $|\varphi(y)| \to \infty$ as $y \to 0$ and

$$\lim_{y \to 0} \int |u(x, y)|^2 \, dx = +\infty.$$

Thus, there is no uniform bound for the integral under the limit sign in terms of $\|u\|^2_{H^{1/2}(\mathbb{R}^2)}$, and the function $u(x,y)$ has no finite trace in $L_2(\mathbb{R})$ at $y = 0$.

It follows from Theorem 1.10.1 that *if $s > m+1/2$, where m is a positive integer, and $a(D)$ is an mth-order partial differential operator with constant (for simplicity) coefficients, then, for $u \in H^s(\mathbb{R}^n)$, the expression $a(D)u$ has a trace on $\mathbb{R}^{n-1}$, which belongs to $H^{s-m-1/2}(\mathbb{R}^{n-1})$, and the operator of passing to this trace is bounded in the corresponding norms.* In particular, this relates to the traces of derivatives with respect to x_n.

1.11 Mollifications and Shifts

Let ω be a function in $\mathcal{D}$. If $u \in H^s(\mathbb{R}^n)$ for some s, then, of course, $u \in \mathcal{S}'$, so that the convolution $u * \omega$ is well defined, at least in the sense of distributions. It is known that $u * \omega$ is an infinitely differentiable function and

$$\text{supp}(u * \omega) \subset \text{supp}\,u + \text{supp}\,\omega. \tag{1.11.1}$$

Theorem 1.11.1. *For $u \in H^s(\mathbb{R}^n)$ and $\omega \in \mathcal{D}$,*

$$F[u * \omega] = F[u]F[\omega]. \tag{1.11.2}$$

Indeed, this relation is known to hold for $u \in \mathcal{S}'$ (see, e.g., Theorem 5.3.1 in [18]). Suppose that

$$\omega(x) \geq 0, \quad \omega(x) = 1 \text{ near } 0, \quad \text{and} \quad \int \omega(x)\,dx = 1. \tag{1.11.3}$$

For $h > 0$, we set

$$\omega_h(x) = h^{-n}\omega\Big(\frac{x}{h}\Big). \tag{1.11.4}$$

These functions form a delta-shaped family as $h \to 0$ (see, e.g., [18, Sec. 1.8]). The convolution $u * \omega_h$ is called the *mollification* of the function (or distribution) u.

Theorem 1.11.2. *For $u \in H^s(\mathbb{R}^n)$,*

$$u * \omega_h \to u \quad \text{in } H^s(\mathbb{R}^n) \quad \text{as } h \to 0. \tag{1.11.5}$$

Proof. It is easy to check that $F[\omega_h](\xi) = F[\omega](h\xi)$. The function on the right-hand side is bounded uniformly in h, and it uniformly converges to $F[\omega](0) = 1$ in any ball $O_R = O_R(0)$ as $h \to 0$. Therefore, $F[u]F[\omega_h] \to F[u]$ in the space $\widehat{H}^s(\mathbb{R}^n)$, which implies (1.11.5). $\square$

Theorem 1.11.3. *For any $s \in \mathbb{R}$ and any $t \in \mathbb{R}^n$, the shift $v_t(x) = v(x+t)$ of any function $v(x) \in H^s(\mathbb{R}^n)$ belongs to this space and tends there to $v(x)$ as $t \to 0$.*

Indeed, we have

$$\widehat{v_t}(\xi) = e^{i\xi \cdot t}\widehat{v}(\xi). \tag{1.11.6}$$

Furthermore,

$$\|v_t - v\|_{H^s(\mathbb{R}^n)}^2 = \int (1 + |\xi|^2)^s |e^{i\xi \cdot t} - 1|^2 |\widehat{v}(\xi)|^2 \, d\xi.$$

This integral can be split into two integrals, over the large ball $O_R(0)$ and over its complement. The latter integral is small at large R, and the former, at fixed R, tends to 0 as $t \to 0$. □

1.12 A Compactness Theorem

In this section we prove a Kondrashov–Rellich type theorem (cf. [217] and [309]), which will have important consequences for spaces on bounded domains and on compact manifolds.

Theorem 1.12.1. *Suppose that $s < \sigma$, X is a bounded set in $H^\sigma(\mathbb{R}^n)$, and the supports of all elements of X are contained in a fixed compact subset K of $\mathbb{R}^n$. Then X is precompact in $H^s(\mathbb{R}^n)$, i.e., any sequence $\{u_l\}$ of elements of X contains a subsequence converging in $H^s(\mathbb{R}^n)$.*

Proof. Since each function u_l can be approximated in $H^\sigma(\mathbb{R}^n)$ by its mollifications with any accuracy, we can consider, without loss of generality, a sequence of functions $u_l \in \mathcal{D}$. Suppose that the norms of u_l in $H^\sigma(\mathbb{R}^n)$ are bounded and their supports are contained in a fixed compact set K' (see (1.11.1)). We put $v_l = F[u_l]$.

Let us show that the functions u_l are uniformly bounded and equicontinuous in any fixed ball O_R. For this purpose, take a function $\varphi \in \mathcal{D}$ equal to 1 identically in a neighborhood of K'. Since $u_l(x) = u_l(x)\varphi(x)$, it follows that

$$v_l(\xi) = (2\pi)^{-n} \int v_l(\eta)\psi(\xi - \eta) \, d\eta,$$

where $\psi = F[\varphi]$. Therefore, by the Schwarz inequality,

$$|v_l(\xi)|^2 \leq c\|u_l\|_{H^\sigma(\mathbb{R}^n)}^2 \int (1 + |\eta|^2)^{-\sigma} |\psi(\xi - \eta)|^2 d\eta$$

and

$$|v_l(\xi) - v_l(\widetilde{\xi})|^2 \leq c\|u_l\|_{H^\sigma(\mathbb{R}^n)}^2 \int (1 + |\eta|^2)^{-\sigma} |\psi(\xi - \eta) - \psi(\widetilde{\xi} - \eta)|^2 d\eta,$$

where c does not depend on ξ and l. Now we use the rapid convergence of $\psi(\eta)$ to zero as $\eta \to \infty$ and the uniform continuity of this function on any compact set.

The last inequalities obviously imply the uniform boundedness and equicontinuity of the functions v_l in any ball O_R.

Next,

$$\|u_l - u_m\|^2_{H^s(\mathbb{R}^n)} \leq \int_{O_R} (1+|\xi|^2)^s |v_l(\xi) - v_m(\xi)|^2$$

$$+ 2 \int_{\mathbb{R}^n \setminus O_R} (1+|\xi|^2)^{s-\sigma}(1+|\xi|^2)^\sigma [|v_l(\xi)|^2 + |v_m(\xi)|^2] \, d\xi.$$

In the second integrand, we have $(1+|\xi|^2)^{s-\sigma} \leq (1+R^2)^{s-\sigma}$. Therefore, given $\varepsilon > 0$, we can choose R so large that this integral is less than $\varepsilon/2$ for all l and m. Then, applying Arzelà's well-known theorem on a sequence of uniformly bounded equicontinuous functions to the family of functions $v_l(\xi)$ on the ball $\bar{O}_R$, we can choose a sequence of indices l_k so that the first integral with $l = l_k$ and $m = l_{k'}$ is less than $\varepsilon/2$ for sufficiently large k and k'.

It follows easily that the sequence $\{u_l\}$ contains a fundamental subsequence in the space $H^s(\mathbb{R}^n)$. This subsequence converges in $H^s(\mathbb{R}^n)$, since this space is complete. □

1.13 Changes of Coordinates

Consider a transformation of coordinates $x = x(y)$. In more detail, let

$$x_j = x_j(y_1, \ldots, y_n) \quad (j = 1, \ldots, n). \tag{1.13.1}$$

Suppose that these functions are infinitely differentiable and uniformly bounded on $\mathbb{R}^n$ together with each of their derivatives. Suppose also that the absolute value of the corresponding Jacobian is uniformly bounded below by a positive number. Then the inverse transformation has similar properties.

Theorem 1.13.1. *All spaces $H^s(\mathbb{R}^n)$ are invariant with respect to such transformations.*

Proof. For $s \geq 0$, we can simply write expressions for the derivatives of the function $u(x(y))$ and see that its norms of fixed order defined by (1.1.7) or (1.7.1) is dominated by the norms of the same order of the function $u(x)$. For negative s, we use the duality between the spaces $H^s(\mathbb{R}^n)$ and $H^{-s}(\mathbb{R}^n)$. □

We also mention that if functions (1.13.1) have finite smoothness, say, belong to $C_b^m(\mathbb{R}^n)$, then only the spaces $H^s(\mathbb{R}^n)$ with $|s| \leq m$ are invariant.

1.14 Discrete Norms and Discrete Representation of Functions in $H^s(\mathbb{R}^n)$

Many authors use norms of yet another type on the spaces $H^s(\mathbb{R}^n)$ and more general spaces. For example, in [376], such norms are taken for initial norms on these spaces (and on more general spaces).

To describe them, we introduce the following partition of unity on $\mathbb{R}^n_\xi$. Let $\psi_0(\xi)$ be a compactly supported infinitely differentiable function with value 1 at $|\xi| \le 1/2$ and 0 at $|\xi| \ge 1$. We set

$$\psi(\xi) = \psi_0(\xi) - \psi_0(2\xi). \tag{1.14.1}$$

The support of this function is contained in the set $\{\xi: 1/4 \le |\xi| \le 1\}$. Let

$$\psi_j(\xi) = \psi(\xi/2^j) \quad (j = 1, 2, \ldots). \tag{1.14.2}$$

Then

$$\psi_j(\xi) = \psi_0(\xi/2^j) - \psi_0(\xi/2^{j-1}) \quad (j = 1, 2, \ldots). \tag{1.14.3}$$

The support of $\psi_j(\xi)$ is contained in $\{2^{j-2} \le |\xi| \le 2^j\}$. Note also that

$$\sum_0^m \psi_j(\xi) = \psi_0(\xi/2^m) \to 1 \quad (m \to \infty); \tag{1.14.4}$$

moreover, on any bounded domain, the sum on the left-hand side simply equals 1 for sufficiently large m. Therefore,

$$\sum_0^\infty \psi_j(\xi) = 1. \tag{1.14.5}$$

This equality is also valid in the Schwartz space $\mathcal{S}'(\mathbb{R}^n)$. Indeed, given a test function $\chi(\xi)$ from $\mathcal{S}$, we split the integral

$$\int [1 - \psi_0(\xi/2^m)]\chi(\xi)\,d\xi$$

into two integrals, over a ball $O_R(0)$ of large radius R and over the complement to this ball. The latter integral is small for large R, because $|\chi(\xi)|$ is rapidly decreasing at infinity and $|1 - \psi_0|$ is uniformly bounded. The former integral vanishes for sufficiently large m if R is fixed.

We denote the Fourier preimage of each function $\psi_j(\xi)$ by $\varphi_j(x)$. These functions belong to $\mathcal{S}(\mathbb{R}^n_x)$. It follows from (1.14.5) that

$$\sum_0^\infty \varphi_j(x) = \delta(x) \tag{1.14.6}$$

in $\mathcal{S}'(\mathbb{R}^n)$. This, in turn, implies the first assertion in the following proposition.

Proposition 1.14.1. *For $u \in S'(\mathbb{R}^n)$,*

$$u = \sum_0^\infty \varphi_j * u \tag{1.14.7}$$

in the sense of convergence in this space.

For any $s \in \mathbb{R}$ and $u \in H^s(\mathbb{R}^n)$, the same equality holds in the sense of convergence in $H^s(\mathbb{R}^n)$.

The second assertion is again proved by applying the Fourier transform and splitting the integral

$$\int |1 - \psi_0(\xi/2^m)|^2 |(Fu)(\xi)|^2 (1 + |\xi|^2)^s \, d\xi$$

into integrals over a ball of large radius R and over its complement.

For any $s \in \mathbb{R}$, we set

$$\|u\|_{H^s(\mathbb{R}^n),\varphi_0} = \left(\sum_0^\infty 2^{2js} \|\psi_j Fu\|_{L_2(\mathbb{R}^n)}^2 \right)^{1/2}. \tag{1.14.8}$$

An elementary verification shows that this is a norm. We check only the triangle inequality. For the L_2-norms $\|\cdot\|$, we have

$$\|\psi_j F(u+v)\|^2 \le \|\psi_j Fu\|^2 + 2\|\psi_j Fu\|\|\psi_j Fv\| + \|\psi_j Fv\|^2,$$

and the Schwarz inequality implies

$$2\sum 2^{2js} \|\psi_j Fu\|\|\psi_j Fv\| \le 2\left(\sum 2^{2js}\|\psi_j Fu\|^2\right)^{1/2}\left(\sum 2^{2js}\|\psi_j Fv\|^2\right)^{1/2}.$$

Thus,

$$\left(\sum 2^{2js}\|\psi_j(Fu+Fv)\|^2\right)^{1/2} \le \left(\sum 2^{2js}\|\psi_j Fu\|^2\right)^{1/2} + \left(\sum 2^{2js}\|\psi_j Fv\|^2\right)^{1/2}.$$

We shall omit similar verifications in Section 10.

Proposition 1.14.2. *The norm* (1.14.8) *is equivalent to* $\|u\|_{H^s(\mathbb{R}^n)}$.

This assertion obviously follows from the fact that, on the support of each function ψ_j, the fraction $2^{js}/|\xi|^s$ is bounded below and above by positive constants not depending on ξ and j.

The right-hand side of (1.14.8) coincides with

$$\left(\sum_0^\infty 2^{2js} \|\varphi_j * u\|_{L_2(\mathbb{R}^n)}^2 \right)^{1/2} \tag{1.14.9}$$

up to a constant factor.

Of course, similar norms can be defined by using any compactly supported infinitely differentiable function ψ_0 identically equal to 1 in a neighborhood of the origin. Moreover, as shown in [320], there is in fact much more arbitrariness in the choice of ψ_0, and this arbitrariness has important applications. We add that it turns out to be convenient and important that the norms under the summation sign do not depend on s. We shall return to this in Section 10.

1.15 Embedding of the Spaces $H^s(\mathbb{R}^n)$ in $L_p(\mathbb{R}^n)$

Here we prove Theorem 1.2.4 (with the nonstrict inequality replaced by the strict one), using the notation introduced in the preceding subsection.

Let $\chi(\xi)$ be a compactly supported smooth function identically equal to 1 on the support of $\psi(\xi)$. Then the function $\chi_j(\xi) = \chi(\xi/2^j)$ equals 1 on the support of $\psi_j(\xi) = \psi(\xi/2^j)$. Let us denote the Fourier preimages of the functions $\chi(\xi)$ and $\chi_j(\xi)$ by $\omega(x)$ and $\omega_j(x)$, respectively. We have

$$\omega_j * \varphi_j * u = \varphi_j * u \qquad (1.15.1)$$

and, for any compactly supported smooth function u,

$$u = \sum_0^\infty \omega_j * \varphi_j * u \qquad (1.15.2)$$

in the sense of distributions. But all summands are functions from S and, of course, belong to $L_p(\mathbb{R}^n)$.

Let us estimate the L_r-norms of the functions ω_j for $r > 1$. These functions are related to ω by the formula

$$\omega_j(x) = 2^{jn}\omega(2^j x).$$

Hence

$$\|\omega_j\|_{L_r(\mathbb{R}^n)} = \left(\int |\omega_j(x)|^r \, dx \right)^{1/r} = 2^{jn(1-1/r)}\|\omega\|_{L_r(\mathbb{R}^n)}$$

and, therefore,

$$\|\omega_j\|_{L_r(\mathbb{R}^n)} \leq C_1 2^{jn/r'}, \qquad (1.15.3)$$

where

$$\frac{1}{r} + \frac{1}{r'} = 1. \qquad (1.15.4)$$

Now we need the *Young inequality*, which we shall obtain in Section 13.2. *Suppose that $v \in L_r(\mathbb{R}^n)$ and $w \in L_2(\mathbb{R}^n)$ (we interchange p and q and set $q = 2$). Then $v * w \in L_p(\mathbb{R}^n)$ and*

$$\|v * w\|_{L_p(\mathbb{R}^n)} \leq C_2\|v\|_{L_r(\mathbb{R}^n)}\|w\|_{L_2(\mathbb{R}^n)} \qquad (1.15.5)$$

if

$$\frac{1}{p} = \frac{1}{2} - \frac{1}{r'}. \tag{1.15.6}$$

It follows that *if*

$$s = \frac{n}{2} - \frac{n}{p} > 0, \tag{1.15.7}$$

then

$$\|\omega_j * \varphi_j * u\|_{L_p(\mathbb{R}^n)} \le C_3 2^{js} \|\varphi_j * u\|_{L_2(\mathbb{R}^n)}. \tag{1.15.8}$$

But if the left-hand side of (1.15.7) is equal to $s - \varepsilon$ with positive ε, then the factor $2^{-j\varepsilon}$ can be inserted into the right-hand side of the last inequality, and the Schwarz inequality yields (see Proposition 1.14.2)

$$\sum_0^\infty \|\omega_j * \varphi_j * u\|_{L_p(\mathbb{R}^n)} \le C_4 \Big(\sum_0^\infty 2^{2js} \|\varphi_j * u\|_{L_2(\mathbb{R}^n)}^2 \Big)^{1/2} \le C_5 \|u\|_{H^s(\mathbb{R}^n)},$$

which is surely sufficient for the convergence of the series (1.15.2) in $L_p(\mathbb{R}^n)$ and for the norm of its sum in $L_p(\mathbb{R}^n)$ to be dominated by the norm of the function u in $H^s(\mathbb{R}^n)$.

The proof in the case $s = n/2 - n/p$ is a little more complicated. It can be found in [60] or [376]; see also Section 14.4 in our book.

2 The Spaces $H^s(M)$ on a Closed Smooth Manifold M

2.1 Closed Smooth Manifolds

Recall that a closed n-dimensional manifold M of class C^∞ is a "good enough" topological space covered by coordinate neighborhoods O. Each neighborhood O is associated with a chart, that is, a domain U in $\mathbb{R}^n$ together with a homeomorphic mapping $x = \kappa(t)$ of U onto O. Cartesian coordinates $t = (t_1, \dots, t_n)$ on U are called local coordinates on O. If coordinate neighborhoods O_1 and O_2 intersect, then the preimages of their intersection on the charts U_1 and U_2 are open sets, and the map $\kappa_1^{-1}\kappa_2$ between them is a C_b^∞ diffeomorphism. The charts form an atlas of the manifold. The closedness of the manifold M means that M is compact (and has no boundary), that is, any cover of M by open sets (in particular, by coordinate neighborhoods) has a finite subcover. An equivalent condition: any sequence of points in M contains a subsequence converging to some point of M.

A function $f(x)$ on M is said to belong to the space $C^\infty(M)$ (or $C^m(M)$, or $C^{m+\theta}(M)$) if each of the functions $f(\kappa(t))$ on the charts of coordinate neighborhoods is of class C^∞ (respectively, of class C^m or $C^{m+\theta}$). Here and in what follows, $m \in \mathbb{Z}_+$ and $0 < \theta < 1$.

Norms on these spaces are introduced by using a sufficiently fine partition of unity on M, namely, a system of functions $\{\varphi_k\}_1^K$ with the properties

$$\varphi_k \in C^\infty(M), \quad \varphi_k(x) \geq 0, \quad \mathrm{supp}\, \varphi_k \subset O_k, \quad \text{and} \quad \sum_1^K \varphi_k(x) \equiv 1. \qquad (2.1.1)$$

Here O_k are coordinate neighborhoods which form a cover of the manifold M, and the partition of unity is subordinate to this cover. We set

$$\|\varphi\|_{C^s(M)} = \max_k \|\varphi_k \varphi\|_{C_b^s(\mathbb{R}^n)}, \qquad (2.1.2)$$

where $s = m$ or $m + \theta$ and the norms on the right-hand side are calculated in local coordinates. We regard the functions $\varphi_k \varphi$ as being *pulled back* to $\mathbb{R}^n$ (and extended by zero outside their supports). On $C^\infty(M)$ the countable set of norms (2.1.2) with $s = m \in \mathbb{Z}_+$ is used. Various choices of partitions of unity, coordinate neighborhoods, and local coordinates lead to equivalent norms. The spaces $C^m(M)$ and $C^{m+\vartheta}(M)$ are the completions of $C^\infty(M)$ with respect to the corresponding norms.

The space $C^{m,1}(M)$ with nonnegative integer m consists of functions locally (in local coordinates) continuously differentiable up to order m whose mth derivatives satisfy the Lipschitz condition, so that their $(m+1)$th derivatives exist almost everywhere and are bounded.

The space $\mathcal{E}'(M)$ of distributions on M is defined as the space of continuous linear functionals on $\mathcal{E}(M) = C^\infty(M)$.

Now we discuss integration over M. We define a *density* on M as a set of infinitely differentiable positive functions $\rho_U(t)$ on the charts U such that if coordinate neighborhoods O and O' have nonempty intersection, then at the points of the charts U and U' corresponding to points of this intersection we have

$$\rho_U(t)\, dt = \rho_{U'}(t')\, dt'. \qquad (2.1.3)$$

This means that

$$\rho_U[t(t')] = \rho_{U'}(t') \left| \det \frac{\partial t(t')}{\partial t'} \right|^{-1}, \qquad (2.1.4)$$

where $\partial t(t')/\partial t'$ is the Jacobian matrix. For a density we use the notation

$$dx = \{\rho_U(t)\, dt\}.$$

If the manifold is oriented, then we assume the determinants in (2.1.4) to be positive and omit the absolute value sign.

Now we define the *integral* of a function $f(x)$ over M by the formula

$$\int_M f(x)\, dx = \sum_1^K \int (f\varphi_k)[\kappa^{(k)}(t)] \rho_{U_k}(t)\, dt. \qquad (2.1.5)$$

This definition involves the partition of unity consisting of the functions φ_k supported on the coordinate neighborhoods O_k, the local coordinates in O_k, and the density $\{\rho_U(t)\,dt\}$ given on M. However, thanks to (2.1.4), the integral depends only on the choice of the density.

In particular, if the support of f is contained in a coordinate neighborhood O with a chart U and a mapping $\kappa\colon U \to O$, then

$$\int_M f(x)\,dx = \int_U f(\kappa(t))\rho_U(t)\,dt; \qquad (2.1.6)$$

cf. [132, Part II, Chap. I, Sec. 1] and [183].

Now we can identify each continuous function $f(x)$ on M with a distribution f belonging to $\mathcal{E}'(M)$ by the relation

$$\langle f, \varphi \rangle_M = \int_M f(x)\varphi(x)\,dx, \quad \varphi \in \mathcal{E}(M). \qquad (2.1.7)$$

We endow M with the *inner product* defined by

$$(f,g)_M = \int_M f(x)\overline{g(x)}\,dx. \qquad (2.1.8)$$

Next, we define the *space* $L_2(M)$ as the Hilbert space obtained by, e.g., completing the space $C(M)$ of continuous functions with respect to the corresponding norm $\|u\|_{0,M} = (u,u)_M^{1/2}$. We can also consider the spaces $L_p(M)$, $1 \le p < \infty$, with norms

$$\|u\|_{L_p(M)} = \left(\int_M |u|^p\,dx \right)^{1/p}. \qquad (2.1.9)$$

Suppose now that M is endowed with a smooth Riemannian metric, which is determined by a covariant metric tensor $(g_{i,j})$. Then the corresponding density on M is defined by $dx = \{\sqrt{|\det(g_{i,j}(t))|}\,dt\}$, and the volume (area if $n = 2$) of the manifold M is expressed as

$$\int_M dx = \sum_1^K \int \varphi_k[\kappa^{(k)}(t)]\rho_{U_k}(t)\,dt. \qquad (2.1.10)$$

Of particular importance is the case of a smooth closed n-dimensional surface in $\mathbb{R}^N$. Consider a surface defined locally by

$$x_j = x_j(t_1,\dots,t_n) \quad (j = 1,\dots,N). \qquad (2.1.11)$$

Then the metric tensor on M determined by the Euclidean metric on $\mathbb{R}^N$ has the form

$$g_{i,j}(t) = \partial_i x(t) \cdot \partial_j x(t), \qquad (2.1.12)$$

where $\partial_i x(t)$ are the derivatives of functions (2.1.11) with respect to the local co-ordinates t_i; for argumentation, see, e.g., [406, Vol. 2, Chap. 12, Sec. 4]. The matrix (2.1.12) is positive definite, because this is a Gram matrix; thus, its determinant is positive.

For comparison, we recall the following expression for the area of a parametrized piece of a 2-dimensional surface in 3-dimensional space, which is taught in calculus courses:

$$S = \iint \sqrt{EG - F^2}\, dt_1 dt_2. \tag{2.1.13}$$

Here $E = \partial_1 x(t) \cdot \partial_1 x(t)$, $F = \partial_1 x(t) \cdot \partial_2 x(t)$, and $G = \partial_2 x(t) \cdot \partial_2 x(t)$.

Remark. Given a density ρ, each point of M has a neighborhood with local coordinates t such that $\rho_U(t) \equiv 1$.

Indeed, let O be any coordinate neighborhood of the point under consideration with local coordinates $\widetilde{t}_j$ ($j = 1, \ldots, n$), and let $\widetilde{U}$ be the corresponding chart. Decreasing $\widetilde{U}$, we can assume it to be a right circular cylinder with axis parallel to the $\widetilde{t}_n$-axis. We set

$$t_1 = \widetilde{t}_1, \ \ldots, \ t_{n-1} = \widetilde{t}_{n-1}, \ t_n = \int_{\widetilde{t}_{n,0}}^{\widetilde{t}_n} \rho_{\widetilde{U}}(\widetilde{t}_1, \ldots, \widetilde{t}_{n-1}, t)\, dt. \tag{2.1.14}$$

Then $\partial t / \partial \widetilde{t}$ is a lower triangular matrix with main diagonal $(1, \ldots, 1, \rho_{\widetilde{U}}(\widetilde{t}))$, and

$$dt = \left| \det(\partial t / \partial \widetilde{t}) \right| d\widetilde{t} = \rho_{\widetilde{U}}(\widetilde{t})\, d\widetilde{t},$$

so that $\rho_U(t)$ is indeed identically equal to 1.

Covering the manifold by neighborhoods with such local coordinates and choosing a finite subcover, we obtain a *special "small" atlas of charts matched with the given density so that all ρ_U are identically equal to 1.* When these charts are used, the expression (2.1.5) for the integral takes the form

$$\int_M f(x)\, dx = \sum_1^K \int (f\varphi_k)[\kappa^{(k)}(t)]\, dt. \tag{2.1.15}$$

2.2 The Spaces $H^s(M)$

The *Bessel potential space* $H^s(M)$ on M, $s \in \mathbb{R}$, is defined as follows. Let $\{O_k\}_1^K$ be a cover of M by coordinate neighborhoods, and let $\{\psi_k\}_1^K$ be a system of functions with the properties

$$\psi_k \in C^\infty(M), \quad \psi_k(x) \geq 0, \quad \text{supp}\, \psi_k \subset O_k, \quad \text{and} \quad \sum \psi_k(x) > 0 \text{ on } M.$$
$$(2.2.1)$$

For any $s \in \mathbb{R}$ and any $u \in C^\infty(M)$, we set

$$\|u\|_{H^s(M)} = \left[\sum \|u\psi_k\|^2_{H^s(\mathbb{R}^n)} \right]^{1/2}, \tag{2.2.2}$$

where the function $u\psi_k$ is assumed to be extended from the corresponding chart on $\mathbb{R}^n$ by zero outside the support of this function and the norm is calculated in $\mathbb{R}^n$. The space $H^s(M)$ is defined as the completion of $C^\infty(M)$ with respect to this norm.

Obviously, this is a Hilbert space with inner product

$$(u, v)_{s,M} = \sum (u\psi_k, v\psi_k)_{s,\mathbb{R}^n}. \tag{2.2.3}$$

We shall omit the subscript $s = 0$.

All norms obtained for various choices of coordinate neighborhoods, local coordinates, and systems of functions ψ_k are equivalent. In particular, for the system of ψ_k we can take a partition of unity:

$$\sum \psi_k(x) \equiv 1. \tag{2.2.4}$$

In this case, we shall usually write φ_k instead of ψ_k. Another useful choice is a system satisfying the condition

$$\sum \psi_k^2(x) \equiv 1. \tag{2.2.5}$$

In this case, using a small atlas of charts matched with the given density, we obtain

$$(u, v)_M = \int_M u\bar{v}\, dx, \tag{2.2.6}$$

and $H^0(M)$ isometrically coincides with $L_2(M)$.

We also set

$$H^\infty(M) = \bigcap H^s(M) \quad \text{and} \quad H^{-\infty}(M) = \bigcup H^s(M). \tag{2.2.7}$$

Remark. For the standard torus $\mathbb{T}^n = [0, 2\pi]^n$, there is another convenient way of defining norms on the spaces H^s, thanks to the existence of a unique system of 2π-periodic coordinates. Each smooth function $u(x)$ can be expanded in an exponential Fourier series:

$$u(x) = \sum_{\alpha \in \mathbb{Z}^n} c_\alpha e^{i\alpha \cdot x}, \quad c_\alpha = (2\pi)^{-n} \int_{\mathbb{T}^n} u(y) e^{-i\alpha \cdot y} dy. \tag{2.2.8}$$

Such an expansion can also be written formally for distributions u; in this case, the Fourier coefficients c_α are understood as the actions of u on the test functions

$(2\pi)^{-n}e^{-i\alpha \cdot y}$. A norm on $H^s(\mathbb{T}^n)$ ($s \in \mathbb{R}$) can be defined as

$$\left(\sum (1 + |\alpha|^2)^s |c_\alpha|^2\right)^{1/2}. \tag{2.2.9}$$

This norm is equivalent to the usual one for reasons that will be explained in Section 13.8.1.

2.3 Basic Properties of the Spaces $H^s(M)$

Below we list the basic properties of $H^s(M)$. The corresponding assertions are easy to derive from our definitions and from similar assertions for $\mathbb{R}^n$; therefore, we omit or only briefly outline the explanations.

1. If $u \in H^s(M)$ and φ is a function in $C^\infty(M)$ supported in a coordinate neighborhood, then the pullback of φu to $\mathbb{R}^n$ belongs to $H^s(\mathbb{R}^n)$.

A similar statement holds for transfers of functions from charts to the manifold.

2. *For $\sigma > s$, the space $H^\sigma(M)$ is continuously embedded in $H^s(M)$. In particular, if $s \geq 0$, then the elements of $H^s(M)$ are square integrable functions.*

3. Theorem 2.3.1. *For $\sigma > s$, the embedding operator $H^\sigma(M) \subset H^s(M)$ is compact.*

Proof. Let $\{u_l\}$ be a bounded sequence in $H^\sigma(M)$. Then $\{\varphi_k u_l\}$ is a bounded sequence in $H^\sigma(M)$ for any function φ_k from a partition of unity subordinated to the cover of the manifold by coordinate neighborhoods O_k ($k = 1, \ldots, N$). Pulling back $\varphi_k u_l$ to $\mathbb{R}^n$, we obtain a bounded sequence of functions in $H^\sigma(\mathbb{R}^n)$ for each k, and the supports of these functions (usual or generalized) are contained in a fixed compact set. By Theorem 1.12.1, there exists a subsequence which is fundamental in $H^s(\mathbb{R}^n)$. The corresponding functions on M form a fundamental sequence in $H^s(M)$. Passing from $k = 1$ to $k = 2, \ldots, N$, we can choose a subsequence of functions u_{l_m} so rare that the sequence $\varphi_k u_{l_m}$ is fundamental in $H^s(M)$ for each k. Then the sequence $\{u_{l_m}\}$ is fundamental in $H^s(M)$. Since the space $H^s(M)$ is complete, it follows that $\{u_{l_m}\}$ converges in this space. $\square$

4. *The space $C^\infty(M)$ is dense in all $H^s(M)$. The space $C^m(M)$ is dense in $H^s(M)$ for $s \leq m$. The space $H^\sigma(M)$ is dense in $H^s(M)$ for $\sigma > s$.*

5. Theorem 2.3.2. *For $s \geq n/2 + \theta$, where $0 < \theta < 1$, any function in $H^s(M)$ (possibly changed on a set of Lebesgue measure zero) is continuous and locally satisfies the Hölder condition of order ϑ in local coordinates for any positive $\vartheta < \theta$. Moreover, the space $H^s(M)$ is continuously embedded in $C^\vartheta(M)$,*

$$\|u\|_{C^\vartheta(M)} \leq C \|u\|_{H^s(M)}, \tag{2.3.1}$$

and this embedding is compact.

A more complete statement is as follows.

Theorem 2.3.3. *For* $s > n/2 + t$, *where* $t > 0$, *the space* $H^s(M)$ *is continuously and compactly embedded in* $C^t(M)$. *If* $s = n/2 + t$, *this is also true for noninteger* t.

In particular, any function in $H^\infty(M)$, possibly corrected on a set of measure zero, is infinitely differentiable, and, conversely, all infinitely differentiable functions on M belong to $H^\infty(M)$. Thus, the stock of functions in $H^\infty(M)$ coincides with that in $C^\infty(M)$.

Let us explain why the embedding is compact. Let $s > s' > n/2 + t$. By Theorem 2.3.1, the space $H^s(M)$ is compactly embedded in $H^{s'}(M)$, and $H^{s'}(M)$ is continuously embedded in $C^t(M)$ (this follows from a similar result for $\mathbb{R}^n$). The composition of a compact and a continuous mapping is compact.

Theorem 2.3.4. *For* $n/2 > s$ *and*

$$s \geq \frac{n}{2} - \frac{n}{p}, \quad 2 < p < \infty, \tag{2.3.2}$$

the space $H^s(M)$ *is continuously embedded in the space* $L_p(M)$ *with norm* (2.1.9). *If the inequality for* s *in* (2.3.2) *is strict, then the embedding is compact.*

6. The space $H^{-m}(M)$ with positive integer m consists of those distributions in $\mathcal{E}'(M)$ which are locally, in local coordinates, linear combinations of derivatives (in the sense of distributions) of functions in $L_2(M)$ up to order m.

Locality is understood in the sense that if $u \in H^{-m}(M)$ and φ is a function in $C^\infty(M)$ supported in a coordinate neighborhood, then the pullback of φu to $\mathbb{R}^n$ has the specified structure.

In particular, we see that $H^{-\infty}(M)$ coincides with $\mathcal{E}'(M)$.

7. Theorem 2.3.5. *The space of continuous linear functionals on* $H^s(M)$ *can be identified with* (i) $H^s(M)$ *by using the form* $(u, v)_{s,M}$ *and* (ii) $H^{-s}(M)$ *by using the form* $(u, v)_M$ *extended to* $H^s(M) \times H^{-s}(M)$.

Proof. The first assertion is obvious; let us verify the second. It is convenient to use local coordinates matched with the given density on M. Suppose that the system of functions ψ_k satisfies conditions (2.2.1) and (2.2.5). Then, according to (2.2.3), we have

$$(u, v)_M = \sum (\psi_k u, \psi_k v)_{0, \mathbb{R}^n}.$$

The generalized Schwarz inequality on $\mathbb{R}^n$ implies the *generalized Schwarz inequality on* M:

$$|(u, v)_M| \leq C \|u\|_{H^s(M)} \|v\|_{H^{-s}(M)}, \tag{2.3.3}$$

where the constant C does not depend on u and v. By virtue of this inequality, the form $(u, v)_M$ can be extended to a bounded form on $H^s(M) \times H^{-s}(M)$. Each element $v \in H^{-s}(M)$ determines a continuous linear functional $(u, v)_M$ on $H^s(M)$. It remains to show that this is a general form of such a functional.

Let $\{\varphi_k\}$ be a partition of unity subordinate to a cover of the manifold by coordinate neighborhoods O_k, and let $\{\psi_k\}$ be another system of infinitely differentiable functions on M supported in O_k and such that $\psi_k = 1$ in a neighborhood of the support of φ_k. Let $f(u)$ be a continuous linear functional on $H^s(M)$. Then

$$f(u) = \sum f(\varphi_k u) = \sum f(\psi_k \varphi_k u),$$

and the $f(\varphi_k u)$ can be treated here as continuous linear functionals on $H^s(\mathbb{R}^n)$. Hence there exist elements $v_k \in H^{-s}(\mathbb{R}^n)$ for which

$$f(\psi_k \varphi_k u) = (\psi_k \varphi_k u, v_k)_{\mathbb{R}^n} = (\varphi_k u, \psi_k v_k)_{\mathbb{R}^n} = (\varphi_k u, \psi_k v_k)_M = (u, \varphi_k v_k)_M.$$

It follows that

$$f(u) = (u, v)_M,$$

where $v = \sum \varphi_k v_k \in H^{-s}(M)$. $\square$

Remark. It would be easier to prove this theorem if we had isometric isomorphisms $\Lambda^t \colon H^s(M) \to H^{s-t}(M)$. But we do not have them so far, they are constructed in the framework of the theory of elliptic pseudodifferential operators.

8. Theorem 2.3.6. *If m is a nonnegative integer, the operator of multiplication by a function in $C^{m,1}(M)$ is bounded in $H^{m+1}(M)$. If $s = m + \theta$, where $0 < \theta < 1$, then the operator of multiplication by a function from $C^{m+\vartheta}(M)$ with $\theta < \vartheta < 1$ is bounded in $H^s(M)$. In $H^s(M)$ with negative s, the operator of multiplication by a function $a(x)$ is defined as the adjoint of the operator of multiplication by $\overline{a(x)}$ in $H^{|s|}(M)$ with respect to the form $(u,v)_M$, and it is bounded if the latter operator is bounded in $H^{|s|}(M)$.*

9. *Given $\tau < s < \sigma$ and any $\varepsilon > 0$, there exists a C_ε such that*

$$\|u\|_{H^s(M)} \le \varepsilon \|u\|_{H^\sigma(M)} + C_\varepsilon \|u\|_{H^\tau(M)} \tag{2.3.4}$$

for any function in $H^\sigma(M)$.

10. Let $n > 1$, and let Γ be an infinite closed smooth $(n-1)$-dimensional submanifold in M. This means that any point of Γ is contained in a coordinate neighborhood O in M in which Γ is determined by the equation $t_n = 0$ in some local coordinates; moreover, $t' = (t_1, \ldots, t_{n-1})$ are local coordinates in $O' = O \cap \Gamma$.

Theorem 2.3.7. *For $s > 1/2$, the map taking each function in $H^s(M)$ to its trace on Γ is defined, and this is a bounded operator from $H^s(M)$ to $H^{s-1/2}(\Gamma)$.*

The trace operator is defined in the same way as in Section 1.10. First, we verify the inequality

$$\|v\|_{H^{s-1/2}(\Gamma)} \le C\|u\|_{H^s(M)} \tag{2.3.5}$$

for usual traces v on Γ (restrictions to Γ) of infinitely differentiable functions $u(x)$ on M with a constant C not depending on u; this is done by considering a suitable partition of unity on M and applying Theorem 1.10.1. After that, approximating a function $u \in H^s(M)$ by a sequence of functions $u_k \in C^\infty(M)$ converging to u in $H^s(M)$, we see that the traces of u_k on Γ converge in $H^{s-1/2}(\Gamma)$ to a function v not depending on the choice of the u_k; this is the trace of the function u on Γ, and it satisfies inequality (2.3.5).

The trace operator has a bounded right inverse. We shall derive this from Theorem 3.3.1.

2.4 Manifolds of Finite Smoothness

If the coordinate diffeomorphisms are of finite smoothness (say, C_b^m), then the spaces $H^s(M)$ are invariantly defined only for $|s| \leq m$.

3 The Spaces $H^s(\mathbb{R}_+^n)$

3.1 Definitions

By $\mathbb{R}_+^n$ and $\mathbb{R}_-^n$ we denote the half-spaces in $\mathbb{R}^n$ consisting of points $x = (x', x_n)$ with $x_n > 0$ and $x_n < 0$, respectively. Their closures are denoted by $\overline{\mathbb{R}_\pm^n}$.

The spaces $C_b^\infty(\overline{\mathbb{R}_+^n})$, $C_b^m(\overline{\mathbb{R}_+^n})$, $C_b^{m+\theta}(\overline{\mathbb{R}_+^n}) = C_b^{m,\theta}(\overline{\mathbb{R}_+^n})$, and $C_b^{m,1}(\overline{\mathbb{R}_+^n})$, where m is a nonnegative integer and $0 < \theta < 1$, and norms on these spaces are defined in the same way as in Section 1.2 with the only difference that the points x and y are taken in $\overline{\mathbb{R}_+^n}$. The linear manifold $C_0^\infty(\mathbb{R}_+^n)$ is contained in $C_b^\infty(\mathbb{R}_+^n)$ and consists of infinitely differentiable functions compactly supported in $\mathbb{R}_+^n$.

The standard inner product in $L_2(\mathbb{R}_+^n)$ is

$$(u, v)_{\mathbb{R}_+^n} = \int_{\mathbb{R}_+^n} u(x)\overline{v(x)}\,dx, \tag{3.1.1}$$

and the corresponding form

$$\langle u, v \rangle_{\mathbb{R}_+^n} = (u, \bar{v})_{\mathbb{R}_+^n} \tag{3.1.2}$$

is used to define distributions on $\mathbb{R}_+^n$.

Definition 1. Let s be any real number. A function (or a distribution) u belongs to $H^s(\mathbb{R}_+^n)$ if it is the restriction to $\mathbb{R}_+^n$ of a function (for $s < 0$, a distribution) $w \in H^s(\mathbb{R}^n)$. The norm $\|u\|_{H^s(\mathbb{R}_+^n)}$ is defined as the greatest lower bound of the norms $\|w\|_{H^s(\mathbb{R}^n)}$ of those $w \in H^s(\mathbb{R}^n)$ whose restrictions to $\mathbb{R}_+^n$ coincide with u.

For nonnegative s, two other definitions can be given.

Definition 2. Suppose that s is nonnegative. The space $H^s(\mathbb{R}^n_+)$ consists of square integrable functions $u(x)$ whose derivatives $D^\alpha u(x)$ (in the sense of distributions) of order $|\alpha| \le s$ on $\mathbb{R}^n_+$ are square integrable as well, and the norms are defined by

$$\|u\|'^2_{\dot{H}^m(\mathbb{R}^n_+)} = \sum_{|\alpha| \le m} \int_{\mathbb{R}^n_+} |D^\alpha u(x)|^2 \, dx \tag{3.1.3}$$

for integer $s = m$ and by

$$\|u\|'^2_{\dot{H}^{m+\theta}(\mathbb{R}^n_+)} = \|u\|'^2_{\dot{H}^m(\mathbb{R}^n_+)} + \sum_{|\alpha|=m} \int_{\mathbb{R}^n_+} \int_{\mathbb{R}^n_+} \frac{|D^\alpha u(x) - D^\alpha u(y)|^2}{|x-y|^{n+2\theta}} \, dx \, dy \tag{3.1.4}$$

for noninteger $s = m + \theta, 0 < \theta < 1$.

Definition 3. Suppose that s is nonnegative. The space $H^s(\mathbb{R}^n_+)$ is defined as the completion of $C_0^\infty(\overline{\mathbb{R}^n_+})$ with respect to the norm (3.1.3) for integer $s = m$ and the norm (3.1.4) for noninteger $s = m + \theta, 0 < \theta < 1$.

Similar definitions are given for the spaces $H^s(\mathbb{R}^n)$.

Theorem 3.1.1. *Definitions 1–3 are equivalent for $s \ge 0$.*

Proof. The proof of this theorem consists of several parts. We consider in detail only the case of a positive integer s. In the case of fractional s, either the same argument applies or additional technical considerations are needed, which we only outline. Note that there is yet another approach to the definition of spaces with fractional s, which involves tools of interpolation theory; it will be mentioned in Section 13.4.

First, we verify the equivalence of the close, second and third, definitions. After that, we prove the equivalence of these definitions to the first one.

For convenience, we temporarily denote the space $H^s(\mathbb{R}^n_+)$ in the sense of Definition k, k = 1, 2, 3, by $H^s_k(\mathbb{R}^n_+)$. We use such a notation only in this proof; in what follows, the subscripts have a quite different meaning.

1°. $H^s_3(\mathbb{R}^n_+) \subset H^s_2(\mathbb{R}^n_+)$, *and the norms on these spaces coincide.*

Indeed, if u is the limit of a sequence of functions $u_k \in C_0^\infty(\overline{\mathbb{R}^n_+})$ in the norm (3.1.3) or (3.1.4), then an elementary verification shows that, for any α with $|\alpha| \le s$, the derivative $D^\alpha u(x)$ in the sense of distributions on $\mathbb{R}^n_+$ is the limit of the sequence $D^\alpha u_k(x)$ in $L_2(\mathbb{R}^n_+)$ (to be more precise, in $H^{s-|\alpha|}_3(\mathbb{R}^n_+)$). Therefore, $D^\alpha u \in L_2(\mathbb{R}^n)$; we see that $H^s_3(\mathbb{R}^n_+) \subset H^s_2(\mathbb{R}^n_+)$ and that the norms in these two spaces of the elements of the former coincide.

2°. $H^s_2(\mathbb{R}^n_+) \subset H^s_3(\mathbb{R}^n_+)$, *and the norms of these spaces coincide.*

First, we make the following observation. In the space $H^s_2(\mathbb{R}^n_+)$, the elements with compact supports contained in $\overline{\mathbb{R}^n_+}$ are dense. Indeed, the elements of this space can be multiplied by functions from $C_0^\infty(\overline{\mathbb{R}^n_+})$, because the derivatives of products with such functions are evaluated by Leibniz's product rule. Let $\theta(x)$ be an infinitely differentiable nonnegative function on $\overline{\mathbb{R}^n_+}$ equal to 1 for $|x| \leq 1$ and 0 for $|x| \geq 2$ and taking values between 0 and 1. For $R > 0$, we set $\theta_R(x) = \theta(x/R)$. If u belongs to $H^s_2(\mathbb{R}^n_+)$, then so does $\theta_R u$, and, as can be shown by a simple calculation, $\theta_R u$ tends in this space to u as $R \to \infty$.

Now let $u \in H^s_2(\mathbb{R}^n_+)$ be a function compactly supported in $\overline{\mathbb{R}^n_+}$. We construct a sequence of functions in $C_0^\infty(\overline{\mathbb{R}^n_+})$ converging to u in this space, and we do this in several steps. First, we extend u by zero to $x_n < 0$ and set

$$u_h(x) = u(x', x_n + h) \quad (h > 0).$$

It is easy to see that, for the restrictions of these functions to $\mathbb{R}^n_+$, we have

$$D^\alpha u_h(x) = (D^\alpha u)(x', x_n + h);$$

therefore, the functions u_h belong to $H^s_2(\mathbb{R}^n_+)$. Note that these restrictions tend to $u(x)$ in $H^s_2(\mathbb{R}^n_+)$ as $h \to 0$. This follows from the continuity in the mean of functions in $L_2(\mathbb{R}^n_+)$.

Let $\psi(x_n)$ be an infinitely differentiable function equal to 0 at $x_n < -2/3$ and 1 at $x_n > -1/3$. We set

$$v_h(x) = \psi(x_n/h)u_h(x).$$

As is easy to verify, these functions belong to $H^s(\mathbb{R}^n)$, and their restrictions to $\mathbb{R}^n_+$ coincide with those of the functions $u_h(x)$; therefore, $v_h|_{\mathbb{R}^n_+} \to u$ as $h \to 0$ in $H^s_2(\mathbb{R}^n_+)$.

Now we approximate $v_h(x)$ by the mollification (see Section 1.11)

$$w_h(x) = v_h(x) * \omega_{h/2}(x).$$

The restriction of this function to $\overline{\mathbb{R}^n_+}$ is an infinitely differentiable function approaching $u(x)$ as $h \to 0$ in $H^s_2(\mathbb{R}^n_+)$. For the required sequence we can take $w_{1/l}(x)|_{\mathbb{R}^n_+}$.

$3°$. $H^s_1(\mathbb{R}^n_+)$ *is continuously embedded in* $H^s_3(\mathbb{R}^n_+)$.

Let u be the restriction to $\mathbb{R}^n_+$ of a function $w \in H^s(\mathbb{R}^n)$. The latter can be approximated in $H^s(\mathbb{R}^n)$ by compactly supported infinitely differentiable functions w_k in the norm (1.1.7) or (1.7.1) for integer or noninteger s, respectively. Then the restrictions u_k of these functions to $\mathbb{R}^n_+$ form a fundamental sequence in $H^s_1(\mathbb{R}^n_+)$. It converges to u in the norm (3.1.3) or (3.1.4). Therefore, any function u in the space $H^s_1(\mathbb{R}^n_+)$ belongs to $H^s_3(\mathbb{R}^n_+)$. Moreover,

$$\|u\|'_{H^s(\mathbb{R}^n_+)} \leq \|w\|'_{H^s(\mathbb{R}^n)}.$$

Passing to the lower bound over w with $w|_{\mathbb{R}^n_+} = u$ on the right-hand side and using the equivalence of the norms $\|w\|_s$ and $\|w\|'_s$ on $\mathbb{R}^n$, we obtain

$$\|u\|'_{H^s(\mathbb{R}^n_+)} \leq C\|u\|_{H^s(\mathbb{R}^n_+)}.$$

This proves the continuous embedding of $H^s_1(\mathbb{R}^n_+)$ in $H^s_3(\mathbb{R}^n)$.

$4°$. *The space $H^s_3(\mathbb{R}^n_+)$ is continuously embedded in $H^s_1(\mathbb{R}^n_+)$.*

To prove this, we need the following theorem.

Theorem 3.1.2. *For any positive integer N, there exists a bounded linear operator $\mathcal{E}_N$ from $H^s_3(\mathbb{R}^n_+)$ to $H^s(\mathbb{R}^n)$, $0 \le s \le N$, which takes each function $u \in H^s_3(\mathbb{R}^n_+)$ to a function $w \in H^s(\mathbb{R}^n)$ whose restriction to $\mathbb{R}^n_+$ coincides with u.*

This *extension operator* will be specified in the proof. It follows from the existence of this operator that any function u from $H^s_3(\mathbb{R}^n_+)$ belongs to $H^s_1(\mathbb{R}^n_+)$ and

$$\|u\|_{H^s(\mathbb{R}^n_+)} \le C\|u\|'_{H^s(\mathbb{R}^n_+)}$$

with a constant C not depending on u, because the left-hand side of this inequality does not exceed the norm $\|w\|_{H^s(\mathbb{R}^n)}$ of any extension w, in particular, of the extension mentioned in Theorem 3.1.2, whose norm is dominated by $\|w\|'_{H^s(\mathbb{R}^n)}$ and, hence, by $\|u\|'_{H^s(\mathbb{R}^n_+)}$.

This completes the proof of Theorem 3.1.1. □

Proof of Theorem 3.1.2. The simplest proof of this theorem employs the following operator proposed by Hestenes [178] for a smooth extension of a smooth function (see also [56]):

$$w(x',x_n) = (\mathcal{E}_N u)(x',x_n) = \begin{cases} u(x',x_n) & \text{if } x_n > 0, \\ \sum_1^N \lambda_\nu u\left(x',-\frac{1}{\nu}x_n\right) & \text{if } x_n < 0, \end{cases} \tag{3.1.5}$$

where the numbers λ_ν are determined by the system of equations

$$\sum_1^N \left(-\frac{1}{\nu}\right)^j \lambda_\nu = 1 \quad (j = 0,\ldots,N-1). \tag{3.1.6}$$

Let us check that this operator has the required properties. First, consider the case of a nonnegative integer $s = m \le N$.

If $u \in C^\infty_0(\overline{\mathbb{R}^n_+})$, then it is easy to show that the extended function belongs to $C^\infty_0(\overline{\mathbb{R}^n})$ on $\overline{\mathbb{R}^n}$ and has equal one-sided derivatives at $x_n = 0$ up to order $m-1$. It is also easy to show that such a function $w(x)$ belongs to $H^m(\mathbb{R}^n)$ and

$$\|w\|'_{H^m(\mathbb{R}^n)} \le C\|u\|'_{H^m(\mathbb{R}^n_+)} \tag{3.1.7}$$

with a constant C not depending on u.

Next, if $\{u_j\}$ is a sequence of functions in $C^\infty_0(\overline{\mathbb{R}^n_+})$ converging to a given function u in the norm $\|\cdot\|'_{m,\mathbb{R}^n_+}$, then the corresponding sequence $\{w_j\}$ converges to a function $w \in H^m(\mathbb{R}^n)$, and it is easy to verify that (3.1.5) remains valid for these u and w.

Now consider the case where $0 < s < 1$ and $N = 1$. In this case, the function w is even with respect to x_n. We have to estimate the three integrals of

$$\frac{|w(x) - w(y)|^2}{|x - y|^{n+2s}}.$$

over $\mathbb{R}^n_- \times \mathbb{R}^n_-$, $\mathbb{R}^n_- \times \mathbb{R}^n_+$, and $\mathbb{R}^n_+ \times \mathbb{R}^n_-$ in terms of the integral of the same expression over $\mathbb{R}^n_+ \times \mathbb{R}^n_+$. There is no problem with the first integral. The second is reduced to the integral

$$\int\limits_{\mathbb{R}^n_+} \int\limits_{\mathbb{R}^n_+} \frac{|u(x) - u(y)|^2}{[(x_1 - y_1)^2 + \ldots + (x_{n-1} - y_{n-1})^2 + (x_n + y_n)^2]^{n/2+s}} \, dx \, dy$$

by substituting $-x_n$ for x_n. Replacing y_n by $-y_n$ in the denominator, we decrease the denominator and increase the integral. This gives the required estimate for this integral. The third integral is estimated in a similar way.

We restrict ourselves to these cases. The general case of fractional s was considered, e.g., in [350] (in a fairly involved notation, for anisotropic spaces, i.e., for spaces of functions with different smoothness with respect to different variables). $\qquad \square$

Remarks on the extension operator.

1. The numbers $1/\nu$ can be replaced by any pairwise different positive numbers.

2. The sum in (3.1.5) can be multiplied by any function $\psi(x_n)$ in $C_0^\infty(\overline{\mathbb{R}}_-)$ identically equal to 1 near the origin.

3. For fixed N, the extension operator $\mathcal{E}_N$ introduced above can be used for a finite range of s: $0 \le s \le N$. If we wish to consider a larger range of s, we must change the operator (increase N). There exist other extension operators, in particular, an operator constructed by Seeley, which does not depend on $s \ge 0$ [336]; see Section 19 in this book. Moreover, a universal extension operator has been constructed for all s (including negatives), substantially more general spaces, and substantially more general (Lipschitz) domains. As mentioned above, such an operator was proposed by Rychkov [320]; see Section 10 below. In particular, the following theorem based on Definition 1 is valid. Its proof is given in Section 10.

Theorem 3.1.3. *There exists a universal extension operator $\mathcal{E}$ which acts boundedly from $H^s(\mathbb{R}^n_+)$ to $H^s(\mathbb{R}^n)$ for all real s and satisfies the condition $\mathcal{E}u|_{\mathbb{R}^n_+} = u$.*

3.2 Properties of the Spaces $H^s(\mathbb{R}^n_+)$

The properties stated below are obvious from definitions or can easily be derived from the corresponding properties of the spaces $H^s(\mathbb{R}^n)$ by using extensions and restrictions.

1. *The space $H^0(\mathbb{R}^n_+)$ coincides with $L_2(\mathbb{R}^n_+)$. For $\sigma > s$, the space $H^\sigma(\mathbb{R}^n_+)$ is continuously and densely embedded in $H^s(\mathbb{R}^n_+)$.*

2. Theorem 3.2.1. *For $s > n/2 + t$, where $t > 0$, the space $H^s(\mathbb{R}^n_+)$ is continuously embedded in $C^t_b(\mathbb{R}^n_+)$. If $s = n/2 + t$, this is also true for noninteger t.*

Theorem 3.2.2. *For $n/2 > s$ and*

$$s \geq \frac{n}{2} - \frac{n}{p}, \quad 2 < p < \infty, \tag{3.2.1}$$

the space $H^s(\mathbb{R}^n_+)$ is continuously embedded in the space $L_p(\mathbb{R}^n_+)$ with norm

$$\|u\|_{L_p(\mathbb{R}^n_+)} = \left(\int\limits_{\mathbb{R}^n_+} |u(x)|^p \, dx \right)^{1/p}. \tag{3.2.2}$$

3. *For positive integer m, the space $H^{-m}(\mathbb{R}^n_+)$ consists of linear combinations of derivatives of order up to m (in the sense of distributions on $\mathbb{R}^n_+$) of functions in $L_2(\mathbb{R}^n_+)$. For $0 < \theta < 1$, the space $H^{-m+\theta}(\mathbb{R}^n_+)$ consists of linear combinations of derivatives of order up to m (in the sense of distributions on $\mathbb{R}^n_+$) of functions in $H^\theta(\mathbb{R}^n_+)$.*

4. Theorem 3.2.3. *Let m be a nonnegative integer. The operator of multiplication by a function $a(x)$ is bounded in $H^s(\mathbb{R}^n_+)$ if $s = m + 1$ and $a \in C^{m,1}_b(\mathbb{R}^n_+)$ or $s = m + \theta, 0 < \theta < 1$, and $a \in C^{m+\vartheta}_b(\mathbb{R}^n_+)$, where $\vartheta \in (\theta, 1)$. Moreover, in the case of the half-space, an estimate of the form (1.9.1) is valid.*

5. *Given $\sigma > s > 0$, for any $\varepsilon > 0$, there exists a $C_\varepsilon > 0$ such that all functions in $H^\sigma(\mathbb{R}^n_+)$ satisfy the inequality*

$$\|u\|_{s,\mathbb{R}^n_+} \leq \varepsilon \|u\|_{\sigma,\mathbb{R}^n_+} + C_\varepsilon \|u\|_{0,\mathbb{R}^n_+}. \tag{3.2.3}$$

6. Theorem 3.2.4. *For $s > 1/2$, any function $u(x)$ in $H^s(\mathbb{R}^n_+)$ has boundary value $(\gamma^+ u)(x') = u(x', 0)$ on the boundary $\mathbb{R}^{n-1}$ of the half-space $\mathbb{R}^n_+$ belonging to $H^{s-1/2}(\mathbb{R}^{n-1})$; moreover, the map which takes each function u to its boundary value is a bounded operator from $H^s(\mathbb{R}^n_+)$ to $H^{s-1/2}(\mathbb{R}^{n-1})$.*

The boundary value of a function is defined similarly to the trace in the case of $\mathbb{R}^n$. Boundary values are often also called traces.

In the theory of partial differential equations, there are tools for overcoming the barrier $s = 1/2$; see the Neumann variational problem in Section 8.

7. *All $H^s(\mathbb{R}^n_+)$ with $s \geq 0$ are Hilbert spaces. It is easy to write out the inner products corresponding to the norms (3.1.3) and (3.1.4). The space dual to $H^s(\mathbb{R}^n_+)$, $s \geq 0$, can therefore be identified with this space.*

There is also another way for realizing the dual of $H^s(\mathbb{R}^n_+)$, which uses an extension of the inner product in $L_2(\mathbb{R}^n_+)$, but to explain it, we must introduce new spaces (see Section 4).

3.3 Boundary Values

We supplement Theorem 3.2.4 with the following important assertion about a right inverse of the trace operator.

Theorem 3.3.1. *There exists a linear operator not depending on s which acts boundedly from $H^{s-1/2}(\mathbb{R}^{n-1})$ to $H^s(\mathbb{R}^n_+)$ for $s \geq 1/2$ and, for $s > 1/2$, takes each function $v(x')$ in the former space to a function $u(x)$ in the latter whose boundary value $(\gamma^+ u)(x')$ coincides with $v(x')$.*

Similarly, there exists a linear operator not depending on s which acts boundedly from $H^{s-1/2}(\mathbb{R}^{n-1})$ to $H^s(\mathbb{R}^n)$ for $s \geq 1/2$ and, for $s > 1/2$, takes each function $v(x')$ in the former space to a function $u(x)$ in the latter whose trace $(\gamma u)(x')$ at $x_n = 0$ coincides with $v(x')$.

Such an operator is constructed explicitly. We emphasize that it is bounded for $s = 1/2$, which is impossible in Theorem 3.2.4.

Proof. It suffices to prove the second assertion of the theorem. Let us write the boundedness condition:

$$\|u\|_{H^s(\mathbb{R}^n)} \leq C_s \|v\|_{H^{s-1/2}(\mathbb{R}^{n-1})}. \tag{3.3.1}$$

We set

$$\xi = (\xi', \xi_n), \quad \sqrt{1 + |\xi'|^2} = \langle \xi' \rangle. \tag{3.3.2}$$

Let $s = m + \theta$, $0 \leq \theta < 1$. Note that if $u(x)$ is a function in $H^s(\mathbb{R}^n)$, then its norm in this space is equivalent to the square root of the expression

$$\iint |(F'u)(\xi', x_n)|^2 \langle \xi' \rangle^{2s}\, dx_n\, d\xi' + \iint |D_n^m(F'u)(\xi', x_n)|^2 \langle \xi' \rangle^{2\theta}\, dx_n\, d\xi'$$

$$+ \iiint \frac{|(F'D_n^m u)(\xi', x_n) - (F'D_n^m u)(\xi', y_n)|^2}{|x_n - y_n|^{1+2\theta}}\, dx_n\, dy_n\, d\xi', \tag{3.3.3}$$

where F' is the Fourier transform with respect to x' and $D_n = -i\partial/\partial x_n$. The second term is omitted if $m = 0$ and the third, if $\theta = 0$.

Let $f(t)$ be a function in $H^\infty(\mathbb{R})$ equal to 1 at the origin, e.g.,

$$f(t) = \frac{1}{1 + |t|^2}. \tag{3.3.4}$$

We set (following [350])

$$u(x) = F'^{-1}[(F'v)(\xi')f(x_n \langle \xi' \rangle)]. \tag{3.3.5}$$

Clearly if, say, $v(x')$ is a smooth compactly supported function, then $u(x', 0) = v(x')$. We shall verify that, for this function, the value of (3.3.3) is finite and can be estimated in terms of the squared norm of v in $H^{s-1/2}(\mathbb{R}^{n-1})$.

The substitution of (3.3.5) reduces the first term in (3.3.3) to the form

$$\iint |(F'v)(\xi')|^2 \langle\xi'\rangle^{2s} |f(x_n\langle\xi'\rangle)|^2 \, dx_n \, d\xi'.$$

Making the change $x_n\langle\xi'\rangle = t$ in this term, we obtain

$$\int |(F'v)(\xi')|^2 \langle\xi'\rangle^{2s-1} \, d\xi' \int |f(t)|^2 \, dt = C_1 \|v\|^2_{H^{s-1/2}(\mathbb{R}^{n-1})}.$$

Similarly, for the second term in (3.3.3), we have

$$\iint |(F'v)(\xi')|^2 |D_n^m f(x_n\langle\xi'\rangle)|^2 \langle\xi'\rangle^{2\theta} \, dx_n \, d\xi',$$

and this term equals

$$\int |(F'v)(\xi')|^2 \langle\xi'\rangle^{2s-1} \, d\xi' \int |f^{(m)}(t)|^2 \, dt = C_2 \|v\|^2_{H^{s-1/2}(\mathbb{R}^{n-1})}.$$

The third term has the form

$$\iiint |(F'v)(\xi')|^2 \frac{|D_n^m f(x_n\langle\xi'\rangle) - D_n^m f(y_n\langle\xi'\rangle)|^2}{|x_n - y_n|^{1+2\theta}} \, dx_n \, dy_n \, d\xi'.$$

Replacing here $x_n\langle\xi'\rangle$ by t and $y_n\langle\xi'\rangle$ by τ, we see that the third term equals

$$\int |(F'v)(\xi')|^2 \langle\xi'\rangle^{2s-1} \, d\xi' \iint \frac{|f^{(m)}(t) - f^{(m)}(\tau)|^2}{|t-\tau|^{1+2\theta}} \, dt \, d\tau = C_3 \|v\|^2_{H^{s-1/2}(\mathbb{R}^{n-1})}.$$

This proves the desired assertion. $\qquad\qquad\qquad\qquad\qquad\qquad\qquad\qquad\qquad\qquad$ □

More general assertions are as follows. We state them for $\mathbb{R}^n_+$. The first of them follows directly from Theorem 3.2.4.

Theorem 3.3.2. *Let $l+1/2 < s$, where l is a nonnegative integer. Then any function $u(x) \in H^s(\mathbb{R}^n_+)$ has boundary values $D_n^j u(x',0)$ on the hyperplane $x_n = 0$, and they belong to the spaces $H^{s-j-1/2}(\mathbb{R}^{n-1})$ $(j = 0,\dots,l)$. Moreover, the map taking each $u(x)$ to the set*

$$u(x',0),\dots,D_n^l u(x',0) \tag{3.3.6}$$

of these boundary values is a bounded operator from $H^s(\mathbb{R}^n_+)$ to the direct product of the spaces $H^{s-j-1/2}(\mathbb{R}^{n-1})$ $(j = 0,\dots,l)$.

These boundary values are called the *Cauchy data* of order l for the function u on the boundary of the half-space.

Theorem 3.3.3. *For any positive integer l, there exists a linear operator not depending on s which acts boundedly from the direct product of the spaces $H^{s-j-1/2}(\mathbb{R}^{n-1})$ $(j = 0,\dots,l)$ to $H^s(\mathbb{R}^n_+)$ for $s \geq l+1/2$ and is such that, for $s > l+1/2$, the image u of any vector function $(v_0,\dots,v_l)$ has Cauchy data $(v_0,\dots,v_l)$ of order l.*

Proof. It suffices to construct such an operator for the case where only one of the derivatives $D_n^k u(x',0) = v_k(x')$, $0 \le k \le l$, is nonzero and the remaining derivatives vanish. We prove a more general assertion from the book [258]; we use it in its complete form later on, in Section 4.3.

Consider the operators

$$T_j : \mathcal{D}(\mathbb{R}^{n-1}) \to \mathcal{D}(\mathbb{R}^n) \quad (j = 0, 1, \ldots) \tag{3.3.7}$$

defined as follows.

Let $\theta_j(t)$ be a function in $\mathcal{D}(\mathbb{R})$ equal to $t^j/j!$ for $|t| < 1$. We set

$$(T_j \varphi)(x) = F'^{-1}[\langle \xi' \rangle^{-j}(F'\varphi)(\xi')\theta_j(x_n \langle \xi' \rangle)], \tag{3.3.8}$$

where, as above, F' is the Fourier transform with respect to x' ($x' \to \xi'$) and $\langle \xi' \rangle$ is defined in (3.3.2).

Lemma 3.3.4. *The operators T_j have the following properties.*

1°. *If $\varphi \in \mathcal{D}(\mathbb{R}^{n-1})$ and $x' \in \mathbb{R}^{n-1}$, then, for any multi-index $\alpha = (\alpha', \alpha_n)$,*

$$(\partial^\alpha T_j \varphi)(x',0) = \begin{cases} \partial^{\alpha'} \varphi(x') & \text{if } \alpha_n = j, \\ 0 & \text{if } \alpha_n \ne j. \end{cases} \tag{3.3.9}$$

2°. *Each T_j can be extended to a bounded linear operator*

$$T_j : H^{\sigma-j-1/2}(\mathbb{R}^{n-1}) \to H^\sigma(\mathbb{R}^n) \quad \text{for } \sigma \in \mathbb{R}. \tag{3.3.10}$$

Proof. The lemma is proved by an elementary verification. Relations (3.3.9) are obtained by differentiation under the sign of the integral (which determines the inverse Fourier transform). To verify (3.3.10), we calculate the Fourier transform of $T_j \varphi$ (by using the substitution $x_n \langle \xi' \rangle = \tau$):

$$(F(T_j \varphi))(\xi) = \langle \xi' \rangle^{-j-1}(F'\varphi)(\xi')(F_n \theta_j)(\xi_n \langle \xi' \rangle^{-1}),$$

where F_n is the Fourier transform with respect to x_n ($x_n \to \xi_n$). Now, substituting $\xi_n = \tau \langle \xi' \rangle$, we obtain

$$\|T_j \varphi\|^2_{H^\sigma(\mathbb{R}^n)} = C_\sigma \|\varphi\|^2_{H^{\sigma-j-1/2}(\mathbb{R}^{n-1})},$$

where

$$C_\sigma = \int (1+\tau^2)^\sigma |(F_n \theta_j)(\tau)|^2 \, d\tau$$

for any σ.   □

In particular, this proves Theorem 3.3.3.   □

Obviously, all of the above is also true for the spaces on $\mathbb{R}^n_-$.

If a function belongs to $H^s(\mathbb{R}^n)$, $s > 0$, then its Cauchy data of order less than $s - 1/2$ on the two sides of the hyperplane $x_n = 0$ coincide, because these are traces in the sense of Section 1.10.

3.4 Extension by Zero

In the proof of Theorem 3.1.2, a bounded operator from $H^s(\mathbb{R}^n_+)$ to $H^s(\mathbb{R}^n)$ extending functions in the former space to functions in the latter was specified. In the range $s \in [0, 1/2)$, instead of this operator, we can use the following *operator of extension by zero*, which we denote by $\mathcal{E}_0$:

$$\mathcal{E}_0 u(x) = \begin{cases} u(x) & \text{if } x_n > 0, \\ 0 & \text{if } x_n < 0. \end{cases} \tag{3.4.1}$$

This is proved in the following theorem, which also covers the interval $(1/2, 1)$ of values of s, in order to prepare further generalizations.

Theorem 3.4.1. 1. *The operator $\mathcal{E}_0$ acts boundedly from $H^s(\mathbb{R}^n_+)$ to $H^s(\mathbb{R}^n)$ for $0 \le s < 1/2$.*

2. *For $1/2 < s < 1$, this operator acts boundedly from the subspace in $H^s(\mathbb{R}^n_+)$ consisting of functions with $u(x', 0) = 0$ to $H^s(\mathbb{R}^n)$.*

Proof. For $s = 0$, the assertion is obvious; thus, assume that $s > 0$. For $v = \mathcal{E}_0 u$, we have

$$\iint \frac{|v(x) - v(y)|^2}{|x-y|^{n+2s}} \, dx \, dy$$

$$= \int\limits_{\mathbb{R}^n_+} \int\limits_{\mathbb{R}^n_+} \frac{|u(x) - u(y)|^2}{|x-y|^{n+2s}} \, dx \, dy + 2 \int\limits_{\mathbb{R}^n_+} |u(x)|^2 \int\limits_{\mathbb{R}^n_-} \frac{1}{|x-y|^{n+2s}} \, dy \, dx. \tag{3.4.2}$$

The inner integral in the second summand on the right-hand side equals

$$\int_0^\infty \int_{\mathbb{R}^{n-1}} \frac{1}{(|y' - x'|^2 + |y_n - x_n|^2)^{n/2+s}} \, dy' \, dy_n,$$

where $x' = (x_1, \ldots, x_{n-1})$. The change $y' - x'$ for y' shows that it does not depend on x', and we can set $x' = 0$. Now the change $y = x_n z$ shows that this integral equals

$$\frac{C}{x_n^{2s}},$$

where C is a constant. Thus, the second summand on the right-hand side in (3.4.2) equals, up to a constant multiplier,

$$\int\limits_{\mathbb{R}^{n-1}} \int\limits_0^\infty \frac{|u(x', x_n)|^2}{x_n^{2s}} \, dx_n \, dx'. \tag{3.4.3}$$

First, we obtain the desired result in the one-dimensional case.

Theorem 3.4.2. *Let $u(t)$ be a smooth compactly supported function on the half-line $\overline{\mathbb{R}}_+$. Then the following assertions hold.*
   1. *For $0 < s < 1/2$,*

$$\int\limits_0^\infty \frac{|u(t)|^2}{t^{2s}} \, dt \leq C \left[\int\limits_0^\infty |u(t)|^2 dt + \int\limits_0^\infty \int\limits_0^\infty \frac{|u(t) - u(\tau)|^2}{|t - \tau|^{1+2s}} \, dt \, d\tau \right] \tag{3.4.4}$$

with a constant not depending on u.
   2. *For $1/2 < s < 1$, the same inequality holds under the assumption $u(0) = 0$.*

Proof. 1. Let us prove the first assertion of the theorem. This is not very easy.
   Note that

$$t^{-s} u(t) = t^{-1-s} \int\limits_t^{2t} [u(t) - u(\tau)] \, d\tau + t^{-1-s} \int\limits_t^{2t} u(\tau) \, d\tau. \tag{3.4.5}$$

In this proof, by $\|f\|_{I_h}$ we mean the L_2-norm of the function f on the interval $I_h = (0, h)$. Our purpose is to derive an estimate for $\|t^{-s} u(t)\|_{I_{1/2}}$ (see (3.4.10) below); having obtained it, we shall easily complete the proof.
   By virtue of (3.4.5), we have

$$\|t^{-s} u(t)\|_{I_{1/2}} \leq \left\| t^{-1-s} \int\limits_t^{2t} [u(t) - u(\tau)] \, d\tau \right\|_{I_{1/2}} + \left\| t^{-1-s} \int\limits_t^{2t} u(\tau) \, d\tau \right\|_{I_{1/2}}. \tag{3.4.6}$$

In the first term on the right-hand side of (3.4.6), we estimate the integral by using the Schwarz inequality:

$$\left| \int\limits_t^{2t} [u(t) - u(\tau)] \, d\tau \right| \leq t^{1/2} \left(\int\limits_t^{2t} |u(t) - u(\tau)|^2 \, d\tau \right)^{1/2}.$$

Since here $0 \leq \tau - t \leq t$, it follows that

$$\left\| t^{-1-s} \int\limits_t^{2t} [u(t) - u(\tau)] \, d\tau \right\|_{I_{1/2}} \leq \left(\int\limits_0^\infty \int\limits_0^\infty \frac{|u(t) - u(\tau)|^2}{|t - \tau|^{1+2s}} \, dt \, d\tau \right)^{1/2} \tag{3.4.7}$$

(on the right-hand side, we have replaced two integrals by integrals over $(0, \infty)$).

Consider the second term on the right-hand side in (3.4.6). We transform and estimate it as follows. For fixed t, we replace τ by $t\sigma$ in the integral; this reduces the term under consideration to the form

$$\left\| t^{-s} \int\limits_1^2 u(t\sigma)d\sigma \right\|_{I_{1/2}},$$

where we use the norm with respect to t. The norm of the integral does not exceed the integral of the norm of the integrand (this is a consequence of the triangle inequality applied to the L_2-norm of a Riemann integral sum); therefore, the last expression does not exceed

$$\int\limits_1^2 \|t^{-s}u(t\sigma)\|_{I_{1/2}}d\sigma.$$

(This quantity is finite, because $s < 1/2$.) For fixed σ, we transform the norm under the integral sign, replacing $t\sigma$ by θ:

$$\left(\int\limits_0^{1/2} t^{-2s}|u(t\sigma)|^2 dt \right)^{1/2} = \sigma^{s-1/2} \left(\int\limits_0^{\sigma/2} \theta^{-2s}|u(\theta)|^2 d\theta \right)^{1/2}.$$

Now we see that the second term on the right-hand side of (3.4.6) is estimated as

$$\left\| t^{-1-s} \int\limits_t^{2t} u(\tau)d\tau \right\|_{I_{1/2}} \leq \int\limits_1^2 \sigma^{s-1/2}d\sigma \|t^{-s}u(t)\|_{I_1}$$

$$\leq \gamma(s)\|t^{-s}u(t)\|_{I_{1/2}} + \gamma(s)2^s\|u(t)\|_{\mathbb{R}_+}, \qquad (3.4.8)$$

where $\|\cdot\|_{\mathbb{R}_+}$ is the L_2-norm over the half-axis and (this is a key inequality in the proof)

$$\gamma(s) = \int\limits_1^2 \sigma^{s-1/2}d\sigma < \int\limits_1^2 d\sigma = 1. \qquad (3.4.9)$$

Relations (3.4.6)–(3.4.9) imply

$$\|t^{-s}u(t)\|_{I_{1/2}} \leq (1-\gamma(s))^{-1} \left(\int\limits_0^\infty \int\limits_0^\infty \frac{|u(t)-u(\tau)|^2}{|t-\tau|^{1+2s}} dt\, d\tau \right)^{1/2} + C_1\|u\|_{\mathbb{R}_+}. \qquad (3.4.10)$$

It remains to take into account the inequality

$$\|t^{-s}u(t)\|_{\mathbb{R}_+} \leq \|t^{-s}u(t)\|_{I_{1/2}} + 2^{-s}\|u(t)\|_{\mathbb{R}_+}.$$

2. Now let us verify the second assertion of the theorem. Instead of (3.4.5) we write

$$t^{-s}u(t) = t^{-1-s} \int_0^t [u(t) - u(\tau)] \, d\tau + t^{-1-s} \int_0^t u(\tau) \, d\tau. \qquad (3.4.11)$$

Both summands on the right-hand side are square integrable, because $u(0) = 0$. For the first summand, similarly to (3.4.7), we obtain

$$\left\| t^{-1-s} \int_0^t [u(t) - u(\tau)] \, d\tau \right\|_{I_1} \leq \left(\int_0^\infty \int_0^\infty \frac{|u(t) - u(\tau)|^2}{|t - \tau|^{1+2s}} \, dt \, d\tau \right)^{1/2}. \qquad (3.4.12)$$

For the second term on the right-hand side of (3.4.11), similarly to (3.4.8), we obtain

$$\left\| t^{-1-s} \int_0^t u(\tau) \, d\tau \right\|_{I_1} \leq \delta(s) \| t^{-s} u(t) \|_{I_1}, \qquad (3.4.13)$$

where

$$\delta(s) = \int_0^1 \sigma^{s-1/2} \, d\sigma < 1. \qquad (3.4.14)$$

As a result, we arrive at the inequality

$$\| t^{-s} u(t) \|_{I_1} \leq (1 - \delta(s))^{-1} \left(\int_0^\infty \int_0^\infty \frac{|u(t) - u(\tau)|^2}{|t - \tau|^{1+2s}} \, dt \, d\tau \right)^{1/2}, \qquad (3.4.15)$$

which brings us to our goal. $\square$

Corollary 3.4.3. *For $0 < s < 1/2$, integral (3.4.3) satisfies the inequality*

$$\int_{\mathbb{R}^{n-1}} \int_0^\infty \frac{|u(x', x_n)|^2}{x_n^{2s}} \, dx_n \, dx'$$

$$\leq C \left[\int_{\mathbb{R}_+^n} |u(x)|^2 \, dx + \int_{\mathbb{R}^{n-1}} dx' \int_0^\infty \int_0^\infty \frac{|u(x', t) - u(x', \tau)|^2}{|t - \tau|^{1+2s}} \, dt \, d\tau \right]. \qquad (3.4.16)$$

The same inequality with $1/2 < s < 1$ holds for functions $u(x)$ with $u(x', 0) = 0$.

Indeed, this inequality is obtained from our result in the one-dimensional case by integration over x'. In turn, this corollary implies the assertion of Theorem 3.4.1, because

$$\int\limits_{\mathbb{R}^n_+} |u|^2\, dx + \int\limits_{\mathbb{R}^{n-1}} dx' \int\limits_0^\infty \int\limits_0^\infty \frac{|u(x',t) - u(x',\tau)|^2}{|t - \tau|^{1+2s}}\, dt\, d\tau \le \|u\|'^2_{H^s(\mathbb{R}^n_+)}.$$

Here, on the left-hand side, we have the squared anisotropic norm of order $(0,s)$ with respect to (x', x_n). This can be shown by applying the Fourier transform with respect to x'. $\qquad\square$

Now we prove a theorem which gives a complete description of those functions in $H^s(\mathbb{R}^n_+)$ with non-half-integer $s > 0$ which are extended by zero to functions in $H^s(\mathbb{R}^n)$.

Theorem 3.4.4. *Suppose that s is a positive non-half-integer and $u(x)$ is a function in $H^s(\mathbb{R}^n_+)$. Let*

$$v(x) = \begin{cases} u(x) & \text{if } x_n > 0, \\ 0 & \text{if } x_n < 0. \end{cases} \tag{3.4.17}$$

If $0 < s < 1/2$, then this function belongs to $H^s(\mathbb{R}^n)$. If $s > 1/2$, then it belongs to $H^s(\mathbb{R}^n)$ if and only if the Cauchy data of order $[s - 1/2]$ for the function $u(x)$ at $x_n = 0$ vanish. In these cases,

$$\|u\|_{H^s(\mathbb{R}^n_+)} \le \|v\|_{H^s(\mathbb{R}^n)} \le C\|u\|_{H^s(\mathbb{R}^n_+)}. \tag{3.4.18}$$

Proof. The necessity of the conditions follows from the fact that the Cauchy data for the extended function must be the same on both sides of $x_n = 0$. Let us verify sufficiency. Integration by parts shows that the derivatives of the extended function of order less than s in the sense of distributions coincide with the extensions by zero of the corresponding derivatives of the initial function. The proof reduces to estimating the norms of the higher-order derivatives of the extended function in terms of the norms of the derivatives of the same order of the initial function, which is done by using Theorem 3.4.1.

We have already obtained the second inequality in (3.4.18), and the first one follows directly from the definition of the norm on the half-space. $\qquad\square$

If s is a positive half-integer, then a function $u \in H^s(\mathbb{R}^n_+)$ extended by zero to $\mathbb{R}^n_-$ belongs to $H^s(\mathbb{R}^n)$ if and only if

$$\int\limits_{\mathbb{R}^n_+} \frac{|\partial^\alpha u(x)|^2}{|x_n|}\, dx < \infty \qquad (|\alpha| = s - 1/2). \tag{3.4.19}$$

Cf. [168] and [237]. Examples show that condition (3.4.19) is not implied by the finiteness of the norm $\|u\|_{H^s(\mathbb{R}^n_+)}$.

The considerations of this subsection will be substantially supplemented in Section 4.

3.5 Gluing Together Functions in $H^s(\mathbb{R}^n_+)$ and $H^s(\mathbb{R}^n_-)$

Theorem 3.5.1. *Suppose that s is a positive non-half-integer number and functions $u_{\pm}(x)$ belong to $H^s(\mathbb{R}^n_{\pm})$, respectively. Let*

$$u(x) = \begin{cases} u_+(x) & \text{if } x_n > 0, \\ u_-(x) & \text{if } x_n < 0. \end{cases} \tag{3.5.1}$$

If $0 < s < 1/2$, then this function belongs to $H^s(\mathbb{R}^n)$. If $s > 1/2$, then it belongs to $H^s(\mathbb{R}^n)$ if and only if the Cauchy data of order $[s-1/2]$ for the functions $u_{\pm}(x)$ at $x_n = 0$ coincide.

Proof. The first assertion is obvious: we can extend the given functions by zero and sum the extensions. Suppose that $s > 1/2$. The necessity of the conditions in the theorem obviously follows from the fact that any function in $H^s(\mathbb{R}^n)$ must have the same Cauchy data on both sides of $x_n = 0$. Sufficiency is verified as follows. From $u(x)$ we subtract a function in $H^s(\mathbb{R}^n)$ with given Cauchy data, thereby obtaining functions on the half-spaces $\mathbb{R}^n_{\pm}$ which can be extended by zero over $\mathbb{R}^n_{\mp}$ to functions in $H^s(\mathbb{R}^n)$ and summed.   □

3.6 Decomposition of the Space $H^s(\mathbb{R}^n)$ with $|s|<1/2$ into the Sum of Two Subspaces

We return to Theorem 3.4.1. It implies the following assertion.

Corollary 3.6.1. *For $0 \le s < 1/2$, the operator of multiplication by the characteristic function θ_+ of the half-space $\mathbb{R}^n_+$ is bounded in $H^s(\mathbb{R}^n)$.*

Indeed, for $s = 0$, this is obvious. Suppose that $0 < s < 1/2$ and $w \in H^s(\mathbb{R}^n)$. Then the restriction u of w to $\mathbb{R}^n_+$ belongs to $H^s(\mathbb{R}^n_+)$, and what was proved above implies

$$\|\mathcal{E}_0 u\|_{H^s(\mathbb{R}^n)} \le C\|u\|_{H^s(\mathbb{R}^n_+)}.$$

It remains to note that the norm on the right-hand side does not exceed $\|w\|_{H^s(\mathbb{R}^n)}$ and the function $\mathcal{E}_0 u$ coincides with $\theta_+ w$.   □

Of course, the situation for $\mathbb{R}^n_-$ is similar.

Corollary 3.6.2. *For $0 \le s < 1/2$, any function w in $H^s(\mathbb{R}^n)$ can be represented as the sum of two functions $w_{\pm}$ from this space supported on $\overline{\mathbb{R}^n_{\pm}}$, respectively. Moreover, the operators of passing from w to $w_{\pm}$ are bounded.*

Indeed, it suffices to set

$$w_{\pm} = \theta_{\pm} w, \tag{3.6.1}$$

where θ_- is the characteristic function of the half-space $\mathbb{R}^n_-$.

Now we need Theorem 1.11.3. Using it, we derive the following assertion from Corollary 3.6.2.

Corollary 3.6.3. *For $0 \le s < 1/2$, each function $w \in H^s(\mathbb{R}^n)$ can be approximated in this space with any accuracy by the sum of two functions $\widetilde{w}_\pm$ in the same space supported strictly inside $\mathbb{R}^n_\pm$ and, therefore, by the sum of two functions $\varphi_\pm$ in $C_0^\infty(\mathbb{R}^n)$ supported strictly inside these half-spaces.*

Corollary 3.6.4. *The operator of multiplication by θ_+ can be extended to a bounded operator in $H^s(\mathbb{R}^n)$ for $-1/2 < s < 0$.*

Indeed, this space is dual to $H^{|s|}(\mathbb{R}^n)$ with respect to the extension of the form $(u,v)_{0,\mathbb{R}^n}$. For functions φ and ψ in $C_0^\infty(\mathbb{R}^n)$, we have

$$(\theta_+\varphi, \psi)_{\mathbb{R}^n} = (\varphi, \theta_+\psi)_{\mathbb{R}^n}.$$

Approximating a given function v in $H^{|s|}(\mathbb{R}^n)$ by such functions ψ and a given (generalized) function u in $H^s(\mathbb{R}^n)$ by functions φ, we can pass to the limit on the right-hand side. Therefore, the left-hand side has a limit as well, which determines the operator in question:

$$(\theta_+u, v)_{\mathbb{R}^n} = (u, \theta_+v)_{\mathbb{R}^n}. \tag{3.6.2}$$

It is easy to see that this operator is well defined, that is, does not depend on the choice of the approximating sequences. It remains to apply Corollary 3.6.1.

Thus, the operators of multiplication by the characteristic functions $\theta_\pm$ of the half-spaces $\mathbb{R}^n_\pm$ are bounded in $H^s(\mathbb{R}^n)$ for $|s| < 1/2$. We denote them by $P_\pm$. The sum of these operators is equal to the identity operator I. These are mutually complementary projections of $H^s(\mathbb{R}^n)$ onto the subspaces consisting of elements supported in $\overline{\mathbb{R}^n_\pm}$. We denote these subspaces by $\widetilde{H}^s(\mathbb{R}^n_\pm)$. More general spaces $\widetilde{H}^s(\mathbb{R}^n_+)$ will be considered in Section 4.

For $s > 1/2$, the operator of multiplication by θ_+ is no longer bounded in $H^s(\mathbb{R}^n)$, essentially because functions in this space may have nonzero traces at $x_n = 0$. In fact, boundedness is lost already at $s = 1/2$. An example can be found in Eskin's book [149].

The above statements have yet another instructive proof given in the same book [149], which is based on the passage to Fourier transforms. It can be shown that if φ is a function in $C_0^\infty(\mathbb{R}^n)$, then

$$F(P_+\varphi)(\xi) = F[\theta_+\varphi](\xi) = \frac{1}{2\pi i} \int \frac{\psi(\xi', \eta_n)}{\xi_n - i0 - \eta_n} d\eta_n$$

$$= \frac{1}{2\pi i} \lim_{\tau \to -0} \int \frac{\psi(\xi', \eta_n)}{\xi_n + i\tau - \eta_n} d\eta_n, \tag{3.6.3}$$

where $\psi = F\varphi$. We denote the expression on the right-hand side by $\Pi_+\psi$. The operator Π_+ can also be written in the form

$$\Pi_+\psi(\xi) = \frac{1}{2}\psi(\xi) + \frac{1}{2\pi i} \text{p.v.} \int \frac{\psi(\xi', \eta_n)}{\xi_n - \eta_n} d\eta_n. \tag{3.6.4}$$

Here we have a singular integral operator (the denominator has a critical singularity); the letters p.v. (principal value) mean that this operator should be understood in the sense of Cauchy principal value, that is, as the limit of the integral over those η_n for which $|\xi_n - \eta_n| > \varepsilon$ as $\varepsilon \to 0$ (see Section 18.4).

A similar operator $\Pi_- = I - \Pi_+$ for $\mathbb{R}^n_-$ has the same expression with the opposite sign in front of the second term:

$$\Pi_-\psi(\xi) = \frac{1}{2}\psi(\xi) - \frac{1}{2\pi i} \, \text{p. v.} \int \frac{\psi(\xi', \eta_n)}{\xi_n - \eta_n} \, d\eta_n. \qquad (3.6.5)$$

It can be proved that Π_+ is a bounded operator in the space $\widehat{H}^s(\mathbb{R}^n)$ of Fourier images of functions in $H^s(\mathbb{R}^n)$ for $|s| < 1/2$. This assertion is also true for the operator $\Pi_- = I - \Pi_+$. In the same book, an example is given which shows that the assertion is false for $s = 1/2$: the operator Π_+ is not bounded in $H^{1/2}(\mathbb{R}^n)$; its domain is only dense in this space.

4 The Spaces $\widetilde{H}^s(\mathbb{R}^n_+)$ and $\mathring{H}^s(\mathbb{R}^n_+)$

4.1 The Spaces $\widetilde{H}^s(\mathbb{R}^n_+)$

For each real s, the *space* $\widetilde{H}^s(\mathbb{R}^n_+)$ is defined as the subspace of $H^s(\mathbb{R}^n)$ consisting of functions supported in $\overline{\mathbb{R}^n_+}$. The norm on $\widetilde{H}^s(\mathbb{R}^n_+)$ is inherited from $H^s(\mathbb{R}^n)$. Obviously, this subspace satisfies the closedness condition. Indeed, suppose that a sequence of functions u_j in $\widetilde{H}^s(\mathbb{R}^n_+)$ converges to u in this space and φ is a function in $C_0^\infty(\mathbb{R}^n)$ supported on $\mathbb{R}^n_-$. Then $\langle u_j, \varphi \rangle_{\mathbb{R}^n} = 0$, and the generalized Schwarz inequality (we consider φ as a function in $H^{-s}(\mathbb{R}^n)$) implies $\langle u, \varphi \rangle_{\mathbb{R}^n} = 0$.

Using the density of the linear manifold $C_0^\infty(\mathbb{R}^n)$ in $H^s(\mathbb{R}^n)$ and the shift operation (see Section 1.11), we can easily verify that *the set of infinitely differentiable functions compactly supported inside* $\mathbb{R}^n_+$ *is dense in* $\widetilde{H}^s(\mathbb{R}^n_+)$. Therefore, $\widetilde{H}^s(\mathbb{R}^n_+)$ can also be defined as the completion of $C_0^\infty(\mathbb{R}^n_+)$ in $H^s(\mathbb{R}^n)$ (we assume that all functions in $C_0^\infty(\mathbb{R}^n_+)$ are extended by zero to $\overline{\mathbb{R}^n_+}$).

For $s \geq 0$, the elements of the space $\widetilde{H}^s(\mathbb{R}^n_+)$ can also be treated as functions in $H^s(\mathbb{R}^n_+)$ whose extensions by zero to $\mathbb{R}^n_-$ belong to $H^s(\mathbb{R}^n)$, with norm taken from $H^s(\mathbb{R}^n)$. This motivates the notation.

Thus, for $s \geq 0$, we have in fact three equivalent definitions of the space $\widetilde{H}^s(\mathbb{R}^n_+)$.

Of course, the extensions by zero of the functions considered in Theorems 3.4.1 and 3.4.4 are contained in $\widetilde{H}^s(\mathbb{R}^n_+)$. Moreover, the operator of extension by zero is bounded for $|s| < 1/2$. This is seen from, e.g., Corollaries 3.6.1 and 3.6.4: extending a function by zero is equivalent to arbitrarily extending this function to a function in $H^s(\mathbb{R}^n)$ and multiplying the result by θ_+. In the reverse direction, the restriction operator is bounded. Thus, we have proved the following theorem.

Theorem 4.1.1. *The spaces $H^s(\mathbb{R}^n_+)$ and $\widetilde{H}^s(\mathbb{R}^n_+)$ with $|s| < 1/2$ can be identified; the corresponding norms are equivalent.*

For $\mathbb{R}^n_-$, all can be repeated with obvious alterations.

The spaces $H^s(\mathbb{R}^n_\pm)$ are in fact quotient spaces:

$$H^s(\mathbb{R}^n_\pm) = H^s(\mathbb{R}^n)/\widetilde{H}^s(\mathbb{R}^n_\mp). \qquad (4.1.1)$$

Consider the operator

$$\Lambda^t_+ = F^{-1}[i\xi_n + \langle \xi' \rangle]^t F. \qquad (4.1.2)$$

Here a real power z^h of a nonzero complex number z with principal value $\arg z \in (-\pi, \pi]$ is defined as

$$z^h = |z|^h e^{ih \arg z}.$$

The following theorem describes the image of the space $\widetilde{H}^s(\mathbb{R}^n_+)$ under the Fourier transform. We give only brief comments; details can be found in Eskin's book [149].

Theorem 4.1.2. *Let $u(x)$ be a function (or a distribution) in $\widetilde{H}^s(\mathbb{R}^n_+)$, and let $v(\xi)$ be its Fourier image. Then, for almost all ξ', $v(\xi', \xi_n)$ admits an analytic continuation to a regular function $v(\xi', \zeta_n)$ with respect to the last variable $\zeta_n = \xi_n + i\tau$ in the lower half-plane $(\tau < 0)$, which is a continuous function of τ $(\tau \le 0)$ with values in $\widehat{H}^s(\mathbb{R}^n)$. Moreover,*

$$\int |v(\xi', \xi_n + i\tau)|^2 (1 + |\xi|^2 + |\tau|^2)^s \, d\xi \le C_u, \quad \tau \le 0, \qquad (4.1.3)$$

where the constant does not depend on τ.

Conversely, any function v with these properties is the Fourier image of a function from $\widetilde{H}^s(\mathbb{R}^n_+)$.

The upper bound of the left-hand side of (4.1.3) can be taken for a squared norm on $\widetilde{H}^s(\mathbb{R}^n_+)$.

The operator Λ^t_+ defines a continuous bijection between the spaces $\widetilde{H}^s(\mathbb{R}^n_+)$ and $\widetilde{H}^{s-t}(\mathbb{R}^n_+)$ for any real s and t. The inverse map is the operator Λ^{-t}_+.

For $s = 0$, the first paragraph of the statement in the case $n = 1$ is one of the versions of the Paley–Wiener theorem; cf. [18] or [335]. We explain this paragraph as follows. If $u(x)$ is a function in $L_2(\mathbb{R})$ vanishing at $x < 0$ and $v(\xi)$ is its Fourier image, then, for $\tau < 0$, the function

$$F[u(x)e^{x\tau}] = \int_0^\infty e^{-ix(\xi+i\tau)} u(x) \, dx = v(\xi + i\tau) \qquad (4.1.4)$$

can be differentiated with respect to the complex variable $\zeta = \xi + i\tau$, and, by virtue of Parseval's identity, its L_2-norm with respect to ξ is bounded uniformly with respect to τ:

$$\|v(\xi + i\tau)\| = c\|u(x)e^{\tau x}\| \le \text{Const}.$$

It is also obvious that

$$\|v(\xi + i\tau) - v(\xi)\| = c\|u(x)[1 - e^{\tau x}]\| \to 0 \quad \text{as } \tau \uparrow 0.$$

Conversely, if $v(x)$ has the properties specified above, then the norm of its Fourier preimage satisfies the condition

$$\|u(x)e^{x\tau}\| \le \text{Const}$$

for $\tau < 0$, using which we can prove by contradiction that $u(x) = 0$ at almost every $x < 0$.

Extending the result to $n > 1$ involves no difficulties.

The operator Λ_+^t is similar to Λ^t (see Section 1.4); clearly, this is a continuous bijection between $H^s(\mathbb{R}^n)$ and $H^{s-t}(\mathbb{R}^n)$. Moreover, for the Fourier images Fu of functions u from $\widetilde{H}^s(\mathbb{R}^n_+)$, it preserves analyticity with respect to $\zeta_n = \xi_n + i\tau$ in the lower half-plane, which is essentially caused by the vanishing of the Fourier preimage $u(x)$ at $x_n < 0$. This operator also preserves estimates of the form (4.1.3) in view of the obvious equivalence

$$|i\zeta_n + \langle \xi' \rangle| \sim 1 + |\xi'| + |\xi_n| + |\tau|.$$

Yet another theorem about the spaces $\widetilde{H}^{-s}(\mathbb{R}^n_+)$ (of negative order) is given in Section 4.3.

4.2 Duality between the Spaces $H^s(\mathbb{R}^n_+)$ and $\widetilde{H}^{-s}(\mathbb{R}^n_+)$

Theorem 4.2.1. *The spaces $H^s(\mathbb{R}^n_+)$ and $\widetilde{H}^{-s}(\mathbb{R}^n_+)$ with $-\infty < s < \infty$ are dual with respect to the extension of the inner product in $L_2(\mathbb{R}^n_+)$ to their direct product. In particular, these spaces are reflexive.*

Proof. The argument uses Remark 1.6.3.

1. Let u_1 be any element of $H^s(\mathbb{R}^n_+)$, and let $v_1 \in H^s(\mathbb{R}^n)$ be an extension of u_1. Take a function u_2, which we first assume to be infinitely differentiable and compactly supported inside $\mathbb{R}^n_+$. The form

$$(u_1, u_2)_{\mathbb{R}^n_+} := (v_1, u_2)_{\mathbb{R}^n} \tag{4.2.1}$$

is defined and does not depend on the choice of the extension v_1. Here u_2 is regarded as an element of $H^{-s}(\mathbb{R}^n)$. This form can be extended to functions $u_2 \in \widetilde{H}^{-s}(\mathbb{R}^n_+)$, because the set of C^∞ functions compactly supported inside $\mathbb{R}^n_+$ is dense in $\widetilde{H}^{-s}(\mathbb{R}^n_+)$. The generalized Schwartz inequality implies

$$|(v_1, u_2)_{\mathbb{R}^n}| \leq \|v_1\|_{H^s(\mathbb{R}^n)} \|u_2\|_{H^{-s}(\mathbb{R}^n)} = \|v_1\|_{H^s(\mathbb{R}^n)} \|u_2\|_{\widetilde{H}^{-s}(\mathbb{R}^n_+)}.$$

Passing to the lower bound of the norm with respect to v_1, we obtain

$$|(v_1, u_2)_{\mathbb{R}^n}| \leq \|u_1\|_{H^s(\mathbb{R}^n_+)} \|u_2\|_{\widetilde{H}^{-s}(\mathbb{R}^n_+)}. \tag{4.2.2}$$

We have introduced the required duality, obtained a generalized Schwartz inequality in $\mathbb{R}^n_+$, and shown that each element u_2 of $\widetilde{H}^{-s}(\mathbb{R}^n_+)$ determines a continuous linear functional on $H^s(\mathbb{R}^n_+)$ by (4.2.1); moreover, the norm of this functional does not exceed $\|u_2\|_{\widetilde{H}^{-s}(\mathbb{R}^n_+)}$.

Conversely, let $f(u_1)$ be a continuous linear functional on $H^s(\mathbb{R}^n_+)$. Taking any $v_1 \in H^s(\mathbb{R}^n)$, we obtain the continuous linear functional

$$g(v_1) = f(v_1|_{\mathbb{R}^n_+})$$

on $H^s(\mathbb{R}^n)$, because the operation of restriction to $\mathbb{R}^n_+$ maps continuously $H^s(\mathbb{R}^n)$ to $H^s(\mathbb{R}^n_+)$. But the functionals on $H^s(\mathbb{R}^n)$ can be realized by using elements of $H^{-s}(\mathbb{R}^n)$:

$$g(v_1) = (v_1, u_2)_{\mathbb{R}^n}.$$

The given functional $f(u_1)$ vanishes if v_1 is an infinitely differentiable function with compact support in $\mathbb{R}^n_-$. Therefore, $\operatorname{supp} u_2 \subset \overline{\mathbb{R}^n_+}$ and $u_2 \in \widetilde{H}^{-s}(\mathbb{R}^n_+)$. We have represented the functional $f(u_1)$ in the form (4.2.1). It does not depend on the choice of the extension v_1 of u_1.

Furthermore,

$$\|u_2\|_{\widetilde{H}^{-s}(\mathbb{R}^n_+)} = \|u_2\|_{H^{-s}(\mathbb{R}^n)} = \sup_{v_1 \neq 0} \frac{|(v_1, u_2)_{\mathbb{R}^n}|}{\|v_1\|_{H^s(\mathbb{R}^n)}}.$$

Let $v_1|_{\mathbb{R}^n_+} = u_1$; then the right-hand side equals

$$\sup_{v_1 \neq 0} \frac{|f(u_1)|}{\|v_1\|_{H^s(\mathbb{R}^n)}} \leq \sup_{u_1 \neq 0} \frac{|f(u_1)|}{\|u_1\|_{H^s(\mathbb{R}^n_+)}}.$$

Here the right-hand side is the norm of the functional under consideration. We have obtained the reverse inequality: the norm $\|u_2\|_{\widetilde{H}^{-s}(\mathbb{R}^n_+)}$ does not exceed the norm of the corresponding functional. Thus, these norms coincide.

2. Let u_1 be any element of $H^s(\mathbb{R}^n_+)$. Then the form (4.2.1) determines a continuous antilinear functional of $u_2 \in \widetilde{H}^{-s}(\mathbb{R}^n_+)$. Here v_1 is any extension of u_1 belonging to $H^s(\mathbb{R}^n)$; its choice does not affect the functional. The norm of this functional is at most $\|u_1\|_{H^s(\mathbb{R}^n_+)}$. This can be verified as at the beginning of the proof, by using the Schwartz inequality and minimizing over extensions.

Now let $h(u_2)$ be a continuous antilinear functional on $\widetilde{H}^{-s}(\mathbb{R}^n_+)$. By the Hahn–Banach theorem, it can be extended to the entire space $H^{-s}(\mathbb{R}^n)$ without increasing its norm and represented in the form

$$h(u_2) = (v_1, u_2)_{\mathbb{R}^n},$$

where $v_1 \in H^s(\mathbb{R}^n)$. Moreover,

$$\|v_1\|_{H^s(\mathbb{R}^n)} = \sup \frac{|(v_1, u_2)_{\mathbb{R}^n}|}{\|u_2\|_{H^{-s}(\mathbb{R}^n)}},$$

where the upper bound is over all nonzero $u_2 \in H^{-s}(\mathbb{R}^n)$. The numerator equals the absolute value of the functional under consideration at u_2. Therefore, on the right-hand side, we have the norm of the extended functional. But the extension does not increase the norm; hence it suffices to take the upper bound only over $u_2 \in \widetilde{H}^{-s}(\mathbb{R}^n_+)$. If v_1 is an extension of $u_1 \in H^s(\mathbb{R}^n_+)$, then we can minimize the left-hand side and obtain a bound for $\|u_1\|_{H^s(\mathbb{R}^n_+)}$ in terms of the norm of the functional. Therefore, these norms coincide. $\square$

Remarks. 1. The duality can also be written as follows. If u_1 belongs to the space $H^s(\mathbb{R}^n_+)$ and u_2 belongs to $\widetilde{H}^{-s}(\mathbb{R}^n_+)$, then in the former space we approximate the function u_1 by a sequence of functions $\varphi_k \in C_0^\infty(\overline{\mathbb{R}^n_+})$, and in the latter we approximate u_2 by a sequence of functions $\psi_k \in C_0^\infty(\mathbb{R}^n_+)$. We may set

$$(u_1, u_2)_{\mathbb{R}^n_+} = \lim_{k\to\infty} (\varphi_k, \psi_k)_{0, \mathbb{R}^n_+}. \tag{4.2.3}$$

2. Using the universal extension operator $\mathcal{E}$ (which is constructed in Section 10), we shall also be able to write

$$(u_1, u_2)_{\mathbb{R}^n_+} = (\mathcal{E}u_1, u_2)_{\mathbb{R}^n}. \tag{4.2.4}$$

4.3 The spaces $\mathring{H}^s(\mathbb{R}^n_+)$

The subspace $\mathring{H}^s(\mathbb{R}^n_+)$ of $H^s(\mathbb{R}^n_+)$ is defined as the completion of the linear manifold $\mathcal{D}(\mathbb{R}^n_+) = C_0^\infty(\mathbb{R}^n_+)$ in $H^s(\mathbb{R}^n_+)$.

For $s > 0$, the elements of this space have zero Cauchy data of order $[s] - 1/2$ (if this number is nonnegative), because this is so for the approximating functions in $C_0^\infty(\mathbb{R}^n_+)$.

We want to find out what elements of $H^s(\mathbb{R}^n_+)$ belong to this subspace. In particular, it is interesting to compare $\mathring{H}^s(\mathbb{R}^n_+)$ with $H^s(\mathbb{R}^n_+)$ itself and with $\widetilde{H}^s(\mathbb{R}^n_+)$. Comparing $\mathring{H}^s(\mathbb{R}^n_+)$ with $\widetilde{H}^s(\mathbb{R}^n_+)$, we must identify a function u on $\mathbb{R}^n_+$ with its extension $\widetilde{u}$ to $\mathbb{R}^n$ by zero outside $\mathbb{R}^n_+$. The spaces $\mathring{H}^s(\mathbb{R}^n_+)$ and $\widetilde{H}^s(\mathbb{R}^n_+)$ are two completions of $C_0^\infty(\mathbb{R}^n_+)$, but with respect to different norms, on $H^s(\mathbb{R}^n_+)$ and on $H^s(\mathbb{R}^n)$.

Before investigating these questions, we give an auxiliary theorem about the spaces $H^{-s}(\mathbb{R}^n)$ of negative order, which is of independent interest. It is borrowed from the book [258].

Let $\mathbb{R}^{n-1}_0$ denote the boundary hyperplane of $\mathbb{R}^n_+$ determined by the equation $x_n = 0$, and let $H_0^{-s}(\mathbb{R}^n)$ be the subspace in $H^{-s}(\mathbb{R}^n)$ consisting of all elements supported in this hyperplane. The following theorem describes this subspace, which obviously coincides with $\widetilde{H}^{-s}(\mathbb{R}^n_-) \cap \widetilde{H}^{-s}(\mathbb{R}^n_+)$.

Theorem 4.3.1. *For $s \leq 1/2$, the only element of the space $H_0^{-s}(\mathbb{R}^n)$ is zero, and for $s > 1/2$, this space consists of all distributions of the form*

$$f = \sum_{0 \leq j < s-1/2} g_j \otimes \delta^{(j)}(x_n), \quad \text{where } g_j \in H^{-s+j+1/2}(\mathbb{R}^{n-1}). \tag{4.3.1}$$

Here $\delta^{(j)}(x_n)$ is the jth derivative of the delta-function $\delta(x_n)$. From (4.3.1) it is seen, in particular, that f treated as a "distribution in x_n" is concentrated at the single point 0.

We prove this theorem only in the case of a half-integer $s = k + 1/2$, $k \in \mathbb{Z}_+$. The case $s \in (k+1/2, k+3/2)$ is a little simpler (the last step of the proof is not needed), and we leave it to the reader. The inequality $j < s - 1/2$ now has the form $j < k$.

Proof of Theorem 4.3.1 for $s = k + 1/2$. First note that all distributions of the form (4.3.1) belong to $H_0^{-s}(\mathbb{R}^n)$. Indeed, they are concentrated in the hyperplane $x_n = 0$, and it is easy to verify that their Fourier images belong to $\widetilde{H}^{-s}(\mathbb{R}^n)$.

Now we make the following observation. If $f \in H_0^{-s}(\mathbb{R}^n)$, $s = k + 1/2$, and ρ is a function in $\mathcal{D}(\mathbb{R}^n)$ with zero Cauchy data *up to* order k at $x_n = 0$, then $\langle f, \rho \rangle = 0$. Indeed, let

$$\rho_\pm(x) = \begin{cases} \rho(x) & \text{at } x \in \mathbb{R}^n_\pm, \\ 0 & \text{outside } \mathbb{R}^n_\pm. \end{cases}$$

Then $\rho_\pm \in \widetilde{H}^\sigma(\mathbb{R}^n_\pm)$ for $\sigma \in (s, s+1)$. Therefore, $\rho_\pm$ can be approximated by functions from $H^\sigma(\mathbb{R}^n)$ supported inside $\mathbb{R}^n_\pm$, which are annihilated by the distribution f. Thus, f annihilates ρ.

Now, given a distribution $f \in H_0^{-s}(\mathbb{R}^n)$, $s = k + 1/2$, we construct distributions

$$g_j \in H^{-s+j+1/2}(\mathbb{R}^{n-1}) \quad (j = 0, \ldots, k)$$

such that

$$\left\langle f - \sum_0^k g_j \otimes \delta^{(j)}, \varphi \right\rangle = \langle f, \rho \rangle \tag{4.3.2}$$

for any $\varphi \in \mathcal{D}(\mathbb{R}^n)$, where ρ is a function in $C_0^\infty(\mathbb{R}^n)$ with zero Cauchy data up to order k at $x_n = 0$. In view of the above observation, the right-hand side of (4.3.2) vanishes; thus, we shall obtain (4.3.1) but with an "extra" summand ($j = k$). We shall show at the end of the proof that this summand vanishes.

To construct the g_j, we use operators (3.3.7) and Lemma 3.3.4. We define g_j by

$$\langle g_j, \phi \rangle = (-1)^j \langle f, T_j \phi \rangle \quad (\phi \in \mathcal{D}(\mathbb{R}^{n-1})). \tag{4.3.3}$$

Then we have relation (4.3.2) with

$$\rho = \varphi - \sum_0^k T_j \psi_j, \quad \psi_j = \partial_n^j \varphi(x',0), \tag{4.3.4}$$

so that ρ has zero Cauchy data up to order k at $x_n = 0$. Moreover,

$$|\langle g_j, \phi \rangle| \le \|f\|_{H^{-s}(\mathbb{R}^n)} \|T_j \phi\|_{H^s(\mathbb{R}^n)} \le C\|f\|_{H^{-s}(\mathbb{R}^n)} \|\phi\|_{H^{s-j-1/2}(\mathbb{R}^{n-1})},$$

whence $\|g_j\|_{H^{-s+j+1/2}(\mathbb{R}^{n-1})} \le C\|f\|_{H^{-s}(\mathbb{R}^n)}$. We have obtained (4.3.1) with an extra (last) summand.

It remains to show that g_k in this last summand vanishes. Suppose that this is not true.

As shown at the first step of the proof, the remaining summands, together with the left-hand side, belong to $H^{-k-1/2}(\mathbb{R}^n)$. Therefore, so does the summand under consideration. Let us write its squared norm in terms of the Fourier transform:

$$\iint \frac{|(F'g_k)(\xi')|^2 \xi_n^{2k}}{(1+|\xi'|^2+\xi_n^2)^{k+1/2}} \, d\xi' \, d\xi_n.$$

Since

$$1 + |\xi'|^2 + \xi_n^2 \le (1+|\xi'|^2)(1+\xi_n^2),$$

it follows that the last integral is not less than

$$\int \frac{|(F'g_k)(\xi')|^2}{(1+|\xi'|^2)^{k+1/2}} \, d\xi' \int \frac{\xi_n^{2k}}{(1+\xi_n^2)^{k+1/2}} \, d\xi_n.$$

But here the last integral diverges; therefore, $(F'g_k)(\xi') = 0$ almost everywhere, i.e., the function g_k is indeed zero.   □

Note that if $g(u)$ is a continuous antilinear functional on $H^s(\mathbb{R}^n_+)$, then it determines the continuous antilinear functional on $H^s(\mathbb{R}^n)$ by the formula $f(u) = g(u|_{\mathbb{R}^n_+})$. If g is supported on $\mathbb{R}^n_0$, then so is f. Therefore, Theorem 4.3.1 gives also a representation of functionals on $H^s(\mathbb{R}^n_+)$ supported on the boundary.

Example. The space $\widetilde{H}^{-1-\sigma}(\mathbb{R}^n_+)$, $|\sigma| < 1/2$, is dual to $H^{1+\sigma}(\mathbb{R}^n_+)$ and contains functionals concentrated in the boundary hyperplane $\mathbb{R}^{n-1}$. As is seen from the theorem proved above, such antilinear functionals can be represented as $(g, u^+)_{\mathbb{R}^{n-1}}$, where $u^+ \in H^{1/2+\sigma}(\mathbb{R}^{n-1})$ and $g \in H^{-1/2-\sigma}(\mathbb{R}^{n-1})$.

Now we pass to the spaces $\mathring{H}^s(\mathbb{R}^n_+)$.

Theorem 4.3.2. *The spaces $\mathring{H}^s(\mathbb{R}^n_+)$ and $H^s(\mathbb{R}^n_+)$ coincide for $s \le 1/2$ and do not coincide for $s > 1/2$.*

In other words, the linear manifold $\mathcal{D}(\mathbb{R}^n_+)$ is dense in $H^s(\mathbb{R}^n_+)$ if and only if $s \le 1/2$.

Proof. It is easy to check this for $s < 1/2$. Indeed, if $0 < s < 1/2$, then any function in $H^s(\mathbb{R}^n_+)$ can be extended by zero to a function in $\widetilde{H}^s(\mathbb{R}^n_+)$. In the latter space, C^∞

functions supported on $\mathbb{R}^n_+$ are dense; therefore, they are also dense in $H^s(\mathbb{R}^n_+)$. As a consequence, functions belonging to $\mathcal{D}(\mathbb{R}^n_+)$ are dense in $H^s(\mathbb{R}^n_+)$ with negative s too. For $s > 1/2$, this cannot be true, because a function in $H^s(\mathbb{R}^n_+)$ may have a nonzero boundary value for such s.

For $s = 1/2$, there is no bounded operator of extension by zero. However, if the closure of $\mathcal{D}(\mathbb{R}^n_+)$ in the Hilbert space $H^{1/2}(\mathbb{R}^n_+)$ did not coincide with this space, then the dual space $\widetilde{H}^{-1/2}(\mathbb{R}^n_+)$ would contain a nonzero functional concentrated in the boundary hyperplane. But, according to Theorem 4.3.1, there are no such functionals. $\quad\square$

Now we compare the spaces $\widetilde{H}^s(\mathbb{R}^n_+)$ and $\overset{\circ}{H}{}^s(\mathbb{R}^n_+)$ with positive s. First note that the former space is embedded in the latter. Indeed, the norm of a function in $H^s(\mathbb{R}^n)$ is not smaller than the norm of its restriction to $\mathbb{R}^n_+$. Therefore, if a function $\widetilde{u}$ is the limit of a sequence of functions $\widetilde{u}_k \in C_0^\infty(\mathbb{R}^n_+)$ in $H^s(\mathbb{R}^n)$, then u is the limit of the sequence of functions u_k in $H^s(\mathbb{R}^n_+)$.

Using results of Section 3.4, we can now give a new, "intrinsic," definition of the space $\widetilde{H}^s(\mathbb{R}^n_+)$.

Let $s = m + \sigma > 0$, where $m = [s]$. The *space* $\widetilde{H}^s(\mathbb{R}^n_+)$ consists of functions u on $\mathbb{R}^n_+$ that belong to $\overset{\circ}{H}{}^s(\mathbb{R}^n_+)$ and have finite norm

$$\left(\|u(x)\|^2_{H^s(\mathbb{R}^n_+)} + \sum_{|\alpha|=m} \|x_n^{-\sigma} D^\alpha u(x)\|^2_{L_2(\mathbb{R}^n_+)} \right)^{1/2}. \tag{4.3.5}$$

Theorem 4.3.3. *This is a complete space. Its new definition is equivalent to those given in Section* 4.1. *For non-half-integer* s, $\widetilde{H}^s(\mathbb{R}^n_+)$ *coincides with* $\overset{\circ}{H}{}^s(\mathbb{R}^n_+)$, *and the norms of these spaces are equivalent. For half-integer* s, *this is a linear manifold in* $H^s(\mathbb{R}^n_+)$ *with a stronger topology. It is continuously embedded in* $\overset{\circ}{H}{}^s(\mathbb{R}^n_+)$ *but not closed.*

Here, for non-half-integer s, the norm (4.3.5) of a function u is equivalent to the norm of its extension $\widetilde{u}$ in $H^s(\mathbb{R}^n)$ (and, by virtue of the theorem, is not very interesting). For half-integer s, the convergence of a sequence of functions in $\widetilde{H}^s(\mathbb{R}^n_+)$ in the norm (4.3.5) implies the norm convergence of this sequence in $H^s(\mathbb{R}^n_+)$, but not vice versa.

As a consequence, for non-half-integer s, $\widetilde{H}^s(\mathbb{R}^n_+)$ turns out to be a (closed) subspace in $H^s(\mathbb{R}^n_+)$, while for half-integer s, this is not so.

Comparing these results with those of Section 3.4, we obtain the following corollary.

Corollary 4.3.4. *For non-half-integer* $s > 1/2$, *the space* $\overset{\circ}{H}{}^s(\mathbb{R}^n_+)$ *coincides with the subspace of those functions in* $H^s(\mathbb{R}^n_+)$ *whose extensions by zero belong to* $\widetilde{H}^s(\mathbb{R}^n_+)$. *Moreover, the norms of such a function in* $\overset{\circ}{H}{}^s(\mathbb{R}^n_+)$ *and its extension in* $\widetilde{H}^s(\mathbb{R}^n_+)$ *are equivalent.*

Corollary 4.3.5. *For non-half-integer* $s > 1/2$, $\widetilde{H}^s(\mathbb{R}^n_+)$ *can be regarded as a (closed) subspace of* $H^s(\mathbb{R}^n_+)$.

For $0 < s < 1/2$, these spaces can be identified: see Theorem 4.1.1.

For half-integer s, the restrictions of functions in $\widetilde{H}^s(\mathbb{R}^n_+)$ to $\mathbb{R}^n_+$ are contained in $\overset{\circ}{H}{}^s(\mathbb{R}^n_+)$. The definition of the norm on $H^s(\mathbb{R}^n_+)$ as a lower bound implies the continuity of the corresponding embedding.

4.4 The Spaces $\widetilde{H}^{-s}(\mathbb{R}^n_+)$ and $H^{-s}(\mathbb{R}^n_+)$ with Non-Half-Integer $s > 1/2$

We begin with the following remark.

Remark 4.4.1. Suppose that a Banach space B is the direct sum of (closed) subspaces B_1 and B_2 (with the same norm). If f is a functional in the dual space B^*, then the restriction of this functional to B_1 belongs to the space B_1^* dual to B_1. Conversely, if $g \in B_1^*$, then, extending g to B_2 by zero and then to B by linearity, we obtain a functional in B^*. In this sense, we obtain a (nonstandard) embedding

$$B_1^* \subset B^*. \tag{4.4.1}$$

Here the procedure for extending a functional g to B is fixed.

We need to emphasize that the complement B_2 of B_1 is not uniquely determined, and a functional on B_1 can be extended to B in many different ways. Hence there is no standard embedding of the left-hand side in the right-hand one.

Next, we can regard the space B^* as the direct sum of the subspaces B_1^* and B_2^*, assuming that the functionals in the former vanish on B_2 and the functional in the latter vanish on B_1.

In particular, in the nonstandard sense specified above, for non-half-integer $s > 1/2$, we have

$$H^{-s}(\mathbb{R}^n_+) \subset \widetilde{H}^{-s}(\mathbb{R}^n_+): \tag{4.4.2}$$

we extend a functional G on $\widetilde{H}^s(\mathbb{R}^n_+)$ from the space on the left-hand side by zero to the orthogonal complement of $\widetilde{H}^s(\mathbb{R}^n_+)$ in the Hilbert space $H^s(\mathbb{R}^n_+)$. The inclusion is strict, because the space on the right-hand side contains functionals supported on the boundary hyperplane. For $0 \le s < 1/2$, these spaces can be identified.

Similar considerations apply to the spaces $H^s(\Omega)$ and $\widetilde{H}^s(\Omega)$ on a domain Ω; we discuss it below in Section 5.1. In Section 11 we consider another decomposition of the space $H^1(\Omega)$ on a Lipschitz domain Ω.

5 The Spaces H^s on Smooth Bounded Domains and Manifolds with Boundary

In this section, by Ω we denote a smooth bounded domain in $\mathbb{R}^n$, i.e., a bounded domain with smooth $(n-1)$-dimensional boundary, by Γ we denote its boundary, and by M, a smooth compact n-dimensional manifold with smooth $(n-1)$-dimensional boundary ∂M. A bounded domain Ω with smooth boundary is a special case of such a manifold. Another important special case is a smooth part of the (smooth) boundary of a bounded $(n+1)$-dimensional domain or, more generally, a hypersurface with smooth boundary in $\mathbb{R}^{n+1}$.

Let us clarify what we mean by the smoothness of the boundary of Ω. Near each boundary point x_0, the boundary is the graph of a C^∞ function $x_n = \phi(x')$ defined on some $(n-1)$-dimensional ball $\{x' : |x'| \le R\}$, $R > 0$, in the appropriately rotated initial Cartesian coordinate system. As the point x passes through the boundary, the difference $x_n - \phi(x')$ changes sign. If we redenote $x_n - \phi(x')$ by x_n, then the boundary will be (locally) transformed into the hyperplane $\{x : x_n = 0\}$; we refer to this change as the *boundary rectification*.

5.1 The Spaces $H^s(\Omega)$

All definitions and main statements in the case of a smooth bounded domain are very similar to those in the case of the half-space $\mathbb{R}^n_+$. It suffices to replace $\mathbb{R}^n_+$ by Ω and $\mathbb{R}^n_-$ by the complement $\mathbb{R}^n \setminus \overline{\Omega}$ of the closure $\overline{\Omega}$ of the domain. The proofs largely are reduced to applications of results for the half-space. We dwell on the main statements, omitting their obvious verification.

The spaces $C^\infty(\overline{\Omega})$, $C^m(\overline{\Omega})$, $C^{m+\theta}(\overline{\Omega}) = C^{m,\theta}(\overline{\Omega})$, and $C^{m,1}(\overline{\Omega})$, where m is a nonnegative integer and $0 < \theta < 1$, are most conveniently defined as consisting of the restrictions to $\overline{\Omega}$ of functions belonging to the corresponding function spaces on $\mathbb{R}^n$ with the inf norms.

The *space* $H^s(\Omega)$ is defined for each s as the space of functions u which are restrictions to Ω of elements of $H^s(\mathbb{R}^n)$ (for $s < 0$, in the sense of distributions) with norm

$$\|u\|_{H^s(\Omega)} = \inf \|v\|_{H^s(\mathbb{R}^n)} : v \in H^s(\mathbb{R}^n), \quad v|_\Omega = u. \tag{5.1.1}$$

For $s \ge 0$, $H^s(\Omega)$ can also be defined as the space of square integrable functions on Ω which have square integrable derivatives in the sense of distributions and finite Sobolev–Slobodetskii norms. For integer $s = m$, this is the norm defined by

$$\|u\|'^2_{H^m(\Omega)} = \sum_{|\alpha| \le m} \|D^\alpha u\|^2_{L_2(\Omega)}. \tag{5.1.2}$$

In the sum on the right-hand side, we can retain only $\alpha = (0,\dots,0)$, $(m,0,\dots,0)$, $\dots, (0,\dots,0,m)$. For noninteger $s = m + \theta$, where $m = [s]$ and $0 < \theta < 1$, we have

$$\|u\|'^2_{H^{m+\theta}(\Omega)} = \|u\|'^2_{H^m(\Omega)} + \sum_{|\alpha|=m} \int_\Omega \int_\Omega \frac{|D^\alpha u(x) - D^\alpha u(y)|^2}{|x-y|^{n+2\theta}} \, dx \, dy. \qquad (5.1.3)$$

Theorem 5.1.1. *For $s \geq 0$, the definitions given above are equivalent.*

The proof of this theorem is similar to that of Theorem 3.1.1 but uses an *extension operator* from $H^s(\Omega)$ to $H^s(\mathbb{R}^n)$ bounded with respect to the norms $\|\cdot\|'_{H^s}$, $0 \leq s \leq' N$. Such an operator (for which we use the same name—the Hestenes operator) is constructed by using partitions of unity as follows.

The compactness of the boundary makes it possible to define a finite smooth partition of unity

$$\sum_{j=0}^m \varphi_j(x) \equiv 1 \qquad (5.1.4)$$

in a neighborhood of the closure of Ω so that the support of the function φ_0 is contained strictly inside the domain and each of the remaining functions φ_j is supported on a "boundary strip"; moreover, in some neighborhood U_j of this support, it is possible to pass to coordinates rectifying the boundary. Therefore, in this neighborhood, we can use the already familiar Hestenes operator for the half-space defined by (3.1.5); we denote it by $\mathcal{E}_N^{(j)}$. Next, for each $j = 1, \ldots, m$, let ψ_j be a C^∞ function supported on U_j and equal to 1 in a smaller neighborhood of the support of φ_j. We set

$$\mathcal{E}_N u = \varphi_0 u + \sum_{j=1}^m \psi_j \mathcal{E}_N^{(j)}(\varphi_j u); \qquad (5.1.5)$$

the right-hand side is written in the initial coordinates.

Theorem 5.1.2. *The operator $\mathcal{E}_N$ acts boundedly from $H^s(\Omega)$ to $H^s(\mathbb{R}^n)$ for $0 \leq s \leq N$ and takes any function u in the former space to a function in the latter whose restriction to Ω coincides with u.*

Using the Rychkov universal extension operator $\mathcal{E}^{(j)}$ (see Section 10) instead of the Hestenes operator $\mathcal{E}_N^{(j)}$ for the half-space, we obtain the universal extension operator for the domain Ω defined by

$$\mathcal{E} u = \varphi_0 u + \sum_{j=1}^m \psi_j \mathcal{E}^{(j)}(\varphi_j u). \qquad (5.1.6)$$

Theorem 5.1.3. *The operator $\mathcal{E}$ acts boundedly from $H^s(\Omega)$ to $H^s(\mathbb{R}^n)$ for any real s and takes any function u in the former space to a function in the latter whose restriction to Ω coincides with u.*

Actually, Rychkov constructed a universal extension operator for domains having very general (Lipschitz) boundaries. Moreover, this operator applies to more general spaces. All this will be discussed in Sections 10 and 14.

All relevant spaces in $\mathbb{R}^n \setminus \overline{\Omega}$, beginning with $C_b^\infty(\mathbb{R}^n \setminus \overline{\Omega})$, are naturally defined in a similar way. The reader can easily formulate the definitions and transfer the statements given below (except those concerning compactness) to these spaces. The unbounded domain $\mathbb{R}^n \setminus \overline{\Omega}$ has compact boundary. We do not consider unbounded domains more general than those of the form $\mathbb{R}^n \setminus \overline{\Omega}$ and $\mathbb{R}_+^n$.

Of course, we could consider more general unbounded domains and noncompact manifolds. But this book is oriented to the construction of the theory of elliptic equations and boundary value problems on compact manifolds and in bounded domains; thus, we wish to avoid considering questions related to conditions on functions at infinity and, above all, to the behavior of boundaries at infinity and weight spaces, which is often natural to consider on unbounded domains and noncompact manifolds.

Further properties of the spaces $H^s(\Omega)$ largely follow from the first definition of these spaces and are similar to properties of the spaces $H^s(\mathbb{R}^n)$ or $H^s(\mathbb{R}_+^n)$.

1. For $s < \sigma$, the space $H^\sigma(\Omega)$ is continuously, compactly, and densely embedded in $H^s(\Omega)$. The space $C^\infty(\overline{\Omega})$ is dense in all spaces $H^s(\Omega)$.

The compactness of the embedding is derived from Theorem 1.12.1.

2. For any integer $m > 0$, the space $H^{-m}(\Omega)$ consists of linear combinations of derivatives of order up to m (in the sense of distributions on Ω) of functions in $L_2(\Omega)$. If $0 < \theta < 1$, then the $H^{-m+\theta}(\Omega)$ consists of linear combinations of derivatives of order up to m (in the sense of distributions on Ω) of functions in $H^\theta(\Omega)$.

3. Theorem 5.1.4. For $s > n/2+t$, where $t > 0$, the space $H^s(\Omega)$ is continuously and compactly embedded in $C^t(\overline{\Omega})$. For $s = n/2+t$, this is also true for noninteger t.

Here we assume that the functions under consideration may be corrected on a set of measure zero.

Theorem 5.1.5. For $n/2 > s$ and

$$s \geq \frac{n}{2} - \frac{n}{p}, \quad 2 < p < \infty, \tag{5.1.7}$$

the space $H^s(\Omega)$ is continuously embedded in the space $L_p(\Omega)$ with norm

$$\|u\|_{L_p(\Omega)} = \left(\int_\Omega |u(x)|^p \, dx \right)^{1/p}. \tag{5.1.8}$$

If the inequality for s in (5.1.7) is strict, then the embedding is compact.

4. If $0 < s < \sigma$, then, for any $\varepsilon > 0$, there exists a number C_ε such that the norms of functions $u \in H^\sigma(\Omega)$ satisfy the inequality

$$\|u\|_{H^s(\Omega)} \leq \varepsilon \|u\|_{H^\sigma(\Omega)} + C_\varepsilon \|u\|_{L_2(\Omega)}. \tag{5.1.9}$$

5. Theorem 5.1.6. *Let m be a nonnegative integer. If a function $a(x)$ belongs to $C_b^{m,1}(\overline{\Omega})$, then the operator of multiplication by $a(x)$ is bounded in $H^{m+1}(\Omega)$. For $s = m + \theta$, $0 < \theta < 1$, the operator of multiplication by a function $a(x)$ is bounded in $H^s(\Omega)$ provided that this function belongs to the space $C_b^{m+\vartheta}(\overline{\Omega}) = C_b^{m,\vartheta}(\overline{\Omega})$, where $\vartheta \in (\theta, 1)$.*

Remark. By duality, similar assertions hold for multipliers on the spaces $\widetilde{H}^{-s}(\Omega)$ defined a little later.

6. Let us state theorems about boundary values.

Theorem 5.1.7. *If $s > 1/2$, then there is a bounded operator which takes each function u in $H^s(\Omega)$ to its boundary value $v = \gamma^+ u$ in $H^{s-1/2}(\Gamma)$ coinciding with the usual boundary value in the case of continuous functions.*

As previously, first, the inequality

$$\|v\|_{H^{s-1/2}(\Gamma)} \leq C\|u\|_{H^s(\Omega)} \tag{5.1.10}$$

is verified for smooth functions, which makes it possible to define boundary values for functions in $H^s(\Omega)$; then, this inequality is extended to functions in $H^s(\Omega)$.

Theorem 5.1.8. *There exists a linear operator not depending on s which acts boundedly from $H^{s-1/2}(\Gamma)$ to $H^s(\mathbb{R}^n)$ for any $s \geq 1/2$ and, for $s > 1/2$, takes any function v in the former space to a function u in the latter whose boundary value $\gamma^+ u$ on Γ coincides with v.*

If $|\alpha| < s - 1/2$, then the derivative $D^\alpha u(x)$ of a function $u(x) \in H^s(\Omega)$ has a boundary value in $H^{s-|\alpha|-1/2}(\Gamma)$, and the operator of passing to this boundary value is bounded. We can consider Cauchy data. Let $\nu(x)$ be the unit inner normal vector to Γ at the point x. Given a C^∞ function $u(x)$ on $\overline{\Omega}$, we can define consecutive derivatives of order k along ν on the boundary for $k = 0$, 1, and so on; these are the Cauchy data. The kth derivative along the normal is a linear combination with smooth coefficients of derivatives of order at most k in the initial coordinates. Therefore, the following analogue of Theorem 3.3.2 is valid.

Theorem 5.1.9. *Given any nonnegative integer $l < s - 1/2$, for each function $u \in H^s(\Omega)$, its Cauchy data $\partial_\nu^k u$ $(k = 0, \dots, l)$ of order l on Γ are defined, and the map taking each function to its Cauchy data is a bounded operator from $H^s(\Omega)$ to the direct product of the spaces $H^{s-k-1/2}(\Gamma)$.*

In principle, it is possible to construct a function with given Cauchy data on a nonplanar boundary, but the explicit construction is not transparent enough, and we shall not dwell on this here. The existence of such a function follows from a simple result of the theory of higher-order elliptic equations, namely, from the solvability of an appropriate Dirichlet problem. We explain this in Section 7.1 (see Problem 3 there).

7. Theorem 5.1.10. *For* $|s| < 1/2$, *the operator* $\mathcal{E}_0$ *of extension by zero acts boundedly from* $H^s(\Omega)$ *to* $H^s(\mathbb{R}^n)$ *(and from* $H^s(\mathbb{R}^n \setminus \overline{\Omega})$ *to* $H^s(\mathbb{R}^n)$*).*

8. Two functions $u^{\pm}(x)$ in $H^s(\Omega)$ and $H^s(\mathbb{R}^n \setminus \overline{\Omega})$ with $0 < s < 1/2$ can be "glued together" into a single function in $H^s(\mathbb{R}^n)$. This is also true for non-half-integer $s > 1/2$, provided that these functions have the same Cauchy data of order $[s - 1/2]$ on Γ.

Now we pass to analogues of definitions and results of Sections 3 and 4. We define the *space* $\widetilde{H}^s(\Omega)$ for any $s \in \mathbb{R}$ as the subspace of $H^s(\mathbb{R}^n)$ consisting of elements supported on $\overline{\Omega}$ and endowed with the norm inherited from $H^s(\mathbb{R}^n)$. This space can also be equivalently defined as the completion of the linear manifold of functions in $C_0^{\infty}(\Omega)$ extended by zero outside Ω in $H^s(\mathbb{R}^n)$. For $s \geq 0$, it can also be defined as the set of functions in $H^s(\Omega)$ whose extensions by zero belong to $H^s(\mathbb{R}^n)$ endowed with the norm taken from the latter space. The definitions of $\widetilde{H}^s(\mathbb{R}^n \setminus \overline{\Omega})$ are similar.

Theorem 5.1.11. *For* $|s| < 1/2$, *the spaces* $H^s(\Omega)$ *and* $\widetilde{H}^s(\Omega)$ *can be identified by using the boundedness of the operator of extension by zero for these* s; *the norms are equivalent.*

The space $H^s(\Omega)$ is a quotient space:

$$H^s(\Omega) = H^s(\mathbb{R}^n) / \widetilde{H}^s(\mathbb{R}^n \setminus \overline{\Omega}).$$

Theorem 5.1.12. *For any* s, *the spaces* $H^s(\Omega)$ *and* $\widetilde{H}^{-s}(\Omega)$ *are dual with respect to the extension of the standard inner product in* $L_2(\Omega)$ *to the direct product of these spaces.*

More precisely, the form mentioned in the theorem is

$$(u_1, u_2)_{\Omega} = \lim_{k \to \infty} (v_1, \psi_k)_{\mathbb{R}^n} = \lim_{k \to \infty} (\varphi_k, \psi_k)_{\Omega}, \qquad (5.1.11)$$

where v_1 is the extension of $u_1 \in H^s(\Omega)$ to a function in $H^s(\mathbb{R}^n)$, $\{\psi_k\}$ is a sequence of functions in $C_0^{\infty}(\mathbb{R}^n)$ supported inside Ω which converges to u_2 in $\widetilde{H}^{-s}(\Omega)$, and $\{\varphi_k\}$ is a sequence of functions in $C^{\infty}(\overline{\Omega})$ converging to u_1 in $H^s(\Omega)$. Using the universal extension operator, we can write

$$(u_1, u_2)_{\Omega} = (\mathcal{E} u_1, u_2)_{\mathbb{R}^n}. \qquad (5.1.12)$$

Theorem 5.1.13. *For* $s \leq 1/2$, *the subspace of* $H^{-s}(\mathbb{R}^n)$ *consisting of elements supported on* Γ *consists only of zero, and for* $s > 1/2$, *this subspace consists of distributions of the form*

$$f = \sum_{0 \leq j < s - 1/2} g_j \otimes \delta^{(j)}(t), \quad \text{where } g_j \in H^{-s+j+1/2}(\Gamma) \qquad (5.1.13)$$

and $\delta^{(j)}(t)$ *is the distribution acting on smooth test functions* $u(x)$ *by the rule*

$$\langle \delta^{(j)}, u\rangle = (-1)^j \partial_\nu^j u(x)|_\Gamma.$$

The distributions of the form (5.1.13) belong to $\widetilde{H}^{-s}(\Omega)$ and act on the boundary values of functions from the space $H^s(\Omega)$ as functionals on this space. Thus, for functions $u \in H^{1+\sigma}(\Omega)$, $|\sigma| < 1/2$, the corresponding antilinear functionals have the form $(g, u^+)_\Gamma$, where $g \in H^{-1/2-\sigma}(\Gamma)$.

The *space* $\mathring{H}^s(\Omega)$ with any $s \in \mathbb{R}$ is defined as the closure in $H^s(\Omega)$ of the linear manifold $C_0^\infty(\Omega)$.

Theorem 5.1.14. *The space $\mathring{H}^s(\Omega)$ coincides with $H^s(\Omega)$ if and only if $s \le 1/2$. For non-half integers $s > 1/2$, it coincides with the subspace in $H^s(\Omega)$ consisting of functions having zero Cauchy data of order less than $s - 1/2$.*

Choose a positive function $\rho(x)$ on Ω equivalent to the distance $d(x)$ from a point x to the boundary Γ near Γ (in the sense that the ratio of $\rho(x)$ to $d(x)$ is contained between two positive constants).

Theorem 5.1.15. *Let $s > 0$. For non-half-integer s, the spaces $\mathring{H}^s(\Omega)$ and $\widetilde{H}^s(\Omega)$ can be identified (assuming a function u on Ω to be identified with its extension $\widetilde{u}$ to $\mathbb{R}^n$ by zero outside Ω). For half-integer $s = m + 1/2$, $m = [s]$, $\widetilde{H}^s(\Omega)$ is a nonclosed linear subspace in $\mathring{H}^s(\Omega)$ with the stronger topology determined by the norm*

$$\left(\|u\|_{H^s(\Omega)}^2 + \sum_{|\alpha|=m} \|\rho^{-1/2} D^\alpha u\|_{L_2(\Omega)}^2\right)^{1/2}. \tag{5.1.14}$$

This norm is equivalent to the original norm on $\widetilde{H}^s(\Omega)$ for the extensions $\widetilde{u}$ of functions u.

We could introduce this norm also for non-half-integer s, but, as is seen from the theorem, this is not very interesting.

Remark 5.1.16. In particular, for non-half-integer $s > 1/2$ (and only for such $s > 1/2$), the space $\widetilde{H}^s(\Omega) = \mathring{H}^s(\Omega)$ is a (closed) subspace in $H^s(\Omega)$. For the dual spaces, we can arrange a (nonstandard) embedding

$$H^{-s}(\Omega) \subset \widetilde{H}^{-s}(\Omega), \tag{5.1.15}$$

if we agree to extend functionals in the space on the lef-hand side to the orthogonal complement of $\widetilde{H}^s(\Omega)$ in $H^s(\Omega)$ by zero.

The inclusion is strict, because the space on the right-hand side contains functionals supported on the boundary. These are sums of the form

$$\sum_{0 \le j < s-1/2} (h_j, \partial_\nu^j u^+)_\Gamma, \tag{5.1.16}$$

where $h_j \in H^{-s+j+1/2}(\Gamma)$ and $\partial_\nu^j u^+$ is the trace on Γ of the jth derivative along the normal of the function u.

For $0 \le s < 1/2$, the spaces in (5.1.15) can be identified.

5.2 The Spaces $H^s(M)$

Now consider the spaces $H^s(M)$ on a smooth compact n-dimensional manifold M with smooth boundary. The interior points of M have coordinate neighborhoods contained inside M with local coordinates introduced in the same way as on a manifold without boundary. These neighborhoods are diffeomorphic to an open ball in $\mathbb{R}^n$. The points of the boundary ∂M have coordinate half-neighborhoods, which are diffeomorphic to the union of an open half-ball and the planar part of its boundary. In such half-neighborhoods, local coordinates $x = (x', x_n)$ in which $x_n \geq 0$ and $|x| < 1$ can be introduced. The points with coordinates $(x', 0)$ belong to ∂M, and the remaining points are interior points of M. All transformations of local coordinates in the intersection of any two neighborhoods or half-neighborhoods are nondegenerate and infinitely differentiable. By virtue of compactness, the closure $\overline{M}$ of the manifold admits a finite cover by such neighborhoods and half-neighborhoods. There exists a finite partition of unity on $\overline{M}$ consisting of C^∞ functions and subordinate to this cover. This makes it possible to introduce norms on $H^s(M)$ in a standard way by using similar norms on the space $\mathbb{R}^n$ and the half-space $\mathbb{R}^n_+$. Choosing another cover, partition of unity, or local coordinates, we obtain an equivalent norm.

Next, usually we can assume that M is a part of a closed smooth manifold M_0 of the same dimension and, moreover, that the boundary coordinate half-neighborhoods are parts of coordinate neighborhoods on M_0. Thus, we can talk about an extension of functions in $H^s(M)$ to functions in $H^s(M_0)$. In particular, $\widetilde{H}^s(M)$ is then defined as the subspace of $H^s(M_0)$ consisting of elements supported on $\overline{M}$.

Other properties of the spaces $H^s(M)$ on a smooth manifold with smooth boundary are quite similar to those of the spaces $H^s(\Omega)$ on a smooth bounded domain, and we shall not repeat the formulations.

Cf. Remark 4.4.1.

Chapter 2
Elliptic Equations
and Elliptic Boundary Value Problems

In this chapter, we prove main theorems of the general theory of "smooth" elliptic equations and problems. Generality is somewhat minimized; in particular, we dwell on scalar equations and problems. More general facts (with similar proofs) are only stated or mentioned. We rarely touch on pseudodifferential operators in this chapter; their theory will be expounded in [26], where the material of this chapter will be substantially supplemented.

In Section 6, the basic ideas of the general theory are explained in the case of a scalar elliptic partial differential equation on a closed C^∞ manifold. In this case, there are no boundary conditions. A particular case is an equation (in fact, on the torus) with periodic boundary conditions. The coefficients in the equation are assumed to be C^∞.

Our main theorems are on the equivalence of ellipticity and the Fredholm property in the spaces H^s, on the smoothness of solutions of equations with smooth right-hand sides, and on the unique solvability of elliptic equations with a parameter (we consider only the case of a linear dependence on the parameter). Their proofs use the method of freezing coefficients. The necessary material related to abstract Fredholm operators is collected in Section 18.1. In particular, all needed definitions are given there.

In Section 7, similar theorems are proved for general scalar boundary value problems in a smooth bounded domain.

In these two sections, 6 and 7, we also outline an explanation of the spectral theory of elliptic equations and problems. Some preliminaries related to abstract notions of spectral theory are given in Section 18.3.

Analogues of the main theorems for matrix equations and problems are stated.

Section 8 is devoted to basic variational boundary value problems, namely, the Dirichlet and Neumann problems. We first consider these problems for second-order strongly elliptic equations and then briefly outline their generalization to higher-order systems. These problems, which have a much longer history, are particularly close to applications.

© Springer International Publishing Switzerland 2015
M.S. Agranovich, *Sobolev Spaces, Their Generalizations and Elliptic Problems
in Smooth and Lipschitz Domains*, Springer Monographs in Mathematics,
DOI 10.1007/978-3-319-14648-5_2

The Sobolev spaces are also very useful in the theory of parabolic and hyperbolic equations and boundary value problems. Parabolic problems will be mentioned in Section 7.1 (and Section 17.6), while hyperbolic problems are beyond the scope of this book.

6 Elliptic Equations on a Closed Smooth Manifold

6.1 Definitions

Consider an mth-order linear partial differential operator on $\mathbb{R}^n$:

$$Au(x) = a(x, D)u(x) = \sum_{|\alpha| \leq m} a_\alpha(x) D^\alpha u(x). \tag{6.1.1}$$

Here $u(x)$ is a scalar function. The coefficients are generally complex-valued functions; we suppose them to be infinitely differentiable and uniformly bounded together with all their derivatives:

$$|D^\beta a_\alpha(x)| \leq C_{\alpha,\beta}. \tag{6.1.2}$$

We know that such an operator acts boundedly from $H^s(\mathbb{R}^n)$ to $H^{s-m}(\mathbb{R}^n)$ for any $s \in \mathbb{R}$:

$$\|Au\|_{H^{s-m}(\mathbb{R}^n)} \leq C_s \|u\|_{H^s(\mathbb{R}^n)}. \tag{6.1.3}$$

Its *symbol* is the polynomial $a(x, \xi)$ obtained from $a(x, D)$ by replacing all D_j by real numbers ξ_j. The *principal symbol* $a_0(x, \xi)$ is the leading homogeneous part of the symbol:

$$a_0(x, \xi) = \sum_{|\alpha|=m} a_\alpha(x) \xi^\alpha. \tag{6.1.4}$$

Before the term "principal symbol" became conventional, this quantity was referred to as the *characteristic polynomial*.

The operator A is said to be *elliptic at a point* x if

$$a_0(x, \xi) \neq 0 \quad (0 \neq \xi \in \mathbb{R}^n). \tag{6.1.5}$$

This operator is said to be *elliptic on a set* $X \subseteq \mathbb{R}^n$ if it is elliptic at each point $x \in X$, and *uniformly elliptic on* X if there is a positive constant C such that

$$|a_0(x, \xi)| \geq C|\xi|^m \tag{6.1.6}$$

for all $x \in X$. For example, the Laplace operator

$$\Delta = -(D_1^2 + \ldots + D_n^2) \tag{6.1.7}$$

is uniformly elliptic on $\mathbb{R}^n$. If X is a compact (i.e., closed and bounded) set, then any operator elliptic on X is automatically uniformly elliptic on X, because, by continuity, an inequality of the form (6.1.6) with some constant holds on the compact set of points (x,ξ) with $x \in X$ and $|\xi| = 1$.

Instead of saying that the operator A is elliptic, we may say that the equation $Au = f$ is elliptic.

Now we define ellipticity with parameter. Let Λ be a closed sector, or angle, in the complex plane with vertex at the origin; e.g., this may be a ray starting at the origin (and containing it). The operator A is said to be *elliptic with parameter in Λ at a point x* if

$$a_0(x,\xi) - \lambda \neq 0 \quad (|\xi| + |\lambda| \neq 0, \ \lambda \in \Lambda); \tag{6.1.8}$$

A is said to be *elliptic with parameter in Λ* on a set X if condition (6.1.8) holds at each point $x \in X$, and *uniformly elliptic with parameter in Λ on X* if there is a positive constant C such that

$$|a_0(x,\xi) - \lambda| \geq C(|\xi|^m + |\lambda|) \tag{6.1.9}$$

for all $x \in X$ and $\lambda \in \Lambda$. For example, the operator $-\Delta$ is uniformly elliptic with parameter on $\mathbb{R}^n$ in any closed sector with vertex at the origin not containing the positive half-axis.

Obviously, ellipticity follows from ellipticity with parameter: it suffices to set $\lambda = 0$ in the definition of the latter.

In the situation considered here, instead of A, the operator $A - \lambda I$ may be referred to as elliptic with parameter.

Remark 6.1.1. Obviously, the ellipticity of an operator A implies that the *coefficients of all higher-order derivatives D_j^m in A are nonzero*. Indeed, this follows from (6.1.5) with the vectors ξ one of whose coordinates, ξ_j, is 1 and the remaining coordinates are zero.

Next, *for $n > 2$, ellipticity implies the evenness of the order m.* Indeed, consider, e.g., the equation

$$a_0(x,\xi',\zeta) = 0 \tag{6.1.10}$$

with respect to ζ; here ζ is written instead of ξ_n. By virtue of ellipticity, for real $\xi' \neq 0$, the roots of this equation are nonreal. If $n > 2$, then we can pass continuously from any point $\xi' \neq 0$ to the point $-\xi'$ along the hyperplane $\xi_n = 0$ avoiding the origin. Under such a passage, the numbers of roots ζ in the upper and the lower half-plane are preserved, and *these numbers are equal*, because, when we replace ξ' by $-\xi'$, each root ζ is mapped to the root $-\zeta$.

If the operator A is elliptic with parameter, then the coincidence of the numbers of the roots in the upper and the lower half-plane and evenness are obtained for $n = 2$ in a similar way, because we can pass from any point $(\xi',0) \neq 0$ in $\mathbb{R}^2 \times \Lambda$, $\xi \neq 0$, to $(-\xi',0)$ avoiding $(0,0)$.

If m is even, we set $m = 2l$.

Now consider an mth-order partial differential operator A on a closed manifold M. Generally, it can be written in the form (6.1.1) only locally, in local coordinates.

A convenient exception is the standard torus $\mathbb{T}^n = [0, 2\pi]^n$, on which global 2π-periodic coordinates $x = (x_1, \ldots, x_n)$ can be used.

If the coefficients in A are infinitely differentiable, then this is a bounded operator from $H^s(M)$ to $H^{s-m}(M)$ for any s.

An examination of the behavior of the principal symbol $a_0(x, \xi)$ under transformations of coordinates shows that $a_0(x, \xi)$ can be treated globally as a function on the cotangent bundle T^*M (see [132]). Therefore, on a manifold, the definitions of ellipticity and ellipticity with parameter still make sense; we mean now global ellipticity and ellipticity with parameter everywhere. They are automatically uniform by the compactness of the manifold.

A classical example of an elliptic operator on a manifold is the Beltrami–Laplace operator Δ on a Riemannian manifold M. Let us recall its expression. If the metric is locally written in the form $ds^2 = \sum g_{j,k}\, dx_j\, dx_k$, where $g_{j,k} = g_{k,j}$, $(g^{j,k})$ is the matrix inverse to $(g_{j,k})$, and g denotes the determinant of $(g_{j,k})$, then

$$\Delta = -\frac{1}{\sqrt{g}} \sum D_j(\sqrt{g}\, g^{j,k} D_k). \tag{6.1.11}$$

The basic statements of the theory of elliptic operators on a closed manifold are the equivalence of the ellipticity of an operator A to the Fredholm property of A (in particular, to the presence of a parametrix; see Section 18.1) as an operator from $H^s(M)$ to $H^{s-m}(M)$ for any s and the equivalence of its ellipticity with a parameter in Λ to the invertibility of the operator $A - \lambda I$ for $\lambda \in \Lambda$ with sufficiently large absolute values. Some results are also obtained for uniformly elliptic operators on $\mathbb{R}^n$, but in this case, there is no equivalence to the invertibility or the Fredholm property of A. The reason for this difference is that, for $s_1 < s_2$, the space $H^{s_2}(M)$ is embedded in $H^{s_1}(M)$ compactly (see Theorem 2.3.1), while for $\mathbb{R}^n$ instead of M, this is not true.

6.2 Main Theorems

Let A be an mth-order partial differential operator on M with C^∞ coefficients.

Theorem 6.2.1. *The following conditions are equivalent.*

$1°$. *The operator A is elliptic on M.*

$2°$. *This is a Fredholm operator from $H^s(M)$ to $H^{s-m}(M)$ for any s.*

$3°$. *The operator A has a two-sided parametrix B acting boundedly from $H^{s-m}(M)$ to $H^s(M)$ for which $T_1 = BA - I$ and $T_2 = AB - I$ are bounded operators from $H^s(M)$ to $H^{s+1}(M)$ for any s.*

$4°$. *The a priori estimate*

$$\|u\|_{H^s(M)} \le C_s(\|Au\|_{H^{s-m}(M)} + \|u\|_{H^{s-1}(M)}) \tag{6.2.1}$$

with a constant not depending on u holds.

The most important assertion is the equivalence of $1°$ and $2°$, and the theorem is usually proved by the scheme

$$1° \Rightarrow 3° \Rightarrow 2° \Rightarrow 4° \Rightarrow 1°. \tag{6.2.2}$$

In comparison with the abstract situation considered in Section 18.1, we additionally have the *scales* of spaces $X_s = H^s(M)$ and $Y_s = H^{s-m}(M)$; the latter is obtained from the former by shifting the index. Moreover, the parametrix has stronger properties than in our abstract case: the operators T_1 and T_2 are smoothing, i.e., increase smoothness; namely, they take functions in $H^s(M)$ to functions in $H^{s+1}(M)$ (in particular, these operators are compact in each $H^s(M)$). We refer to such parametrices as *qualified*.

We also mention that the above a priori estimate turns out to be *two-sided*: its right-hand side is dominated by the left-hand side. This shows that the spaces H^s are adequate to the operators under consideration.

As mentioned in Proposition 18.1.6, if the kernel of A is trivial, then the term $\|u\|_{s-1}$ on the right-hand side of the estimate can be omitted. In the general case, this norm can be replaced by the norm of any order lower than m.

In [26], a similar theorem will be proved for more general pseudodifferential elliptic operators by means of the *calculus* of these operators, which is constructed in advance. All analytical work is concentrated in the construction of this calculus. Instead, here we give an outline of the proof by a classical method of the theory of elliptic equations, which was developed long before this calculus. Its key ingredients are localization by using a partition of unity on M and "freezing coefficients." This method is known as the *method of freezing coefficients*.

Proof of Theorem 6.2.1. Let us verify that $1° \Rightarrow 3°$. We describe the construction of the right parametrix in detail.

Step 1. Consider the operator A on $\mathbb{R}^n$. We remove the lower-order terms and freeze the coefficients at a point x_0, obtaining the homogeneous operator

$$A_0 = a_0(x_0, D) = \sum_{|\alpha|=m} a_\alpha D^\alpha \tag{6.2.3}$$

with constant coefficients. We set

$$B_0 = F^{-1} \frac{|\xi|^m}{(|\xi|^m + 1)a_0(x_0, \xi)} F, \tag{6.2.4}$$

where F is the Fourier transform in the sense of distributions. The numerator cancels the singularity in the denominator at the origin. The fraction $|\xi|^m/(|\xi|^m + 1)$ tends to 1 as $|\xi| \to \infty$. Obviously, we have $A_0 B_0 = I + T_0$, where

$$T_0 = -F^{-1} \frac{1}{|\xi|^m + 1} F$$

is a smoothing operator (it increases the order of smoothness by m, acting boundedly from $H^s(\mathbb{R}^n)$ to $H^{s+m}(\mathbb{R}^n)$). Of course, this operator is not compact in $H^s(\mathbb{R}^n)$, but its compactness is not required. We refer to the operator B_0 as a *quasi-parametrix* for A_0.

Step 2. Consider the operator $a(x,D)$ on $\mathbb{R}^n$ with any lower-order terms (of order less than m) and leading coefficients close to constants equal to their values at x_0. We write it in the form

$$a(x,D) = a_0(x_0,D) + a_1(x,D) + a_2(x,D),   \qquad (6.2.5)$$

where $a_1(x,D)$ is the sum of the lower-order terms of $a(x,D)$ and $a_2(x,D)$ is obtained from the leading part of $a(x,D)$ by subtracting $a_0(x_0,D)$. Now we shall show that

$$a(x,D)B_0 = I + T_1 + T_2,   \qquad (6.2.6)$$

where T_1 is again a smoothing operator and T_2 is an operator with norm less than 1 for fixed s, provided that the values of the leading coefficients in $a(x,D)$ are close enough to their values at x_0. The operator $I + T_2$ is invertible, and the required quasi-parametrix for $a(x,D)$ is obtained in the form

$$B_0(I+T_2)^{-1}.   \qquad (6.2.7)$$

Let us verify the assertion concerning the right-hand side of (6.2.6). As we have seen, $a_0(x_0,D)B_0$ is the sum of the identity and a smoothing operator. Consider $a_2(x,D)B_0$. This operator consists of products of the coefficients in $a_2(x,D)$ (whose absolute values have small upper bounds) and operators of order zero. According to Corollary 1.9.4, this operator decomposes into the sum of a smoothing operator and the operator T_2, whose norm is small if the coefficients in $a_2(x,D)$ are close enough to zero. Gathering all smoothing terms in T_1, we obtain the quasi-parametrix (6.2.7).

Step 3. Take two systems of infinitely differentiable functions on the manifold, $\{\varphi_j(x)\}$ and $\{\psi_j(x)\}$. The former is a sufficiently fine finite partition of unity: $\sum \varphi_j = 1$. The latter consists of functions $\psi_j(x)$ such that, for each j, $\psi_j(x) = 1$ in a neighborhood of the support of φ_j and the support of ψ_j is contained in some coordinate neighborhood U_j on M. The operator A can be written in the form

$$A = \sum \psi_j A \varphi_j \cdot .   \qquad (6.2.8)$$

Suppose that we can pass to local coordinates in the jth summand and replace the operator A by an elliptic operator A_j on $\mathbb{R}^n$ with leading coefficients close to constants which has the quasi-parametrix B_j (constructed at Step 2). Then

$$A = \sum \psi_j A_j \varphi_j,   \qquad (6.2.9)$$

and we construct a right parametrix for A in the form

$$B = \sum \varphi_k B_k \psi_k \cdot .   \qquad (6.2.10)$$

Obviously,

$$ABf = \sum \psi_j A_j \varphi_j \varphi_k B_k \psi_k f. \qquad (6.2.11)$$

The only nonzero terms in this sum are those in which $\varphi_j \varphi_k \neq 0$. In these terms, we pass from the expression for A in the form A_j to the expression in the form A_k. This can be done if the supports of all φ_j are sufficiently small. The commutator $A_k(\varphi_j \varphi_k) - (\varphi_j \varphi_k) A_k$ is a partial differential operator of order at most $m - 1$. We obtain

$$ABf = \sum \varphi_j \varphi_k f + Tf = f + Tf,$$

where T is an operator increasing the smoothness of functions in $H^{s-m}(M)$ by 1 and, therefore, compact in $H^{s-m}(M)$. We have reached our goal.

Actually, this construction is performed for all s in a finite but arbitrarily long interval simultaneously (this is a little less than promised in the statement of the theorem; the truth of this theorem in full completeness will be proved in [26]).

The construction of a left parametrix is similar; the only difference is in obvious permutations in the expressions written above. We leave this construction to the reader. Both parametrices turn out to be two-sided (see Proposition 18.1.4).

$3° \Rightarrow 2° \Rightarrow 4°$: see Propositions 18.1.3 and 18.1.6.

The implication $4° \Rightarrow 1°$ is proved by contradiction. Suppose that ellipticity is violated at some point (x_0, ξ_0) of the cotangent bundle, where $\xi_0 \neq 0$. Then the a priori estimate turns out to be false: it is disproved by substituting the function $\varphi(x) \exp(\lambda x \cdot \xi_0)$ in local coordinates, where φ is a smooth function that has small support containing x_0 and takes the value 1 near this point and λ is a positive parameter tending to infinity. After reducing the exponential we see that if the support of φ is small enough, then the left-hand side grows somewhat faster than the right-hand one. We leave the verification of this assertion to the reader. □

The index of an elliptic operator does not depend on s. This will be explained in Section 6.3.

Remark 6.2.2. The proof can be changed as follows. Instead of constructing the left parametrix, we can derive an a priori estimate (again in three steps), which will also imply the finite dimensionality of the kernel and the closedness of the range; see Proposition 18.1.7.

Theorem 6.2.3. *Suppose that A is an elliptic operator and $Au = f$, where $u \in H^s(M)$ but $f \in H^{s-m+\tau}(M)$, $\tau > 0$. Then $u \in H^{s+\tau}(M)$.*

We refer to this theorem as the *theorem on the regularity of solutions*, or *on increasing the smoothness of solutions*. It is proved by applying the parametrix to our equation on the left:

$$BAu = u + Tu = Bf. \qquad (6.2.12)$$

We see that $u \in H^{\min(s+1, s+\tau)}(M)$; if $\tau > 1$, then the parametrix is applied again as many times as needed.

Corollary 6.2.4. *The kernel $\operatorname{Ker} A$ of an elliptic operator on M in any space $H^s(M)$ consists of infinitely differentiable functions and, therefore, does not depend on s.*

We add that Theorem 6.2.3 has a local version, which asserts that if the right-hand side has higher smoothness in some domain on the manifold, then the smoothness of the solution is accordingly higher on this domain. To prove this, we apply the parametrix on the left to the equality

$$A\psi u = (A\psi u - \psi A u) + \psi f,$$

where ψ is a smooth function supported in the domain under consideration and taking the value 1 in a smaller subdomain.

Theorem 6.2.5. *If an operator A on M is elliptic with a parameter in a sector Λ, then the equation*

$$(A - \lambda)u = f \tag{6.2.13}$$

with $f \in H^0(M)$ is uniquely solvable in $H^m(M)$ for $\lambda \in \Lambda$ with sufficiently large $|\lambda|$. Moreover, the a priori estimate

$$\|u\|_{H^m(M)} + |\lambda| \|u\|_{H^0(M)} \le C \|f\|_{H^0(M)} \tag{6.2.14}$$

with a constant not depending on λ holds for these λ. The condition of ellipticity with parameter is necessary for the validity of this estimate.

For simplicity, we give here the statement only for $s = m$.

Proof. The proof of this theorem is similar to but simpler than that of Theorem 6.2.1. We consider the solution in the norm $\||u\||_{H^m(M)}$, which depends on the parameter and equals the left-hand side of (6.2.14). Already at the first step, instead of the quasi-parametrix, we obtain the inverse operator in the form

$$B_0 = F^{-1} \frac{1}{a_0(x_0, \xi) - \lambda} F; \tag{6.2.15}$$

it is sufficient to assume $\lambda \in \Lambda$ to be nonzero. At the second step we obtain a right inverse operator for λ with sufficiently large absolute values. Even larger absolute values of λ should be taken at the third step, and again a right inverse operator is obtained.   □

For the remaining values of λ, we obtain the Fredholm property for $A - \lambda$ with index zero.

6.3 Adjoint Operators

Let $(u, v)_M$ denote an inner product in $L_2(M)$, and let A and A^* be partial differential operators of order m related by

$$(Au, v)_M = (u, A^* v)_M \tag{6.3.1}$$

for infinitely differentiable functions u and v. Then these operators are said to be *formally adjoint*, and if $A = A^*$, then A is a *formally self-adjoint operator*. For example, the Beltrami–Laplace operator on a Riemannian manifold is formally self-adjoint. Of course, the adjointness relation between A and A^* depends on the choice of the inner product. If the local coordinates are compatible with the density on M determining the inner product (see Section 2.1), then, in these coordinates,

$$A^* = \sum_{|\alpha| \le m} D^\alpha [\overline{a_\alpha(x)}\cdot] \quad \text{if and only if} \quad A = \sum_{|\alpha| \le m} a_\alpha(x) D^\alpha. \qquad (6.3.2)$$

In particular, this is true for operators on the standard torus with periodic coordinates.

Clearly, $A^{**} = A$.

Suppose that the operator A is elliptic. Then so is the operator A^*, because its principal symbol is the function complex conjugate to the principal symbol of A. Both these operators have finite-dimensional kernels consisting of infinitely differentiable functions. For A and A^* considered as operators from $H^m(M)$ to $H^0(M) = L_2(M)$, relation (6.3.1) remains valid (it is transferred to functions in $H^m(M)$ by the passage to the limit). This means that A and A^* *are adjoint as unbounded operators in the Hilbert space* $H^0(M)$ (and have common domain $H^m(M)$). Their ranges $R(A)$ and $R(A^*)$ are closed in $H^0(M)$.

Proposition 6.3.1. *The following relations hold*:

$$H^0(M) = R(A) \oplus \operatorname{Ker} A^* = R(A^*) \oplus \operatorname{Ker} A. \qquad (6.3.3)$$

Proof. It is seen from (6.3.1) that if a function v belongs to $\operatorname{Ker} A^*$, then it is orthogonal to $R(A)$ in $L_2(M)$. Let us verify the converse. Relation (6.3.1) remains valid for $u \in H^m(M)$ and $v \in H^0(M)$; in this case, we consider A^*v in $H^{-m}(M)$, extending the inner product on the right to $H^m(M) \times H^{-m}(M)$. This means that $A \colon H^m(M) \to H^0(M)$ and $A^* \colon H^0(M) \to H^{-m}(M)$ *are adjoint as operators in Banach spaces* (which are Hilbert here). Suppose that v is orthogonal to $R(A)$. Then $(u, A^*v)_M = 0$ for all $u \in H^m(M)$, so that $A^*v = 0$ in $H^{-m}(M)$; cf. Proposition 18.1.8. But we know that the kernel of an elliptic operator consists of C^∞ functions, in particular, $v \in \operatorname{Ker} A^* \subset H^m(M)$. We have proved the first equality for $H^0(M)$ in (6.3.3); the proof of the second is similar. $\qquad\square$

Corollary 6.3.2. *The codimension of the range of an elliptic operator $A \colon H^m(M) \to H^0(M)$ coincides with the dimension of the kernel of A^*. Therefore, for the index $\varkappa(A)$ of A, the formula*

$$\varkappa(A) = \dim \operatorname{Ker} A - \dim \operatorname{Ker} A^* \qquad (6.3.4)$$

is true, and

$$\varkappa(A^*) = -\varkappa(A). \qquad (6.3.5)$$

Now the index of an operator $A \colon H^s(M) \to H^{s-m}(M)$ can be defined by (6.3.4) for any s. We see that it does not depend on s.

Let us make the following additional remark. For any s, relation (6.3.1) remains valid if $u \in H^s(M)$ and $v \in H^{m-s}(M)$. This means that, *for any s, the operators $A: H^s(M) \to H^{s-m}(M)$ and $A^*: H^{m-s}(M) \to H^{-s}(M)$ are mutually adjoint as operators in Banach spaces*. It follows again that a function belongs to $R(A)$ in $H^{s-m}(M)$ if and only if it is orthogonal to $\operatorname{Ker} A^*$ with respect to the extension of the form $(\cdot, \cdot)_M$. Therefore, the codimension of the range of A in $H^{s-m}(M)$ does not depend on s; thus, the assertion that *the index $\varkappa(A)$ does not depend on s* remains valid under the original definition of index (given in Section 18.1). Moreover, the dimensions of the kernels $\operatorname{Ker} A$ and $\operatorname{Ker} A^*$, whose difference equals the index $\varkappa(A)$, do not depend on s either.

Properties of index will be discussed in more detail in [26]. We shall verify that the index of an operator does not depend on the lower-order terms of this operator and is homotopy invariant, that is, does not change under an ellipticity-preserving continuous variation of the coefficients in the principal symbol (cf. Proposition 18.1.12 (3)). The problem of calculating the index of a general (matrix) elliptic operator was essentially stated in Gel'fand's celebrated paper [164]. This paper has exerted a strong influence on the development of the theory of elliptic equations. In particular, it gave rise to the theory of elliptic pseudodifferential operators. The problem of index calculation was solved in topological terms for operators on a closed manifold by Atiyah and Singer [45].

6.4 Some Spectral Properties of Elliptic Operators

Let A be an mth-order elliptic operator on a manifold M. Consider it as an unbounded operator in $L_2(M)$ with domain $H^m(M)$. Suppose that the resolvent set of A is nonempty. (A necessary condition for this is the vanishing of the index.) Then this is an operator with discrete spectrum, because its resolvent is a bounded operator from $L_2(M)$ to the space $H^m(M)$, which is compactly embedded in $L_2(M)$.

It follows from Theorem 6.2.3 that all generalized eigenfunctions (root functions) of such an operator, that is, all eigenfunctions or eigen- and associated functions, belong to all spaces $H^s(M)$ and, hence, are infinitely smooth. They remain generalized eigenfunctions in all these spaces.

Any formally self-adjoint elliptic operator turns out to be self-adjoint in $L_2(M)$. An example of such an operator is the Beltrami–Laplace operator on a closed Riemannian manifold. Numerous mathematicians, beginning with H. Weyl (1912), studied the asymptotic behavior of eigenvalues of a self-adjoint elliptic operator. Its principal symbol (in the scalar case) is real and of constant sign. If this sign is plus, then all but possibly finitely many eigenvalues of the operator are positive and have a unique accumulation point, $+\infty$. Let us number them with positive integers in nondecreasing order with multiplicities taken into account. Then the following (Weyl's) asymptotic formula is valid:

$$\lambda_j \sim cj^q, \quad q = \frac{m}{n}, \tag{6.4.1}$$

where c is a positive constant expressed in terms of the principal symbol. We do not give this expression here. The difficult problem of estimating the remainder in (6.4.1), i.e., the difference between the left- and right-hand sides, has been studied by many authors. Hörmander showed that an optimal (in the general case) estimate for semibounded scalar (pseudodifferential) operators has the form $O(j^{(m-1)/n})$ (see [184]); under certain conditions, this estimate can be somewhat improved.

Eigenfunctions of a self-adjoint elliptic operator form an orthonormal basis $\{e_j\}$ in the space $L_2(M)$. Any function $f \in L_2(M)$ can be expanded in a series of the form

$$f = \sum c_j e_j, \quad c_j = (f, e_j)_M, \tag{6.4.2}$$

which unconditionally converges in the norm of this space. But the eigenfunctions belong to all spaces $H^s(M)$ and provide unconditional bases in all of them (with account of isomorphisms). Therefore, for $f \in H^s(M)$, the series unconditionally converges in $H^s(M)$.

By a *nearly self-adjoint elliptic operator* (or a *weak perturbation of a self-adjoint elliptic operator*) of order m we mean an operator which differs from a self-adjoint one by a term of order at most $m-1$. According to Theorem 18.3.1, the system of its generalized eigenfunctions is complete in $L_2(M)$ provided that its principal symbol is positive; as a consequence, this system is complete in all $H^s(M)$. Theorem 18.3.2 also applies to non-self-adjoint elliptic operators with $p = n/m$.

The directions of the most rapid decay of the resolvent (if they exist) are precisely the directions of ellipticity with parameter.

The spectral properties of elliptic operators will be discussed in more detail in [26].

The self-adjoint operator $-\Delta + I$ on a Riemannian manifold has positive eigenvalues. Therefore, its real powers $(-\Delta + I)^t$ are defined. This operator isomorphically maps $H^s(M)$ to $H^{s-2t}(M)$ for any s and t. This can be verified by means of interpolation theory (see Section 13.8.1).

6.5 Generalizations

Here we only list some generalizations of the theory of elliptic operators, without going into details.

1. The theory can be generalized to *matrix* operators, which act on vector-valued functions. This generalization is simplest when the leading parts of all elements of the matrix

$$a(x, D) = (a_{j,k}(x, D)) \tag{6.5.1}$$

are considered as having the same order m. In this case, from the symbols of these higher-order parts a (matrix) principal symbol $a_0(x, \xi)$ is composed. Some of its elements may be zero. The ellipticity of the operator (6.5.1) means that the determinant of the principal symbol does not vanish for $\xi \neq 0$. A self-adjoint matrix elliptic operator has Hermitian principal symbol with real nonzero eigenvalues. If there are eigenvalues of both signs, then the eigenvalues of the operator have the accumulation points $\pm\infty$ with asymptotics of type (6.4.1). For an optimal estimate of the remainder (in the case of pseudodifferential operators), see Ivrii [188].

A more general definition of ellipticity is as follows. Suppose that the matrix $a(x,D)$ has size $d\times d$. We fix two systems $m_1,\ldots,m_d$ and $t_1,\ldots,t_d$ of nonnegative integers and assume that the leading order in $a_{j,k}(x,D)$ is $m_j + t_k$. The principal symbol is composed of the symbols of these higher-order parts (some of which may vanish), and the ellipticity condition is again the assumption that the determinant of the principal symbol is nonzero. This property is known as *Douglis–Nirenberg ellipticity* [131]. The corresponding operator can be treated as an operator from the direct product of the spaces $H^{s+t_k}(M)$ to the direct product of the spaces $H^{s-m_j}(M)$. Note that it is not easy to define ellipticity with parameter in this case if the orders $m_j + t_j$ of diagonal elements depend on j.

2. Matrix operators can be considered *on sections of vector bundles*. Roughly speaking, this means that not only the local coordinate system for the independent variables but also the coordinate system in which the vector of the scalar functions under consideration is written change as a point moves along the manifold; see, e.g., [185, vol. 3, Sec. 18].

3. We can define ellipticity with parameter for operators polynomially depending on a parameter; see, e.g., [10] and [34]. The parameter is then considered as having a fixed weight with respect to differentiation. For example, the operator may have the form

$$\sum_{j=0}^{m} \lambda^{m-j} A_j(x,D), \tag{6.5.2}$$

where each A_j is an operator of order j; here the weight of the parameter equals 1, while in Eq. (6.2.13) the weight equals m. If $a_{0,j}(x,\xi)$ is the principal symbol of A_j, then for (6.5.2), ellipticity with parameter in a sector Λ means that

$$\det \sum \lambda^{m-j} a_{0,j}(x,\xi) \neq 0, \quad (\xi,\lambda) \neq 0, \ \lambda \in \Lambda. \tag{6.5.3}$$

Such operators are the subject of the more general spectral theory of *operator pencils*, that is, operators polynomially depending on the spectral parameter; see, e.g., [249].

A relationship between problems elliptic with parameter and parabolic equations is described in the next section.

4. Relaxing the smoothness assumptions, we can consider elliptic operators on spaces $H^s(M)$ with s ranging in a finite interval. We shall have an occasion to use this possibility in Section 12 (see Theorem 12.1.1).

7 Elliptic Boundary Value Problems in Smooth Bounded Domains

In this section we consider only spaces H^s with nonnegative s, i.e., the Sobolev–Slobodetskii L_2-spaces.

7.1 Definitions and Statements of Main Theorems

Let Ω be a bounded domain in $\mathbb{R}^n$ with smooth boundary Γ. Suppose that we have a scalar linear partial differential operator

$$A = a(x, D) = \sum_{|\alpha| \leq 2l} a_\alpha(x) D^\alpha \qquad (7.1.1)$$

on Ω of even order $2l$ with coefficients infinitely differentiable on $\overline{\Omega}$.

Suppose also given l boundary operators

$$B_j = b_j(x, D) = \sum_{|\beta| \leq r_j} b_{j\beta}(x) D^\beta \quad (j = 1, \ldots, l) \qquad (7.1.2)$$

on Γ of nonnegative orders r_j with coefficients infinitely differentiable on Γ. The boundary value problem which we shall consider in this section is

$$a(x, D)u(x) = f(x) \text{ in } \Omega, \qquad (7.1.3)$$

$$b_j(x, D)u(x) = g_j(x) \text{ on } \Gamma \quad (j = 1, \ldots, l). \qquad (7.1.4)$$

With this boundary value problem we associate the operator $\mathfrak{A}u = (f, g_1, \ldots, g_l)$ taking each solution to the corresponding set of right-hand sides.

In the simplest functional setting of the problem, the spaces for the functions u, f, and g_j are chosen as follows:

$$u \in H^s(\Omega), \quad f \in H^{s-2l}(\Omega), \quad g_j \in H^{s-r_j-1/2}(\Gamma). \qquad (7.1.5)$$

Here we assume for simplicity that

$$s \geq 2l, \quad s > \max r_j + 1/2. \qquad (7.1.6)$$

The following proposition must be obvious for those who have read the preceding sections.

Proposition 7.1.1. *The operator* $\mathfrak{A}$ *acts boundedly from the space* $H^s(\Omega)$ *to the direct product*

$$H^s(\Omega, \Gamma) = H^{s-2l}(\Omega) \times \prod_{j=1}^{l} H^{s-r_j-1/2}(\Gamma). \qquad (7.1.7)$$

In other words, the estimate

$$\|f\|_{H^{s-2l}(\Omega)} + \sum_{j=1}^{l} \|g_j\|_{H^{s-r_j-1/2}(\Gamma)} \le C_s \|u\|_{H^s(\Omega)} \qquad (7.1.8)$$

with a constant not depending on u is valid.

Below we give the definition of ellipticity for this problem.

The *first ellipticity condition* is that the operator A is elliptic on $\overline{\Omega}$, that is, for its main symbol $a_0(x,\xi)$, we have

$$a_0(x,\xi) \ne 0 \quad (x \in \overline{\Omega},\, 0 \ne \xi \in \mathbb{R}^n). \qquad (7.1.9)$$

The *second ellipticity condition*, which is called the *regular ellipticity* of A, is that the equation $a_0(x,\xi',\zeta) = 0$ with $\xi' \ne 0$ has the same number of roots ζ in the upper and lower half-planes (and this number equals l). It was mentioned in Remark 6.1.1 that this condition holds automatically if $n > 2$ or $n = 2$ and the operator A is elliptic with parameter. For $n = 2$, it eliminates examples of the type $(D_1 + iD_2)^2$. By continuity, regular ellipticity is preserved by any rotation of the coordinate system, and if it holds at some point x, then it holds everywhere.

The *third condition* is known as the *Lopatinskii condition*. It is imposed on the operators of the problem at each point x_0 of the boundary Γ. Its statement has the simplest form when the origin is transferred to a point x_0 and the coordinate system is rotated so that the $t = x_n$ axis is directed along the inner normal to the boundary at this point. Suppose that the operators of the problem are rewritten in this coordinate system. Consider the following problem on the ray $\mathbb{R}_+ = \{t : t > 0\}$ for fixed $\xi' = \xi_0' \ne 0$:

$$\begin{aligned} a_0(x_0,\xi_0',D_t)v(t) &= 0 \qquad (t > 0), \\ b_{j0}(x_0,\xi_0',D_t)v|_{t=0} &= h_j \quad (j = 1,\dots,l), \end{aligned} \qquad (7.1.10)$$

where $b_{j0}(x,\xi)$ are the principal symbols of the operators b_j:

$$b_{j,0}(x,\xi) = \sum_{|\beta|=r_j} b_{j\beta}(x)\xi^{\beta}. \qquad (7.1.11)$$

Problem (7.1.10) is required to have precisely one solution in $L_2(\mathbb{R}_+)$ for any $\xi_0' \ne 0$ and any numbers h_j.

Note that, under the above assumptions, the space of those solutions of the equation $a_0(x_0,\xi_0',D_t)v(t) = 0$ which belong to $L_2(\mathbb{R}_+)$ consists of functions with absolute value decreasing exponentially as $t \to +\infty$; the dimension of this space is l, and the number of boundary conditions equals this dimension (cf. Remark 7.1.2, 1 below). Problem (7.1.10) is obtained from the original problem by freezing the coefficients at the point x_0, removing the lower-order terms, and applying the formal Fourier transform with respect to the tangent variables.

The Lopatinskii condition first appeared in its full generality in Lopatinskii's paper [242]. It is sometimes called the *Shapiro–Lopatinskii condition*, because similar considerations were performed by Shapiro [340, 341] under more special assumptions. This condition is also referred to by saying that the boundary operators *cover* the given elliptic operator.

If all of the three conditions stated above hold, then the problem is said to be *elliptic*.

The simplest example of an elliptic problem is the Dirichlet problem for the Poisson equation:

$$-\Delta u = f \text{ on } \Omega, \quad u = g \text{ on } \Gamma. \tag{7.1.12}$$

The second example is the Neumann problem for the same equation:

$$-\Delta u = f \text{ on } \Omega, \quad \partial_\nu u = g \text{ on } \Gamma, \tag{7.1.13}$$

where ∂_ν is the inner (for definiteness) normal derivative.

Problem 1. Verify the ellipticity of these problems.

Remark 7.1.2. The Lopatinskii condition can be reformulated as follows.

1. If the operator $a(x, D)$ is regularly elliptic, then, at each $\xi' \neq 0$, the polynomial $a(\zeta) = a_0(x_0, \xi', \zeta)$ can be factorized as

$$a_0(\zeta) = a_0^+(\zeta)a_0^-(\zeta), \tag{7.1.14}$$

where the roots of the polynomials $a_0^+(\zeta)$ and $a_0^-(\zeta)$ belong to the upper and lower half-planes, respectively. The coefficients in these polynomials depend smoothly on (x_0, ξ'). All solutions of the equation $a_0^+(D_t)v(t) = 0$ decrease exponentially in absolute value as $t \to \infty$; these are all solutions of the first equation in (7.1.10) whose absolute values decrease as $t \to \infty$. We set $b_{j,0}(\zeta) = b_{j,0}(x_0, \xi', \zeta)$. Given any $\xi' \neq 0$, the Lopatinskii condition at the point under consideration can be written as follows: *The problem*

$$a_0^+(D_t)v(t) = 0 \quad (t > 0), \quad b_{j,0}(D_t)v(t)|_{t=0} = h_j \quad (j = 1, \dots, l) \tag{7.1.15}$$

is uniquely solvable for any h_j.

It is easy to see from here that the Lopatinskii condition is also equivalent to the condition that *the remainders $\widetilde{b}_{0,j}(\zeta)$ of the division of the polynomials $b_{0,j}(\zeta)$ by $a_0^+(\zeta)$ are linearly independent.* In (7.1.15), the polynomials $b_{j,0}$ can be replaced by $\widetilde{b}_{j,0}$.

2. This also implies the equivalence of the Lopatinskii condition to the uniqueness for problem (7.1.10) or (7.1.15) in $L_2(\mathbb{R}_+)$: if $h_j = 0$ for all j, then the solution belonging to $L_2(\mathbb{R}_+)$ is trivial.

3. On the other hand, the Lopatinskii condition is equivalent to that obtained by replacing zero on the right-hand side of the first equation by any function $f(t) \in L_2(\mathbb{R}_+)$. Indeed, extending this function by zero to $t < 0$, we can easily construct a

solution of the equation $a_0(D_t)v_0(t) = f(t)$ in $L_2(\mathbb{R}_+)$ by using the Fourier transform as

$$v_0(t) = F^{-1}[a_0(\xi)]^{-1}(Ff)(\xi) \, d\xi.$$

Subtracting $v_0(t)$ from the solution $v(t)$ of the equation $a_0(D_t)v(t) = f(t)$, we obtain problem (7.1.10).

The *Dirichlet problem* for a regularly elliptic equation of order $2l$ is the problem with boundary conditions

$$\partial_\nu^j u = g_j \quad (j = 0, \ldots, l-1). \tag{7.1.16}$$

As a consequence of the first remark, we obtain the following generalization of the first example: *The Dirichlet problem for any (scalar) regularly elliptic equation of order $2l$ is elliptic.* Indeed, in this case, the remainders are the polynomials $b_{j,0}(\zeta) = \zeta^j$ themselves.

In particular, the Dirichlet problem for the equation

$$\Delta^l u = f \tag{7.1.17}$$

is elliptic.

The "mainest" theorem of the theory of the elliptic problems is similar to Theorem 6.2.1 and is stated as follows.

Theorem 7.1.3. *The following conditions are equivalent.*

1°. *The boundary value problem* (7.1.3)–(7.1.4) *is elliptic.*

2°. *The operator* $\mathfrak{A} \colon H^s(\Omega) \mapsto H^s(\Omega, \Gamma)$ *is Fredholm.*

3°. *The operator* $\mathfrak{A}$ *has a two-sided parametrix* $\mathfrak{R}$ *acting boundedly from* $H^s(\Omega, \Gamma)$ *to* $H^s(\Omega)$ *and such that the operators* $\mathfrak{A}\mathfrak{R} - \mathcal{I}$ *and* $\mathfrak{R}\mathfrak{A} - I$ *act boundedly from* $H^s(\Omega, \Gamma)$ *to* $H^{s+1}(\Omega, \Gamma)$ *and from* $H^s(\Omega)$ *to* $H^{s+1}(\Omega)$, *respectively. Here* $\mathcal{I}$ *and* I *denote the corresponding identity operators.*

4°. *The a priori estimate*

$$\|u\|_{H^s(\Omega)} \leq C_s' \left[\|f\|_{H^{s-2l}(\Omega)} + \sum_{j=1}^{l} \|g_j\|_{H^{s-r_j-1/2}(\Gamma)} + \|u\|_{H^0(\Omega)} \right] \tag{7.1.18}$$

holds, where the constant does not depend on u.

We again have two scales of spaces. As to the estimate, it is again two-sided: the right-hand side is dominated by the left-hand side (see (7.1.8)). This is yet another evidence that the Sobolev–Slobodetskii spaces are adequate to the problems under consideration. In the case of uniqueness, the last term on the right-hand side in (7.1.18) is unnecessary. The parametrix is again qualified.

The next theorem is on increasing the smoothness of solutions, or on the regularity of solutions.

Theorem 7.1.4. *Suppose that the boundary value problem* (7.1.3)–(7.1.4) *is elliptic and a number s satisfies conditions* (7.1.6). *Suppose also that $\tau > 0$,*

$u \in H^s(\Omega)$, $f \in H^{s-2l+\tau}(\Omega)$,   and $g_j \in H^{s-r_j-1/2+\tau}(\Gamma)$ for $j = 1,\ldots,l$.

Then $u \in H^{s+\tau}(\Omega)$.

In particular, if $f \in C^\infty(\overline{\Omega})$ and $g_j \in C^\infty(\Gamma)$, then $u \in C^\infty(\overline{\Omega})$.

This theorem has a local version, which asserts that if the smoothness of the right-hand sides is enhanced near some point, then the smoothness of the solution near this point is enhanced accordingly.

Next, there are conditions of *ellipticity with parameter*, which guarantee the unique solvability of the problem at large absolute values of the parameter. Below we state the simplest result of this kind. Instead of the equation $a(x, D)u = f$, we consider the equation with parameter

$$a(x, D)u - \lambda u = f \qquad (7.1.19)$$

in Ω. For simplicity, we assume that the boundary conditions do not depend on λ. The parameter λ varies within a closed sector Λ in the complex plane with vertex at the origin. The definition of the ellipticity of a problem with parameter is similar to that of ellipticity. The changes are that the principal symbol $a_0(x, \xi)$ is replaced by $a_0(x, \xi) - \lambda$, it is assumed that this difference does not vanish for $(\xi, \lambda) \neq 0$ with $\xi \in \mathbb{R}^n$ and $\lambda \in \Lambda$, and in the Lopatinskii condition it is assumed that the problem on the ray for the equation

$$[a_0(x_0, \xi_0', D_t) - \lambda]v(t) = 0 \qquad (7.1.20)$$

has a unique solution decreasing as $t \to +\infty$ for $(\xi_0', \lambda) \neq 0$, $\lambda \in \Lambda$. The conditions of ellipticity with parameter contain the ellipticity conditions (which correspond to $\lambda = 0$).

For example, the Dirichlet problem for the equation $\Delta u + \lambda u = f$ is elliptic with parameter along any ray except the positive half-axis. The same is true for the Neumann problem.

Problem 2. Prove this. Prove also that, if a regularly elliptic equation of order $2l$ is elliptic with parameter along some rays, then the Dirichlet problem for this equation is elliptic with parameter along the same rays.

Theorem 7.1.5. *If the boundary value problem (7.1.19), (7.1.4) is elliptic with parameter in Λ, then, for any s satisfying condition (7.1.6), this problem is uniquely solvable for $\lambda \in \Lambda$ with sufficiently large absolute values.*

Moreover, there is an a priori estimate uniform in the parameter. The simplest estimate under the assumptions $s = 2l > \max r_j$ and $g_j = 0$ is

$$\|u\|_{H^{2l}(\Omega)} + |\lambda| \|u\|_{H^0(\Omega)} \leq C\|f\|_{H^0(\Omega)}, \qquad (7.1.21)$$

where the constant does not depend on u and λ. The conditions of ellipticity with parameter are necessary for this estimate to hold.

For other λ, the problem turns out to be Fredholm with index zero.

Note that problems with parameter are obtained from nonstationary mixed problems in the cylindrical domain $\Omega \times (0,\infty)$ with coefficients not depending on time $t \in (0,\infty)$ (above, the letter t had a different meaning). For example, consider the heat equation

$$\Delta U(x,t) - \partial_t U(x,t) = F(x,t) \tag{7.1.22}$$

in such a domain with Dirichlet boundary condition

$$U(x,t) = G(x,t) \quad (x \in \Gamma) \tag{7.1.23}$$

on the lateral surface and homogeneous (for simplicity) initial condition

$$U(x,0) = 0. \tag{7.1.24}$$

The formal Laplace transform

$$u(x,\lambda) = \int_0^\infty U(x,t)e^{-\lambda t}dt \tag{7.1.25}$$

reduces this problem to the problem with parameter for the Laplace operator, which was mentioned before the statement of Theorem 7.1.5. Here Λ is the right half-plane. In fact, this is a way to study nonstationary, "parabolic," problems; see, e.g., [10] and [34]. There is also a direct way—the study of mixed problems with coefficients depending in addition on t by the method of freezing coefficients (see, e.g., Solonnikov's paper [357] and Eidel'man's survey [145]).

Problem 3. Verify that the Dirichlet problem for the equation $\Delta^l u - \lambda u = 0$ is elliptic with parameter along any ray except $\mathbb{R}_+$ or $\mathbb{R}_-$. Derive from this that, for $s \geq 2l$, a bounded operator which maps any set of functions $v_j \in H^{s-j-1/2}(\Gamma)$ $(j = 0,\ldots,l-1)$ to a function $u \in H^s(\Omega)$ with Cauchy data $v_0,\ldots,v_{l-1}$ can be defined independently of s. (Cf. the statement after Theorem 5.1.9.)

Now we outline the proofs of Theorems 7.1.3–7.1.5. To simplify calculations, we assume that $r_j < 2l$ for all j and $s = 2l$. We dwell on only important points and omit some technical details similar to those considered in the preceding section but a little more cumbersome.

7.2 Proofs of Main Theorems

As in Section 6, the proof of Theorem 7.1.3 consists of three steps. We begin with the first one. Consider the problem in the half-space $\mathbb{R}_+^n$ for operators without lower-order terms and assume that the coefficients do not depend on x. The problem has the form

$$a_0(D)u(x) = f(x) \quad (x_n > 0), \tag{7.2.1}$$
$$b_{j,0}(D)u(x)|_{x_n=0} = g_j(x') \quad (j = 1,\ldots,l). \tag{7.2.2}$$

Applying the (formal) Fourier transform F' with respect to the tangent variables and setting $t = x_n$, we obtain the problem

$$a_0(\xi', D_t)v(\xi', t) = h(\xi', t) \quad (t > 0), \tag{7.2.3}$$
$$b_{j,0}(\xi', D_t)v(\xi', t)|_{t=0} = h_j(\xi') \quad (j = 1,\ldots,l), \tag{7.2.4}$$

where $v = F'u$, $h = F'f$, and $h_j = F'g_j$. We denote the space of solutions of the equation $a_0(\xi', D_t)v(t) = 0$ which decrease as $t \to \infty$ by $\mathfrak{M} = \mathfrak{M}(\xi')$ and refer to any basis $\omega_1(t),\ldots,\omega_l(t)$ in this space as a *stable basis*.

7.2.1 A Canonical Basis

In (7.2.3)–(7.2.4) with $h = 0$, we can replace the polynomial $a_0(\xi', \zeta)$ in ζ by $a_0^+(\xi', \zeta)$ and the polynomials $b_j(\xi', \zeta)$ in ζ by the remainders $\widetilde{b}_{j,0}(\xi', \zeta)$ after dividing them by $a_0^+(\xi', \zeta)$ (the variables ξ' play the role of parameters):

$$a_0^+(\xi', D_t)v(\xi', t) = 0 \quad (t > 0), \tag{7.2.5}$$
$$\widetilde{b}_{j,0}(\xi', D_t)v(\xi', t)|_{t=0} = h_j(\xi') \quad (j = 1,\ldots,l). \tag{7.2.6}$$

Without loss of generality, we assume that the leading coefficient in a_0^+ is equal to 1. Note that the functions $\widetilde{b}_{j,0}(\xi', \zeta)$ are positive homogeneous in (ξ', ζ) of degree r_j for $\xi' \neq 0$.

A stable basis can be constructed in the form of the contour integrals

$$\omega_k(\xi', t) = \frac{1}{2\pi} \int_\gamma \frac{e^{i\zeta t}\zeta^{k-1}}{a_0^+(\xi', \zeta)} d\zeta \quad (k = 1,\ldots,l), \tag{7.2.7}$$

where $\gamma = \gamma(\xi')$ is a closed contour in the upper half-plane enclosing all roots ζ of the polynomial $a_0^+(\xi', \zeta)$, i.e., all roots of the polynomial $a_0(\xi', \zeta)$ in this half-plane. Obviously, these are solutions of Eq. (7.2.5) decreasing in absolute value as $t \to +\infty$, and these solutions are linearly independent. The contour may be changed, but locally, near the chosen point ξ'_0, it can be assumed to be independent of ξ'. We define a *canonical basis* of $\Omega_k(\xi', t)$ in $\mathfrak{M}(\xi')$ by the conditions

$$\widetilde{b}_{j,0}(\xi', D_t)\Omega_k(\xi', t)|_{t=0} = \delta_{j,k} \quad (j, k = 1,\ldots,l). \tag{7.2.8}$$

It exists by virtue of the Lopatinskii condition, and each function $\Omega_k(\xi', t)$ can be determined by substituting a linear combination of the functions $\omega_1(\xi', t), \ldots, \omega_l(\xi', t)$ into condition (7.2.8) with fixed k. We obtain a linear system of equations with nonzero determinant, which uniquely determines the coefficients. Moreover, the $\Omega_k(\xi', t)$ have the form (7.2.7) but with the ζ^{k-1} replaced by some polynomials $N_k(\xi',\zeta)$ in ζ. These polynomials are determined not uniquely, but they can be assumed to satisfy the conditions

$$\widetilde{b}_{j,0}(\xi',\zeta)N_k(\xi',\zeta) = \delta_{j,k}\zeta^{l-1}.$$

Clearly, each $N_k(\xi',\zeta)$ is a polynomial of degree at most $l-1$ in ζ, which is positive homogeneous in (ξ',ζ) of degree $l-r_k-1$ for $\xi' \neq 0$. We have proved the following assertion.

Proposition 7.2.1. *The canonical basis has the form*

$$\Omega_k(\xi',t) = \frac{1}{2\pi i}\int_\gamma \frac{e^{i\zeta t}N_k(\xi',\zeta)}{a_0^+(\xi',\zeta)}\,d\zeta \quad (k = 1,\dots,l), \tag{7.2.9}$$

where the $N_k(\xi',\zeta)$ are polynomials in ζ of degree at most $l-1$ in which all coefficients are C^∞ functions with respect to ξ' at $\xi' \neq 0$. Furthermore, the N_k are positive homogeneous of degree $l-r_k-1$ with respect to (ξ',ζ) at these ξ'.

Explicit expressions for N_k are given in [9] in a slightly different notation.

Proposition 7.2.2. *The following estimates with a constant not depending on ξ' are valid:*

$$\int_0^\infty |D_t^j\Omega_k(\xi',t)|^2\,dt \leq C_1|\xi'|^{2(j-r_k)-1} \quad (j = 0,\dots,2l;\ k = 1,\dots,l). \tag{7.2.10}$$

Proof. Suppose that the contour γ does not depend on ξ' with $|\xi'| = 1$, that is, $\gamma(\xi') = \gamma_0$ for such ξ'. Suppose also that, for other ξ', $\gamma(\xi')$ is obtained from γ_0 by a homothety with coefficient $|\xi'|$, that is, $\gamma(\xi') = |\xi'|\gamma_0$. Clearly, this assumption can be made. Then, differentiating the integrand in (7.2.9) j times with respect to t and treating $\zeta/|\xi'|$ as a new variable, we obtain

$$|D_t^j\Omega_k(\xi',t)| \leq C|\xi'|^{j-r_k}e^{-\varepsilon|\xi'|t} \quad (j = 0,\dots,2l,\ k = 1,\dots,l)$$

with positive C and ε not depending on ξ' and t. Therefore,

$$\int_0^\infty |D_t^j\Omega_k(\xi',t)|^2\,dt \leq C|\xi'|^{2(j-r_k)}\int_0^\infty e^{-2\varepsilon|\xi'|t}\,dt.$$

Taking $|\xi'|t$ for a new variable, we obtain the required estimate. $\qquad\square$

7.2.2 The A Priori Estimate

We assume for simplicity that $s = 2l > r_k$. At the first step, we want to obtain the estimate

$$\|u\|^2_{H^{2l}(\mathbb{R}^n_+)} \le C\Big[\|f\|^2_{H^0(\mathbb{R}^n_+)} + \sum_{k=1}^{l} \|g_k\|^2_{H^{2l-r_k-1/2}(\mathbb{R}^{n-1})} + \|u\|^2_{H^0(\mathbb{R}^n_+)}\Big] \qquad (7.2.11)$$

for problem (7.2.1), (7.2.2). For convenience, we have replaced all norms by their squares. First, suppose that $f = 0$. Then the solution of problem (7.2.5), (7.2.6) can be expressed as

$$v(\xi',t) = \sum_{k=1}^{l} \Omega_k(\xi',t)h_k(\xi'). \qquad (7.2.12)$$

Now it suffices to prove the estimate

$$\sum_{j=0}^{2l} |\xi'|^{2(2l-j)} \int_0^\infty |D_t^j v(\xi',t)|^2 dt \le C' \sum_{k=1}^{l} |\xi'|^{2(2l-r_k)-1} |h_k(\xi')|^2 \qquad (7.2.13)$$

with a constant not depending on ξ' and the functions under consideration. Indeed, integrating this estimate with respect to ξ', we shall obtain an estimate which differs from the required estimate (7.2.11) in that it contains seminorms instead of norms. It will remain to add the zeroth-order squared norms of the right-hand sides g_j of the boundary conditions on the right and the squared zeroth norm of the solution on the right and left.

But estimate (7.2.13) follows directly from (7.2.10).

Now suppose that $f \ne 0$. This case is reduced to that considered above as follows. We continue $h(\xi',t)$ by zero to $t < 0$ and set

$$v_0(\xi',t) = F_n^{-1} a_0^{-1}(\xi) F_n h(\xi',t) \qquad (7.2.14)$$

for $\xi' \ne 0$. Twice applying Parseval's identity with respect to the last variable, we obtain

$$|\xi'|^{2(2l-j)} \int_0^\infty |D_t^j v_0(\xi',t)|^2 dt$$

$$\le C_1 \int \frac{|\xi'|^{2(2l-j)} \xi_n^{2j}}{|a_0(\xi)|^2} |(F_n h)(\xi)|^2 d\xi_n \le C_2 \int_0^\infty |h(\xi',t)|^2 dt.$$

This implies the required estimate, because the integration of the integral on the right-hand side with respect to ξ' yields $\|f\|^2_{H^0(\mathbb{R}^n_+)}$.

Thus, we have obtained estimate (7.2.11) and, thereby, completed the first step. Let us rewrite the result for the operators without lower-order terms with coefficients frozen at x_0:

$$\|u\|^2_{H^{2l}(\mathbb{R}^n_+)}$$

$$\leq C\Big[\|a_0(x_0,D)u\|^2_{H^0(\mathbb{R}^n_+)} + \sum_{k=1}^{l}\|b_{k,0}(x_0,D)u\|^2_{H^{2l-r_k-1/2}(\mathbb{R}^{n-1})} + \|u\|^2_{H^0(\mathbb{R}^n_+)}\Big].$$

At the second step we consider the problem in the half-space for operators containing lower-order terms:

$$a(x,D) = a_0(x,D) + a_1(x,D) \quad \text{and} \quad b_k(x,D) = b_{k,0}(x,D) + b_{k,1}(x,D).$$

Here we assume that all coefficients in the higher-order terms $a_0(x,D)$ and $b_{k,0}(x,D)$ are close to those at the boundary point x_0. We obtain

$$\|u\|^2_{H^{2l}(\mathbb{R}^n_+)} \leq C\Big[\|a(x,D)u\|^2_{H^0(\mathbb{R}^n_+)} + \sum_{k=1}^{l}\|b_k(x,D)u\|^2_{H^{2l-r_k-1/2}(\mathbb{R}^{n-1})}$$

$$+ \|u\|^2_{H^0(\mathbb{R}^n_+)} + T_0 + \sum_{k=1}^{l}T_k\Big],$$

where

$$T_0 = \|a_1(x,D)u\|^2_{H^0(\mathbb{R}^n_+)} + \|[a_0(x_0,D) - a_0(x,D)]u\|^2_{H^0(\mathbb{R}^n_+)}$$

and

$$T_k = \|b_{k,1}(x,D)u\|^2_{H^{2l-r_k-1/2}(\mathbb{R}^{n-1})} + \|[b_{k,0}(x_0,D) - b_{k,0}(x,D)]u\|^2_{H^{2l-r_k-1/2}(\mathbb{R}^{n-1})}.$$

To complete the second step, it remains to obtain an estimate of the form

$$C\Big[T_0 + \sum_{k=1}^{l}T_k\Big] \leq \frac{1}{2}\|u\|^2_{H^{2l}(\mathbb{R}^n_+)} + C_1\|u\|^2_{H^0(\mathbb{R}^n_+)}$$

for higher-order coefficients close enough to constants. This can be done thanks to the workpieces which we have prepared: Theorem 3.3.1 about trace on the boundary of the half-space, Theorem 3.2.3 on a bound for the norm of the operator of multiplication by a smooth function on the half-space with using the upper bound for the absolute value of this function, inequality (3.2.3) for intermediate norms on the half-space, and similar results of Section 1 for $\mathbb{R}^{n-1}$ instead of $\mathbb{R}^n$ (Theorem 1.9.2 and Proposition 1.8.1).

At the third step, we multiply a function $u \in H^{2l}(\Omega)$, after it is extended, by the elements of a partition of unity on the closure of a neighborhood of the domain Ω:

$$u = \sum_{0}^{N}\varphi_j u,$$

where the φ_j are smooth functions, the support of φ_0 is contained strictly inside the domain, and each function φ_j with $j > 0$ vanishes outside a small neighborhood of some boundary point. Here we assume that the functions are replaced by their restrictions to Ω. We also assume that, in this neighborhood, it is possible to rectify the boundary and apply results of the second step, because the leading coefficients are close to constants. An estimate for $\varphi_0 u$ is taken from Section 6. Now we can assume that, in the initial coordinates, we have

$$\|u\|^2_{H^{2l}(\Omega)} \le C\Bigg[\sum_{j=0}^{N} \|a(x,D)(\varphi_j u)\|^2_{H^0(\Omega)}$$

$$+ \sum_{j=1}^{N} \sum_{k=1}^{l} \|b_k(x,D)(\varphi_j u)\|^2_{H^{2l-r_j-1/2}(\Gamma)} + \|u\|^2_{H^0(\Omega)}\Bigg].$$

Next, we interchange the differential operators and the functions φ_j; then there arise additional terms, which can be estimated via $\varepsilon\|u\|^2_{H^{2l}(\Omega)}$, where ε is a small coefficient, and $\|u\|^2_{H^0(\Omega)}$ by using estimates for the intermediate norms and the trace theorem for a domain. At the last step we get rid of the functions φ_j by using the boundedness of the operators of multiplication by them.

7.2.3 A Right Parametrix

In deriving the a priori estimate we supposed given a solution. Now we are given right-hand sides and must construct an approximate solution.

We only outline the first step. We set

$$\mathfrak{R}(f,g) = R_0 f + \sum_{1}^{l} R_j(g_j - B_j R_0 f). \qquad (7.2.15)$$

Here

$$R_0 f = F^{-1}\frac{|\xi|^{2l}}{1+|\xi|^{2l}}a_0^{-1}(\xi)F\mathcal{E}f, \qquad (7.2.16)$$

$\mathcal{E}$ is a bounded operator of extension of the function to the entire space, and

$$R_j g_j = F'^{-1}\frac{|\xi'|^{r_j+1}}{1+|\xi'|^{r_j+1}}\Omega_j(\xi',t)F'g_j. \qquad (7.2.17)$$

The fractions are aimed at canceling the singularities of the function $a_0^{-1}(\xi)$ at the point $\xi = 0$ and of the functions in the canonical basis at the point $\xi' = 0$.

The necessity of the algebraic ellipticity conditions is proved by contradiction by substituting special families of functions depending on an additional parameter into the a priori estimate.

In the case of a problem with a parameter, an exact right inverse of the problem operator is constructed in this way. For a problem in the half-space with constant leading coefficients without lower-order terms, such an operator can be constructed explicitly by using the Fourier transform with respect to the tangent variables. At the next two steps, we use the invertibility of an operator close in norm to an invertible one. This proximity is ensured by increasing the absolute value of the parameter λ.

Details can be found, e.g., in [34].

After these proofs appeared, the calculus of boundary value problems was developed [68, 69]. To boundary value problems with, generally, pseudodifferential (rather than differential) operators, matrix operators are assigned, which form an algebra; the operators have matrix symbols, and each elliptic problem has a two-sided parametrix within this algebra. This theory was expounded in the book [311]. There is also calculus of problems with a parameter [175].

7.3 Normal Systems of Boundary Operators and Formally Adjoint Boundary Value Problems. Boundary Value Problems with Homogeneous Boundary Conditions

Consider the system of boundary operators

$$B_j = b_j(x, D) = \sum_{|\beta| \le r_j} b_{j,\beta}(x) D^\beta \quad (j = 1, \dots, k) \tag{7.3.1}$$

of orders r_j. In this subsection, we assume that their coefficients are infinitely differentiable in a neighborhood of the boundary and all r_j are less than $2l$, where $2l$ is the order of the operator $a(x, D)$; we also assume that k does not exceed $2l$. We denote the leading parts of these operators by $B_{j,0} = b_{j,0}(x, D)$.

Definitions. System (7.3.1) is said to be *normal* if the orders r_j are pairwise different and the boundary Γ is noncharacteristic for each of the operators B_j at each point. The latter means that if the B_j are written in coordinates in which the x_n axis is normal to the boundary, then the coefficient of the leading derivative with respect to x_n (of order r_j) is a nonvanishing function.

If there is a ray of ellipticity with parameter for problem (7.1.19), (7.1.4), then the boundary operators of this problem form a normal system. This can be verified by setting $\xi' = 0$ and $\lambda \ne 0$ in the Lopatinskii condition.

A *Dirichlet system* is, by definition, a normal system of $2l$ operators.

An example is the system of consecutive normal derivatives

$$1, D_\nu, \dots, D_\nu^{2l-1}. \tag{7.3.2}$$

Obviously, if $k < 2l$, then any normal system can be completed to a Dirichlet system by adding, e.g., the normal derivatives of missing orders.

Note also that any two Dirichlet systems can be linearly expressed in terms of each other by using matrix partial differential operators in the tangent variables with C^∞ coefficients. For example, suppose that (7.3.1) is a Dirichlet system, $k = 2l$, and $r_j = j - 1$. We decompose the operator $b_j(x, D)$ in powers of the normal derivative:

$$b_j(x, D) = b_{j,0}(x, D') + b_{j,1}(x, D')D_\nu + \ldots + b_{j,j-1}(x)D_\nu^{j-1}. \tag{7.3.3}$$

This is the expression for the operators of system (7.3.1) in terms of the operators of system (7.3.2). Here each $b_{j,k}(x, D')$ is a differential operator of order at most $j - 1 - k$ with C^∞ coefficients containing differentiations only along the tangent directions, in which the last coefficient $b_{j,j-1}(x)$ is a nowhere vanishing numerical function. Thus, the matrix of the passage from system (7.3.2) to (7.3.1) is a lower triangular nonsingular matrix of differential operators. The orders of its elements increase under each shift from right to left and from top to bottom. The inverse matrix has a similar structure. As a consequence, any Dirichlet system can be linearly expressed in terms of any other Dirichlet system in a similar way.

Recall that the operator *formally adjoint* to A is defined as

$$A^*v = \sum_{|\alpha| \le 2l} D^\alpha |\overline{a_\alpha(x)}v(x)]. \tag{7.3.4}$$

This operator is elliptic and, as is easy to verify, regularly elliptic together with A. The operators A and A^* are related by

$$(Au, v)_\Omega = (u, A^*v)_\Omega \tag{7.3.5}$$

for functions $u, v \in \mathcal{D}(\Omega) = C_0^\infty(\Omega)$ (which is verified by integration by parts). Just this relation means that the operators A and A^* are formally adjoint on Ω.

We return to system (7.3.1); now we shall assume that $k = l$, the system is normal, and the problem for the operator A with given boundary operators is elliptic.

Let us complete the system $\{B_1, \ldots, B_l\}$ to a Dirichlet system $\{B_1, \ldots, B_{2l}\}$.

Theorem 7.3.1. *There exists another Dirichlet system* $\{C_1, \ldots, C_{2l}\}$ *such that, for any functions* $u, v \in H^{2l}(\Omega)$, *the Green identity*

$$(Au, v)_\Omega - (u, A^*v)_\Omega = \sum_1^l (B_{l+j}u, C_jv)_\Gamma - \sum_1^l (B_ju, C_{l+j}v)_\Gamma \tag{7.3.6}$$

is valid.

Here all operators are assumed to have C^∞ coefficients. The sum of the orders of B and C in each summand equals $2l - 1$. The nonuniqueness of the construction is contained in the choice of $B_{l+1}, \ldots, B_{2l}$; as soon as these operators are chosen, the second system is determined uniquely, as we shall see later on. Note that if functions $u, v \in H^{2l}(\Omega)$ satisfy the conditions

$$B_1u = \ldots = B_lu = 0, \quad C_1v = \ldots = C_lv = 0, \tag{7.3.7}$$

then the Green identity (7.3.6) takes the form (7.3.5).

If (7.3.6) holds, then the problems with the operators $A, B_1, \ldots, B_l$ and with the operators $A^*, C_1, \ldots, C_l$ are said to be *formally adjoint*. In particular, any problem coinciding with a problem formally adjoint to it is said to be *formally self-adjoint*.

Let us outline the proof of the theorem. We can assume that the functions $u(x)$ and $v(x)$ are infinitely differentiable and have small supports near boundary points. Rectifying the boundary, we assume that the functions are defined on the half-space $\{x \colon x_n > 0\}$. Let us decompose the operator A in powers of the derivative D_n:

$$A = A_0 + A_1 D_n + \ldots + A_{2l} D_n^{2l}.$$

Here the A_j are differential operators with respect to the tangent variables of order at most $2l - j$, and the last coefficient is a nonvanishing function (see Remark 6.1.1). It is easy to trace the procedure of integration by parts, which leads to the relation

$$(Au, v)_{\mathbb{R}_+^n} - (u, A^* v)_{\mathbb{R}_+^n} = \sum_{k=1}^{2l} (D_n^{k-1} u, N_k v)_{\mathbb{R}^{n-1}},$$

where each N_k is a partial differential operator of order $2l - k$ in which the coefficient of D_n^{2l-k} coincides with A_{2l}^* up to sign. Therefore, these operators form a normal system.

We write the sum of boundary terms in the form of the inner product $[Du, Nv]$ of the columns

$$Du = (u, \ldots, D_n^{2l-1} u)' \quad \text{and} \quad Nv = (N_1 v, \ldots, N_{2l} v)'.$$

For a while, we rearrange the $B_j u$ so that their orders decrease and denote the resulting column by $\widetilde{B} u$. We have $Du = \mathcal{B} \widetilde{B} u$, where $\mathcal{B}$ is a lower triangular matrix of partial differential operators with respect to x' whose main diagonal consists of nonvanishing functions, and

$$[Du, Nv] = [\mathcal{B} \widetilde{B} u, Nv] = [\widetilde{B} u, \mathcal{B}^* Nv].$$

The orders of the operators N_k decrease from top to bottom, and $\mathcal{B}^*$ is an upper triangular matrix. The verification that v is here subject to the action of a Dirichlet system and that an appropriate change of notation yields the required formula (7.3.6) is left to the reader.

Theorem 7.3.2. *If boundary value problems with operators $A, B_1, \ldots, B_l$ and A^*, $C_1, \ldots, C_l$ are formally adjoint and one of them is elliptic, then so is the other.*

For a detailed proof we refer the reader to the book [237, Chap. 2, Sec. 2]. (A different, purely algebraic and more formal, proof is given in [330].) Here we only explain the main idea. Performing localization, freezing the coefficients at a boundary point, omitting lower-order terms, and applying the Fourier transform with respect to the tangent variables, we arrive at the following situation. There are two problems on the ray $\mathbb{R}_+ = \{t \colon t > 0\}$ for ordinary differential operators of order $2l$ with constant coefficients:

$$a_0(D)u(t) = f(t) \quad (t > 0), \qquad b_{j,0}u(t)|_{t=0} = 0 \quad (j = 1,\ldots,l),$$
$$a_0^*(D)v(t) = g(t) \quad (t > 0), \qquad c_{j,0}u(t)|_{t=0} = 0 \quad (j = 1,\ldots,l).$$

We assume that the right-hand sides and solutions of these problems belong to $L_2(\mathbb{R}_+)$. For functions u and v satisfying the given boundary conditions, we have

$$\int_0^\infty a_0(D)u(t) \cdot \overline{v(t)}\, dt = \int_0^\infty u(t) \cdot \overline{a_0^*(D)v(t)}\, dt.$$

Since the initial problem is elliptic, the first problem with right-hand side in $L_2(\mathbb{R}_+)$ is uniquely solvable in this space. As to the second problem, it suffices to show that if the function $g(t)$ is identically zero, then this problem has only the trivial solution in this space (see item 2 in Remark 7.1.2). But, for such $g(t)$, the last relation implies

$$\int_0^\infty a_0(D)u(t) \cdot \overline{v(t)}\, dt = 0.$$

Now it suffices to find a solution $u(t)$ of the first problem with $f(t) = v(t)$. We obtain

$$\int_0^\infty |v(t)|^2 dt = 0,$$

whence $v(t) = 0$, as required.

Let $H_B^{2l}(\Omega)$ denote the subspace of functions in $H^{2l}(\Omega)$ which satisfy the homogeneous (i.e., zero) boundary conditions $B_j u = 0$ ($j = 1,\ldots,l$), and let A_B denote the operator $H_B^{2l}(\Omega) \to L_2(\Omega)$ defined by $A_B u = Au$. It corresponds to the problem

$$Au = f \text{ on } \Omega, \quad B_1 u = \ldots = B_l u = 0 \text{ on } \Gamma. \tag{7.3.8}$$

Theorem 7.3.3. *Under the above assumptions, the operator A_B is Fredholm. Its kernel consists of C^∞ functions (possibly corrected on a set of measure zero).*

Proof. This follows from similar properties of the operator corresponding to the initial problem with inhomogeneous boundary conditions. Indeed, the kernel of A_B coincides with that of the operator corresponding to the initial problem; therefore, it is finite-dimensional and consists of infinitely differentiable functions. The a priori estimate, which we already know, takes the form

$$\|u\|_{H^{2l}(\Omega)} \le C[\|A_B u\|_{L_2(\Omega)} + \|u\|_{L_2(\Omega)}]. \tag{7.3.9}$$

This implies the closedness of the range of A_B. The codimension of the range remains finite. $\qquad\square$

Note that the domain $H_B^{2l}(\Omega)$ of A_B is dense in $L_2(\Omega)$. The obtained estimate implies also the closedness of the operator A_B in $L_2(\Omega)$: if $u_j \to u$ and $A_B u_j \to f$ in $L_2(\Omega)$, then $u \in H_B^{2l}(\Omega)$ and $A_B u = f$.

The situation with the operator $(A^*)_C$ corresponding to the adjoint problem is similar.

The operators A_B and $(A^*)_C$ are related by

$$(A_B u, v)_\Omega = (u, (A^*)_C v)_\Omega \tag{7.3.10}$$

on their domains, and they are mutually adjoint in $L_2(\Omega)$. The situation is similar to that considered in Proposition 6.3.1, and the following assertion similar to this proposition is valid.

Corollary 7.3.4. *Under the same assumptions, the orthogonal complement to the range of the operator A_B in $L_2(\Omega)$ coincides with the kernel of $(A^*)_C$, and the orthogonal complement to the range of $(A^*)_C$ coincides with the kernel of A_B.*

7.4 Spectral Boundary Value Problems

The simplest spectral boundary value problem for an elliptic equation with a spectral parameter inside a domain (i.e., contained in the equation) has the form

$$Au(x) = \lambda u(x) \text{ on } \Omega, \tag{7.4.1}$$
$$B_j u(x) = 0 \text{ on } \Gamma \quad (j = 1, \dots, l). \tag{7.4.2}$$

We use the same notation as in the preceding section and assume that the same assumptions hold, that is, the problem under consideration is an elliptic problem for the operator $A = a(x, D)$ of order $2l$ with normal system of boundary conditions of orders $r_j < 2l$. To this problem we assign an unbounded operator A_B in $L_2(\Omega)$ with domain $D(A_B) = H_B^{2l}(\Omega)$. This operator can also be regarded as a bounded Fredholm operator from $H_B^{2l}(\Omega)$ to $L_2(\Omega)$. If the problem is uniquely solvable for some $\lambda = \lambda_0$, then this operator has discrete spectrum, and all generalized eigenfunctions turn out to be infinitely differentiable.

The simplest are self-adjoint problems, and they are particularly important for applications. The corresponding operator A_B is then self-adjoint in $L_2(\Omega)$, i.e.,

$$(A_B u, v)_\Omega = (u, A_B v)_\Omega \quad (u, v \in D(A_B)). \tag{7.4.3}$$

The eigenvalues of A_B are real, and their asymptotic behavior is known. Describing it has turned out to be a very difficult problem and long been taking much effort of many mathematicians. If the principal symbol $a_0(x, \xi)$ is positive, then the eigenvalues (which are positive in this case, at least beginning with some number) numbered by positive integers in nondecreasing order with multiplicities taken into account satisfy the asymptotic relation

$$\lambda_k \sim c k^{2l/n} \tag{7.4.4}$$

as $k \to \infty$, where the constant c is calculated in terms of the principal symbol. In the general case, the optimal estimate of the remainder term in this asymptotics, i.e., the difference between the left- and right-hand sides, has the form $O(k^{(2l-1)/n})$. See Safarov and Vassiliev [321].

In $L_2(\Omega)$, there is an orthonormal basis of eigenfunctions, and it remains a basis in the domain $H_B^{2l}(\Omega)$ of the operator A_B.

If the operator A_B is non-self-adjoint, then $(A^*)_C$ is adjoint to it.

We say that the operator A_B is *nearly self-adjoint* (or is a *weak perturbation of the self-adjoint operator* $\frac{1}{2}(A_B + A_B^*)$) if A and A^* have the same principal symbol and A_B and A_B^* have the same system of boundary operators. In this case, an arbitrarily narrow sector symmetric with respect to $\mathbb{R}_+$ contains the spectra of the operators A_B and A_B^*, possibly except finitely many eigenvalues. (Moreover, all but finitely many eigenvalues belong to some "parabolic neighborhood" of the ray $\mathbb{R}_+$.) We mean the case of a scalar problem with A having positive principal symbol. The eigenvalue asymptotics specified above remains valid. Ellipticity with parameter takes place along any ray except $\mathbb{R}_+$.

If A_B is non-self-adjoint, then there arises the problem of *conditions for the completeness of the system of generalized eigenfunctions* of the operator A_B in $L_2(\Omega)$ and in $D(A_B)$, i.e., for the density of the linear combinations of generalized eigenfunctions in these spaces. A sufficient completeness condition (in these spaces) consists in the existence of rays of ellipticity with parameter such that the angles between any two neighboring rays is less than $2l\pi/n$. In particular, this condition holds for nearly self-adjoint operators.

Completeness also holds in some other spaces, including the intermediate spaces obtained from $L_2(\Omega)$ and $D(A_B)$ by complex interpolation (this will be explained in Section 13). In the self-adjoint case, the eigenfunctions form a basis in these spaces. See also Section 17.2.

Examples. Consider the spectral Dirichlet problem for the Laplace equation

$$-\Delta u = \lambda u \text{ in } \Omega, \quad u = 0 \text{ on } \Gamma. \tag{7.4.5}$$

Let us denote the corresponding operator by $-\Delta_D$. This is a self-adjoint operator with discrete spectrum consisting of positive eigenvalues of finite multiplicity, which tend to $+\infty$ as $ck^{2/n}$. Self-adjointness follows from the Green identity

$$-\int_\Omega \Delta u \cdot \overline{v}\, dx = \int_\Omega \nabla u \cdot \nabla \overline{v}\, dx - \int_\Gamma \partial_\nu u \cdot \overline{v}\, dS, \tag{7.4.6}$$

or, more precisely, from the second Green identity for the Laplace operator

$$\int_\Omega \Delta u \cdot \overline{v}\, dx - \int_\Omega u \cdot \Delta \overline{v}\, dx = \int_\Gamma \partial_\nu u \cdot \overline{v}\, dS - \int_\Gamma u \cdot \partial_\nu \overline{v}\, dS. \tag{7.4.7}$$

Relation (7.4.6) implies also the positivity of all eigenvalues. Indeed, setting $v = u$, we see that if $\lambda \leq 0$ and the boundary value vanishes, then $\nabla u = 0$ almost everywhere and, hence, everywhere in Ω, so that $u \equiv \text{const}$ on Ω, which implies $u \equiv 0$, because $u = 0$ on the boundary.

A similar situation holds in the case of the Neumann spectral problem for the Laplace operator, that is, the problem

$$-\Delta u = \lambda u \text{ in } \Omega, \quad \partial_\nu u = 0 \text{ on } \Gamma. \tag{7.4.8}$$

We denote the corresponding operator in $L_2(\Omega)$ by $-\Delta_N$. This is a self-adjoint operator with discrete spectrum consisting of nonnegative eigenvalues, which tend to $+\infty$ and have the same asymptotics. The number 0 is an eigenvalue, and the corresponding space of eigenfunctions is one-dimensional and contains only constants. Self-adjointness and the absence of negative eigenvalues again follow from the Green identity.

Spectral elliptic boundary value problems with a spectral parameter in boundary conditions are interesting and useful as well. For example, consider the problem

$$Lu = 0 \text{ in } \Omega, \quad \partial_\nu u = \lambda u \text{ on } \Gamma \tag{7.4.9}$$

for a second-order elliptic equation (e.g., for the Laplace equation). This is the *Poincaré–Steklov spectral problem* (cf. [306] and [362]).

Suppose that the Dirichlet problem for the equation $Lu = 0$ is uniquely solvable. Then the Poincaré–Steklov problem reduces to a spectral equation on the boundary Γ. Let D_Γ denote the operator which takes the right-hand side of the Dirichlet boundary condition for solutions of the homogeneous equation to the right-hand side of the Neumann boundary condition:

$$u|_\Gamma \mapsto u \mapsto \partial_\nu u. \tag{7.4.10}$$

The operator D_Γ is known as the *Dirichlet-to-Neumann operator*. It acts boundedly from $H^{s-1/2}(\Gamma)$ to $H^{s-3/2}(\Gamma)$ for $s > 3/2$.

If the Neumann problem is uniquely solvable, then we can introduce the operator N_Γ which maps the right-hand side of the Neumann boundary condition to the right-hand side of the Dirichlet condition. This is the *Neumann-to-Dirichlet operator*. It acts boundedly from $H^{s-3/2}(\Gamma)$ to $H^{s-1/2}(\Gamma)$ and is compact in $H^{s-3/2}(\Gamma)$, because so is the embedding of the latter space in the former. If both problems, Dirichlet and Neumann, are uniquely solvable, then the operators D_Γ and N_Γ are mutually inverse. These operators play an important role in the theory of inverse problems for elliptic equations (see, e.g., [380]).

It has been proved in the theory of pseudodifferential operators that D_Γ is an *elliptic pseudodifferential operator of order* 1. In particular, we can set $s = 3/2$ and consider D_Γ as an *unbounded* operator in $L_2(\Gamma)$ with domain $H^1(\Gamma)$. Problem (7.4.9) reduces to the equation

$$(D_\Gamma - \lambda I)\varphi = 0, \tag{7.4.11}$$

where $\varphi = u|_\Gamma$.

In the case of the equation $-\Delta u = \mu u$ with real μ different from the eigenvalues of the Dirichlet and Neumann problems for the Laplace operator, both operators turn out to be self-adjoint in $L_2(\Gamma) = H^0(\Gamma)$; this follows from the relation

$$\int_\Gamma \partial_\nu u \cdot \overline{v} \, dS = \int_\Gamma u \cdot \partial_\nu \overline{v} \, dS, \qquad (7.4.12)$$

which is a consequence of the second Green identity. The spectrum of D_Γ is discrete and consists of eigenvalues of finite multiplicity, which tend to $+\infty$ as $c' k^{1/(n-1)}$. The eigenfunctions (solutions of the homogeneous equation $(D_\Gamma - \lambda I)\varphi = 0$) belong to $C^\infty(\Gamma)$ (provided that the boundary is smooth). They form an orthonormal basis in $L_2(\Gamma)$, which can be shown to remain a basis in all spaces $H^s(\Gamma)$.

We shall return to similar problems in Section 11 in the context of the theory of second-order strongly elliptic systems in Lipschitz domains.

A more detailed information about elliptic spectral problems and bibliography can found in the surveys [12, 64, 315].

7.5 Generalizations

1. General elliptic boundary value problems can be considered for Douglis–Nirenberg elliptic systems (see, e.g., [9] and [183, 185]). In principle, this does not involve anything essentially new; it is only required to choose adequate spaces for solutions and right-hand sides.

The Dirichlet problem for an even-order elliptic system with leading part homogeneous with respect to differentiation may be nonelliptic and, therefore, non-Fredholm. The first example of a non-Fredholm Dirichlet problem was given by Bitsadze [67]. This is the problem for the 2×2 system in the plane with matrix

$$\begin{pmatrix} \partial_1^2 - \partial_2^2 & 2\partial_1\partial_2 \\ -2\partial_1\partial_2 & \partial_1^2 - \partial_2^2 \end{pmatrix}. \qquad (7.5.1)$$

There are simplified versions of the Lopatinskii condition for the matrix Dirichlet problem, which are essentially due to Lopatinskii [244], in terms of the principal symbol $a_0(x, \xi)$ of the system; see also [146] and [11]. In particular, a necessary and sufficient condition for the ellipticity of the Dirichlet problem is the factorizability of the principal symbol:

$$a_0(x, \xi', \zeta) = a_-(x, \xi', \zeta) a_+(x, \xi', \zeta). \qquad (7.5.2)$$

Here x is a boundary point, the x_n axis is normal to the boundary, and the zeros ζ of the determinants of the matrices a_+ and a_- at $\xi' \neq 0$ belong to the upper and lower half-planes, respectively. There is also another, equivalent, condition; for a second-order system, it consists in that the matrix-integral

$$\int_{\gamma} e^{i\zeta t} a_0^{-1}(x, \xi', \zeta) \, d\zeta \tag{7.5.3}$$

over a contour enclosing all roots ζ of the determinant of $a_0(x, \xi', \zeta)$ in the upper half-plane is nonsingular for $\xi' \neq 0$.

2. Boundary value problems for systems can be considered on a compact smooth manifold with smooth boundary, and vector-valued functions can be replaced by sections of vector bundles; see, e.g., [183, 185].

3. Boundary value problems elliptic with parameter can be considered in a more general setting, with all operators polynomially depending on a parameter; see, e.g., [10] and [34]. Of interest are also the corresponding spectral problems, which have an extensive literature; see, e.g., [249].

4. Relaxing the smoothness assumptions, we can consider boundary value problems in spaces with index s varying in a finite interval.

5. There is an extensive theory of differential elliptic operators on a manifold with conical, edge, and similar singularities and of boundary elliptic problems in a domain with boundary singularities of these types. Outside the singularities, the manifold or the boundary is assumed to be smooth. This theory is substantially more complicated. It introduces and uses spaces related in a special way to the singularities and studies the asymptotic behavior of solutions near the singularities; see the initial paper [216] by Kondrat'ev and books [117] by Dauge, [284] by Nazarov and Plamenevsky, [220, 221] by Kozlov, Maz'ya, and Rossman, and [253] by Maz'ya and Rossman.

8 Strongly Elliptic Equations and Variational Problems

In the theory of elliptic problems, there is a different approach, which was developed before the general theory of elliptic problems outlined in the preceding section. As applied to the simplest problems, it is explained in textbooks on mathematical physics, such as [225] and [349]; see also [265]. This approach is less general in the sense that it requires the given elliptic equation to be associated with an "energy" quadratic form with positive definite real part. This is the strong ellipticity condition. It holds for many equations arising in applications.

In the case of smooth boundary and coefficients, this approach provides a faster way to theorems on the unique solvability or the Fredholm property of problems, but in spaces of lower smoothness (which is interesting by itself), and proving the smoothness of solutions to equations with smooth right-hand sides requires more effort. This case is considered below.

The approach in question is particularly effective in the case of nonsmooth co-efficients (see Remark 8.1.6 below) and a nonsmooth (e.g., Lipschitz) boundary. Second-order strongly elliptic systems in Lipschitz domains are considered in Sections 11, 12 and 16, 17.

In the context of this book, of most interest in this approach is the choice of function spaces.

We begin with a detailed consideration of the Dirichlet and Neumann problems for a second-order scalar equation and then mention generalizations to higher-order systems.

8.1 The Dirichlet and Neumann Problems for a Second-Order Scalar Equation

Let Ω be a bounded domain with boundary Γ. For simplicity, we first assume it to be C^∞. In Ω, consider the second-order scalar equation with leading part written in the *divergence form*

$$Lu = f, \tag{8.1.1}$$

where

$$Lu(x) = -\sum_{j,k=1}^{n} \partial_j a_{j,k}(x)\partial_k u(x) + \sum_{j=1}^{n} b_j(x)\partial_j u(x) + c(x)u(x). \tag{8.1.2}$$

The coefficients are generally complex-valued functions, which are assumed for simplicity to be infinitely differentiable on $\overline{\Omega}$. We can also assume (although this is not necessary) that

$$a_{j,k}(x) = a_{k,j}(x) \quad (j \neq k). \tag{8.1.3}$$

This condition means that, after differentiation, the coefficients of similar terms in the first sum become equal. If, in addition, $a_{j,k} = \overline{a_{k,j}}$ (so that the $a_{j,k}$ are real), $b_j = 0$, and $c = \overline{c}$, then we have a formally self-adjoint equation. The first sum can be rewritten in the form

$$\nabla a(x)\nabla u(x) = \operatorname{div}[a(x)\operatorname{grad} u(x)], \tag{8.1.4}$$

where $a(x)$ is a symmetric (but not necessarily real) $n \times n$ matrix with elements $a_{j,k}(x)$.

On the higher-order part we impose the *strong ellipticity* condition introduced by Vishik in [388]. This is the requirement that the form

$$a(x,\xi) = \sum a_{j,k}(x)\xi_j\xi_k \tag{8.1.5}$$

with real $\xi = (\xi_1,\ldots,\xi_n)$ have positive definite real part, i.e., the form

$$\operatorname{Re} a(x,\xi) = \sum \frac{a_{j,k}(x) + \overline{a_{j,k}(x)}}{2}\,\xi_j\xi_k \tag{8.1.6}$$

be positive definite. To be more precise, we assume that the uniform condition

$$\operatorname{Re} a(x,\xi) \ge C_0|\xi|^2 \quad (x \in \overline{\Omega}), \tag{8.1.7}$$

where C_0 is a positive constant, is satisfied.

Given a complex vector $\zeta = \xi + i\eta$, consider the form $a(x,\zeta)$ defined by

$$a(x,\zeta) = \sum a_{j,k}(x)\zeta_k\overline{\zeta_j}. \tag{8.1.8}$$

Under condition (8.1.3), we have

$$a(x,\zeta) = a(x,\xi) + a(x,\eta),$$

so that inequality (8.1.7) is generalized to complex numbers ζ:

$$\operatorname{Re} a(x,\zeta) \ge C_0|\zeta|^2. \tag{8.1.9}$$

For real coefficients $a_{j,k}(x) = a_{k,j}(x)$, strong ellipticity follows from the ellipticity condition $a(x,\xi) > 0$. Conversely, ellipticity always follows from strong ellipticity. We again emphasize that we assume the strong ellipticity condition to hold here.

The most important problems for Eq. (8.1.1) are the Dirichlet problem with boundary condition

$$u^+(x) = g(x) \text{ on } \Gamma \tag{8.1.10}$$

and the Neumann problem. From now on, we use the superscript + to denote boundary values on Γ. To write the Neumann boundary condition, we introduce the so-called *conormal derivative*. Let $\nu = \nu(x) = (\nu_1(x), \ldots, \nu_n(x))$ be the unit outer normal vector to the boundary Γ at a boundary point x. If the function $u(x)$ is smooth (it suffices to assume that it belongs to $H^s(\Omega)$ with $s > 3/2$), then we set

$$T^+u(x) = \sum_{j=1}^{n} \nu_j(x)a_{j,k}(x)\gamma^+\partial_k u(x) \tag{8.1.11}$$

on Γ. This is what is called the conormal derivative for Eq. (8.1.1). The Neumann boundary condition has the form

$$T^+u(x) = h(x) \text{ on } \Gamma. \tag{8.1.12}$$

The conormal derivative is related to the equation closer than the usual normal derivative (see the Green identity below). In the case of the Laplace equation, it coincides with the normal derivative.

Both problems are elliptic: we have already mentioned the ellipticity of the Dirichlet problem in Section 7.1, and the ellipticity of the Neumann problem is easy to verify (by rotating the coordinate system so that ∂_n becomes the normal derivative).

Moreover, *these problems are elliptic with parameter in the sector of opening $\pi + \varepsilon$ with bisector $\mathbb{R}_-$, where $\varepsilon > 0$ is sufficiently small.* The sector of ellipticity with parameter is larger than the left half-plane because, by continuity, the strong ellipticity condition is preserved under the replacement of $a_{j,k}(x)$ by $e^{i\theta} a_{j,k}(x)$ with sufficiently small $|\theta|$.

Let us introduce the form

$$\Phi_\Omega(u,v) = \int_\Omega \left[\sum_{j,k} a_{j,k} \partial_k u \cdot \partial_j \overline{v} + \sum_j b_j \partial_j u \cdot \overline{v} + c u \overline{v} \right] dx. \qquad (8.1.13)$$

The first sum can be rewritten in the form

$$a(x)\nabla u(x) \cdot \overline{\nabla v(x)} = a(x) \operatorname{grad} u(x) \cdot \operatorname{grad} \overline{v(x)}. \qquad (8.1.14)$$

If $u \in H^2(\Omega)$ and $v \in H^1(\Omega)$, then, integrating by parts, we obtain the *first Green identity*

$$(Lu, v)_\Omega = \Phi_\Omega(u, v) - (T^+ u, v^+)_\Gamma. \qquad (8.1.15)$$

Here $(u, v)_\Omega$ and $(\varphi, \psi)_\Gamma$ are standard inner products in $L_2(\Omega)$ and $L_2(\Gamma)$, respectively. It is convenient for what follows to denote the operator formally adjoint to L by $\widetilde{L}$; it can be written as

$$\widetilde{L}v = - \sum_{j,k=1}^n \partial_j \overline{a_{k,j}(x)} \partial_k v(x) - \sum \overline{b_j(x)} \partial_j v(x)$$

$$+ \left[\overline{c(x)} - \sum \partial_j \overline{b_j(x)} \right] v(x). \qquad (8.1.16)$$

We want to have the first Green identity for $\widetilde{L}$ with the same form $\Phi_\Omega(u, v)$:

$$(u, \widetilde{L}v)_\Omega = \Phi_\Omega(u, v) - (u^+, \widetilde{T}^+ v)_\Gamma. \qquad (8.1.17)$$

Then the corresponding conormal derivative of a function $v \in H^s(\Omega)$, $s > 3/2$, must be

$$\widetilde{T}^+ v(x) = \sum_{j,k} \nu_j(x) \overline{a_{k,j}(x)} \gamma^+ \partial_k v(x) + \sum \nu_j(x) \overline{b_j(x)} v^+(x). \qquad (8.1.18)$$

Relation (8.1.17) is obtained for $u \in H^1(\Omega)$ and $v \in H^2(\Omega)$. The first Green identities (8.1.15) and (8.1.17) with $u, v \in H^2(\Omega)$ imply the *second Green identity*

$$(Lu, v)_\Omega - (u, \widetilde{L}v)_\Omega = (u^+, \widetilde{T}^+ v)_\Gamma - (T^+ u, v^+)_\Gamma. \qquad (8.1.19)$$

For $s > 3/2$, passing to the limit, we can extend relation (8.1.15) to $u \in H^s(\Omega)$ and $v \in H^1(\Omega)$, relation (8.1.17) to $u \in H^1(\Omega)$ and $v \in H^s(\Omega)$, and relation (8.1.19) to $u, v \in H^s(\Omega)$. For this purpose, we approximate functions from $H^s(\Omega)$ by functions from $H^2(\Omega)$.

Although we assume the boundary and the coefficients to be smooth in this section, of great interest are equations with nonsmooth right-hand sides, whose solutions are functions of low smoothness. It is possible to consider them in the context of the weak setting of the Dirichlet and Neumann problems, which we describe below, and weak solutions.

First, consider these problems with homogeneous boundary conditions, i.e., zero right-hand sides.

In this case, the Dirichlet problem is written as

$$(f,v)_\Omega = \Phi_\Omega(u,v). \tag{8.1.20}$$

Here v is any test function,

$$u,v \in \widetilde{H}^1(\Omega), \quad \text{and} \quad f \in H^{-1}(\Omega). \tag{8.1.21}$$

We assume that the form $(f,v)_\Omega$ is extended to the direct product $H^{-1}(\Omega) \times \widetilde{H}^1(\Omega)$. The choice of the space containing u and v is caused by the homogeneity of the Dirichlet condition: recall that $\widetilde{H}^1(\Omega)$ is identified with $\overset{\circ}{H}^1(\Omega)$.

The Neumann problem with homogeneous boundary condition is written in the same form (8.1.20) but with

$$u,v \in H^1(\Omega) \quad \text{and} \quad f \in \widetilde{H}^{-1}(\Omega). \tag{8.1.22}$$

Thus, we need spaces with negative indices. In both cases, f and u, v belong to spaces dual to each other with respect to the (extended) form $(f,v)_\Omega$. Note that the form $\Phi_\Omega(u,v)$ is bounded on $H^1(\Omega)$ and, in particular, on $\widetilde{H}^1(\Omega)$:

$$\begin{aligned}
|\Phi_\Omega(u,v)| &\le C_1 \|u\|_{H^1(\Omega)} \|v\|_{H^1(\Omega)} \quad (u,v \in H^1(\Omega)), \\
|\Phi_\Omega(u,v)| &\le C_1 \|u\|_{\widetilde{H}^1(\Omega)} \|v\|_{\widetilde{H}^1(\Omega)} \quad (u,v \in \widetilde{H}^1(\Omega)).
\end{aligned} \tag{8.1.23}$$

In the case of the Dirichlet problem with homogeneous boundary condition, L is bounded as an operator from the space $\widetilde{H}^1(\Omega)$ of functions u to the space $H^{-1}(\Omega)$ of functions f. Indeed, in this case we have

$$\|f\|_{H^{-1}(\Omega)} \le C_2 \sup_{v \ne 0} \frac{|(f,v)_\Omega|}{\|v\|_{\widetilde{H}^1(\Omega)}} = C_2 \sup_{v \ne 0} \frac{|\Phi_\Omega(u,v)|}{\|v\|_{\widetilde{H}^1(\Omega)}} \le C_3 \|u\|_{\widetilde{H}^1(\Omega)}.$$

In the case of the Neumann problem with homogeneous boundary condition, a similar argument shows that L is a bounded operator from $H^1(\Omega)$ to $\widetilde{H}^{-1}(\Omega)$.

At first sight, it seems strange that the functions u and f belong to spaces of different classes, H and $\widetilde{H}$. But this is natural for problems under consideration, in which the right-hand sides belong to spaces with negative indices dual to the solution spaces; cf. the Lax–Milgram theorem in Section 18.2. Recall also that the spaces $H^s(\Omega)$ and $\widetilde{H}^s(\Omega)$ can be identified if $|s| < 1/2$ (see Section 5.1). In Section 13.8, we shall see that, for this reason, the spaces containing u and f belong to the same interpolation scale.

First, consider the Dirichlet problem in more detail.

Theorem 8.1.1. *For a strongly elliptic operator of the form* (8.1.2), *there exist constants* $C_4 > 0$ *and* $C_5 \geq 0$ *such that all functions in* $\widetilde{H}^1(\Omega)$ *satisfy the inequality*

$$\|u\|^2_{\widetilde{H}^1(\Omega)} \leq C_4 \operatorname{Re} \Phi_\Omega(u, u) + C_5 \|u\|^2_{H^0(\Omega)}. \tag{8.1.24}$$

This inequality is known as the *Gårding inequality*. If it holds, the form $\Phi_\Omega(u, u)$ is said to be *coercive on* $\widetilde{H}^1(\Omega)$, and (8.1.24) is also called the *coercivity condition* for Φ_Ω on $\widetilde{H}^1(\Omega)$.

Proof. In the scalar case, under condition (8.1.9) (which follows, as we saw, from (8.1.7) and (8.1.3)), the proof is quite simple. The substitution of $\zeta_j = \partial_j u(x)$ into (8.1.9) and integration with respect to x yield the inequality

$$\sum \|\partial_j u\|^2_{H^0(\Omega)} \leq C_4 \operatorname{Re} \Phi_{0,\Omega}(u, u),$$

where $\Phi_{0,\Omega}$ is the principal part of the form Φ_Ω:

$$\Phi_{0,\Omega}(u, v) = \int_\Omega a(x) \nabla u(x) \cdot \nabla \overline{v(x)} dx. \tag{8.1.25}$$

If all coefficients b_j vanish, then we immediately obtain the required assertion. If some of the coefficients b_j are nonzero, then we apply (5.1.9). $\qquad\square$

If $\operatorname{Re} c(x)$ is sufficiently large (or $c(x)$ is replaced by $c(x) - \lambda$ and $\mu = -\operatorname{Re}\lambda$ is sufficiently large), then estimate (8.1.24) with $C_5 = 0$ is valid:

$$\|u\|^2_{\widetilde{H}^1(\Omega)} \leq C_4 \operatorname{Re} \Phi_\Omega(u, u). \tag{8.1.26}$$

In what follows, we usually assume for simplicity that this estimate does hold. We call it the *strong Gårding inequality*, or the *strong coercivity condition for* Φ_Ω *on* $\widetilde{H}^1(\Omega)$. As we shall see below, this condition implies the unique solvability (rather than the Fredholm property) of the problem. Relation (8.1.20) implies

$$\|u\|^2_{\widetilde{H}^1(\Omega)} \leq C_6 \|f\|_{H^{-1}(\Omega)} \|u\|_{\widetilde{H}^1(\Omega)};$$

therefore, we have the *a priori estimate*

$$\|u\|_{\widetilde{H}^1(\Omega)} \leq C_6 \|f\|_{H^{-1}(\Omega)}, \tag{8.1.27}$$

which implies uniqueness for the Dirichlet problem.

As mentioned above, the reverse inequality is valid as well. Thus, we have the *two-sided estimate*

$$\|u\|_{\widetilde{H}^1(\Omega)} \leq C_6 \|f\|_{H^{-1}(\Omega)} \leq C_7 \|u\|_{\widetilde{H}^1(\Omega)}. \tag{8.1.28}$$

This shows that the chosen spaces are adequate to the problem under consideration.

Next, estimate (8.1.26) implies the existence of a solution of the problem for any right-hand side f. In the important special case where the equation is formally self-adjoint, which means that

$$\overline{\Phi_\Omega(u,v)} = \Phi_\Omega(v,u), \tag{8.1.29}$$

this is a consequence of F. Riesz' theorem, according to which the form $\Phi_\Omega(u,v)$ has properties of inner product, so that we can represent the functional $(f,v)_\Omega$ in the form $\Phi_\Omega(u,v)$. But even without this assumption, the form $\Phi_\Omega(u,v)$ with a fixed function u is a general continuous antilinear functional on $\widetilde{H}^1(\Omega)$, which uniquely determines u by virtue of the Lax–Milgram theorem (see Section 18.2). Thus, we have proved the following existence and uniqueness theorem.

Theorem 8.1.2. *If inequality* (8.1.26) *holds, then the Dirichlet problem $Lu = f$ in Ω, $u^+ = 0$ has precisely one weak solution in $\widetilde{H}^1(\Omega)$ for any right-hand side $f \in H^{-1}(\Omega)$, and the a priori estimate* (8.1.27) *holds.*

Remark 8.1.3. To have inequality (8.1.26), it is not necessary to assume the presence of a zero-order term in the equation. For example, in the case of the Poisson equation $-\Delta u = f$, we can use the *Friedrichs inequality* (see, e.g., [225] or [358]): for functions vanishing on Γ,

$$\|u\|^2_{L_2(\Omega)} \le C_\Omega \|\nabla u\|^2_{L_2(\Omega)}. \tag{8.1.30}$$

It is well known that the Dirichlet problem in this case is uniquely solvable. The Friedrichs inequality is also valid in the L_p-norms.

To extend the theorem on the unique solvability of the Dirichlet problem to the case of an inhomogeneous boundary condition $u^+ = g \in H^{1/2}(\Gamma)$ (in which the solution is sought in $H^1(\Omega)$), it suffices to subtract a function $u_0 \in H^1(\Omega)$ with given boundary value g from the solution, which yields the problem considered above for the difference $u - u_0$. This is explained in detail in Section 11.1.

Considering the equation with parameter

$$Lu - \lambda u = f \tag{8.1.31}$$

instead of (8.1.1), we obtain, as shown below, the estimate

$$\|u\|_{\widetilde{H}^1(\Omega)} + \mu\|u\|_{H^{-1}(\Omega)} \le C\|f\|_{H^{-1}(\Omega)} \tag{8.1.32}$$

for sufficiently large $\mu = -\text{Re}\,\lambda$, which is uniform in the parameter and resembles (7.1.21). This can be compared with the remark on the ellipticity of problems with parameter before (8.1.13).

Let us derive estimate (8.1.32). Note that if we add $-(\lambda u, u)_\Omega$ to $\Phi_\Omega(u,u)$ under the Re sign, inequality (8.1.26) remains valid. But

$$\Phi_\Omega(u,u) - \lambda(u,u)_\Omega = (f,u)_\Omega.$$

Therefore,
$$\|u\|^2_{\widetilde{H}^1(\Omega)} \le C\|f\|_{H^{-1}(\Omega)}\|u\|_{\widetilde{H}^1(\Omega)},$$

which gives the required estimate of the first term on the left-hand side in (8.1.32). Next, we have
$$\mu\|u\|_{H^{-1}(\Omega)} \le \|f\|_{H^{-1}(\Omega)} + \|Lu\|_{H^{-1}(\Omega)},$$

and the last term in this expression can be estimated via the first term on the left-hand side in (8.1.32), which has just been estimated. This gives the estimate of the second term on the left-hand side in (8.1.32).

Moreover, an estimate of the form (8.1.32) with $|\lambda|$ instead of μ, that is,
$$\|u\|_{\widetilde{H}^1(\Omega)} + |\lambda|\|u\|_{H^{-1}(\Omega)} \le C\|f\|_{H^{-1}(\Omega)}, \tag{8.1.33}$$

holds outside any angle with vertex at the origin which encloses an angle containing the values of the quadratic form $\Phi_\Omega(u,u)$. The proof of this estimate uses the possibility of multiplying the form $\Phi_\Omega(u,u)$ by $e^{i\theta}$ with sufficiently small $|\theta|$.

We proceed to the Neumann problem. In this case, we need an inequality for functions in $H^1(\Omega)$ similar to (8.1.24):
$$\|u\|^2_{H^1(\Omega)} \le C_4 \operatorname{Re}\Phi_\Omega(u,u) + C_5\|u\|^2_{H^0(\Omega)}. \tag{8.1.34}$$

In the case of a scalar equation satisfying condition (8.1.3), which we consider in this section, inequality (8.1.34) is obtained in precisely the same way as above. We say that the *form $\Phi_\Omega(u,v)$ is coercive on $H^1(\Omega)$*. For sufficiently large $\operatorname{Re}c(x)$, this inequality holds with $C_5 = 0$:
$$\|u\|^2_{H^1(\Omega)} \le C_4 \operatorname{Re}\Phi_\Omega(u,u). \tag{8.1.35}$$

We refer to (8.1.35) as the *strong coercivity condition for the form Φ_Ω on $H^1(\Omega)$*. Having (8.1.35), we can repeat the above considerations of the Dirichlet problem for the Neumann problem almost without changes. As a result, we obtain the following theorem.

Theorem 8.1.4. *If inequality (8.1.35) holds, then the Neumann problem $Lu = f$ in Ω, $T^+u = 0$ on Γ has precisely one weak solution in $H^1(\Omega)$ for any right-hand side $f \in \widetilde{H}^{-1}(\Omega)$, and this solution satisfies the a priori estimate*
$$\|u\|_{H^1(\Omega)} \le C_6\|f\|_{\widetilde{H}^{-1}(\Omega)}. \tag{8.1.36}$$

We again have a two-sided estimate. Moreover, an estimate with a parameter similar to (8.1.33) holds.

For the weak setting of the Neumann problem with inhomogeneous boundary condition, we need the complete Green identity
$$(f,v)_\Omega = \Phi_\Omega(u,v) - (h,v^+)_\Gamma \quad (v \in H^1(\Omega)). \tag{8.1.37}$$

Here $v^+ \in H^{1/2}(\Gamma)$; therefore, $h \in H^{-1/2}(\Gamma)$ and $(h, v^+)_\Gamma$ is a continuous antilinear functional on $H^{1/2}(\Gamma)$ and, hence, on $H^1(\Omega)$.

Probably somewhat unexpectedly, for functions $u, v \in H^1(\Omega)$, the Green identities (8.1.15) and (8.1.37) cannot be proved. The point is that, in the framework of the trace theorem which we know, the expression (8.1.11) does not make sense for a function in $H^1(\Omega)$. Moreover, the right-hand side of the equation $Lu = f$ is uniquely determined by u as a distribution only inside the domain Ω. As an element of the space of continuous (anti)linear functionals on $H^1(\Omega)$, this right-hand side may contain a component concentrated on Γ, namely, a functional of the form $(w, v^+)_\Gamma$, where $w \in H^{-1/2}(\Gamma)$. But such a component can be transferred (or not transferred) to the boundary term in (8.1.37).

The way out of this situation generally accepted in the literature (see, e.g., [258]) is to *postulate* the Green identity (8.1.37). *Given $u \in H^1(\Omega)$ and $f \in \widetilde{H}^{-1}(\Omega)$, this identity is taken for the definition of the conormal derivative.* Generally, the conormal derivative is no longer expressed by (8.1.11). We also take relation (8.1.37) for the definition of *a solution of the Neumann problem with inhomogeneous boundary condition, i.e., with given $f \in \widetilde{H}^{-1}(\Omega)$ and the conormal derivative $T^+u = h \in H^{-1/2}(\Gamma)$.* Thus, there is an arbitrariness in the statement of the problem, since f and h are not independent. But if f or T^+u is given, then T^+u or f, respectively, is determined uniquely. In particular, the two Neumann problems with $f = 0$ and with $h = 0$ are of independent significance.

Note also that, in the case where f is known to belong to $L_2(\Omega)$, it is usually this function f extended by zero outside Ω which is considered as the right-hand side of the equation $Lu = f$ in $\widetilde{H}^{-1}(\Omega)$; see [258, p. 117]. Under this convention, the conormal derivative is determined uniquely. In particular, this relates to the case where $f = 0$ on Ω.

We shall return to this point in Section 11. In Section 11.2, we shall propose a method for eliminating the arbitrariness mentioned above.

However, it is useful to know that the functional $(T^+u, v^+)_\Gamma$ can be represented in the form $(f_1, v)_\Omega$ with $f_1 \in \widetilde{H}^{-1}(\Omega)$. Thus, we can always reduce the general Neumann problem to the problem with homogeneous boundary condition, which was considered above, by changing the right-hand side of the equation. Therefore, the unique solvability theorem remains valid in the case of an inhomogeneous boundary condition.

We refer to the conormal derivative defined by (8.1.11) as the *smooth conormal derivative*.

A feature of the general Neumann problem which is of particular interest for us in this section is that its setting allows specifying the right-hand side of the boundary condition in the space with negative index $-1/2$.

Now let us explain the term "variational problem." *This term refers to the case where the problem is formally self-adjoint, or, equivalently, satisfies condition (8.1.29), and the form $\Phi_\Omega(v, v)$ is nonnegative.* Consider the functional

$$\Psi(v) = \text{Re}[\Phi_\Omega(v, v) - 2(f, v)_\Omega]. \qquad (8.1.38)$$

This functional attains its minimum value at a solution. Indeed, we have

$$\Psi(u+v) - \Psi(u) = \Phi_\Omega(v,v) + 2\operatorname{Re}[\Phi_\Omega(u,v) - (f,v)_\Omega],$$

and if u is a solution, then the expression in brackets vanishes, so that the left-hand side is nonnegative.

For this reason, the weak settings of the Dirichlet and Neumann problems considered in this section are also called *variational settings.* Using this name, we do not exclude the case of the absence of formal self-adjointness.

For studying the smoothness of the solution under additional smoothness conditions on the right-hand sides and the boundary, there is Nirenberg's method of difference quotients [290]; see our Section 16.5. For example, it turns out that if the right-hand side of the equation belongs to $L_2(\Omega)$, then the solution belongs to $H^2(\Omega')$ on any interior subdomain Ω', and if the boundary is C^2, then the solution of the Dirichlet problem belongs to $H^2(\Omega)$.

But particularly simple is the proof of the following theorem in the case of smooth coefficients and boundary; it follows from the consistency of usual and variational elliptic theories.

Theorem 8.1.5. *Let $s \geq 2$. Then the solution of the Dirichlet problem with $f \in H^{s-2}(\Omega)$ and $g \in H^{s-1/2}(\Gamma)$ belongs to $H^s(\Omega)$ and is a solution in the sense of general theory* (see Section 7). *Therefore, it belongs to $H^s(\Omega)$. The same is true for the solution of the Neumann problem with $f \in H^{s-2}(\Omega)$ and $h \in H^{s-3/2}(\Gamma)$.*

Proof. It suffices to verify that a usual solution is a variational solution for $s = 2$, which is performed elementarily by integration by parts. Note that, in the case under consideration, f has no components supported on Γ and the conormal derivative is smooth and determined uniquely. Since the variational solution is unique, it follows that this is also a solution in the usual sense, and we can apply the familiar theorem on the smoothness of solutions of usual elliptic problems.   □

Remark 8.1.6. Of independent interest is the question of what results are valid for equations with nonsmooth coefficients. Theorem 8.1.1 remains true when the coefficients in (8.1.2) are only bounded measurable functions. This assumption is also sufficient for proving Theorems 8.1.2 and 8.1.4 if only weak solutions are considered.

Remark 8.1.7. Now consider the Dirichlet problem in the case where strong ellipticity is present but the form $\Phi_\Omega(u,v)$ is not strongly coercive on $\widetilde{H}^1(\Omega)$. Let L be the corresponding operator. Then the operator $L_\tau = L + \tau$ satisfies the strong coercivity condition for sufficiently large τ. The operator $\mathcal{L}_D \colon \widetilde{H}^1(\Omega) \to H^{-1}(\Omega)$ corresponding to the original Dirichlet problem with homogeneous boundary condition turns out to be a weak perturbation of the invertible operator $\mathcal{L}_{\tau,D}$. To be more precise, their difference is a bounded operator in $\widetilde{H}^1(\Omega)$, which is, of course, compact as an operator from $\widetilde{H}^1(\Omega)$ to $H^{-1}(\Omega)$. Therefore, $\mathcal{L}_D$ is a *Fredholm operator with index zero*; see Proposition 18.1.12. Next, by virtue of the second Green identity, the operators $\mathcal{L}_D$ and $\widetilde{\mathcal{L}}_D$, where $\widetilde{\mathcal{L}}_D$ corresponds to the Dirichlet problem for a formally adjoint operator, turn out to be adjoint as operators in Banach (Hilbert in the case under consideration) spaces, i.e.,

$$(\mathcal{L}_D u, v)_\Omega = (u, \widetilde{\mathcal{L}}_D v)_\Omega \quad (u, v \in \widetilde{H}^1(\Omega)). \tag{8.1.39}$$

The operator $\widetilde{\mathcal{L}}_D$ is Fredholm and has index zero as well as $\mathcal{L}_D$, so that, e.g., *the equation $\mathcal{L}_D u = f \in H^{-1}(\Omega)$ is solvable if and only if the right-hand side f satisfies the condition*

$$(f, v)_\Omega = 0 \tag{8.1.40}$$

for all solutions of the homogeneous equation $\widetilde{\mathcal{L}}_D v = 0$; see Proposition 18.1.8.

A similar situation occurs in the case of the Neumann problem if the form $\Phi_\Omega(u, v)$ is coercive but not strongly coercive on $H^1(\Omega)$. In this case, *the operator $\mathcal{L}_N \colon H^1(\Omega) \to \widetilde{H}^{-1}(\Omega)$ corresponding to the Neumann problem with homogeneous boundary condition is Fredholm and has index zero, and the equation $\mathcal{L}_N u = f$ is solvable if and only if f satisfies condition* (8.1.40) *for all solutions of the adjoint homogeneous equation $\widetilde{\mathcal{L}}_N v = 0$.*

Our Theorem 8.1.5 can be extended to the Fredholm situation as well. For example, it suffices to rewrite the equation $\mathcal{L}_D u = f$ in the form $\mathcal{L}_{\tau,D} u = f + \tau u$ with an invertible operator on the left.

We must also mention that an important role in the theory of strongly elliptic equations is played by *surface potentials*. If $E(x, y)$ is a fundamental solution, i.e., a solution of the equation

$$L_x E(x, y) = \delta(x - y), \tag{8.1.41}$$

then the classical *single-layer potential* is defined by

$$u(x) = \mathcal{A}\psi(x) = \int_\Gamma E(x, y)\psi(y)\, dS_y, \tag{8.1.42}$$

and the *double-layer potential* is defined by

$$u(x) = \mathcal{B}\varphi(x) = \int_\Gamma [\partial_{\nu_y} E(x, y)]\varphi(y)\, dS_y \quad (x \notin \Gamma). \tag{8.1.43}$$

Here for simplicity we wrote the last expression only for a formally self-adjoint operator L. The functions ψ and φ must be regular in a certain sense. Both operators, (8.1.42) and (8.1.43), map functions given on the boundary to solutions of the homogeneous equation outside the boundary. Of great importance are also the restrictions of these operators to Γ and the conormal derivatives of these functions on Γ. In the case of a smooth boundary and smooth coefficients in L, the last four operators are pseudodifferential operators on Γ, and we can study them by means of the calculus of these operators. In the nonsmooth case, there is a simplified approach to derive properties of these operators and the so-called hypersingular operator $H = -T^+\mathcal{B}$, which is based on the assumption that the Dirichlet and Neumann problems are uniquely solvable. This approach is explained in Section 12 in a more general situation, namely, for second-order strongly elliptic systems on Lipschitz domains. But the explanation begins with a revision of the definitions of these operators.

Potential-type operators are convenient for solving problems in the case of a homogeneous equation and an inhomogeneous boundary condition; we shall see this in Section 12.

Operators acting on the boundary include also the Neumann-to-Dirichlet operator N taking Neumann data for a solution of the homogeneous equation to Dirichlet data and the inverse Dirichlet-to-Neumann operator D. For uniquely solvable Dirichlet and Neumann problems, the operator D acts boundedly from $H^{1/2}(\Gamma)$ to $H^{-1/2}(\Gamma)$ and is invertible, and the operator N has similar properties in the reverse direction. These two operators, which have already been mentioned in Section 7.4 in a different context, are considered in more detail in Section 11.

8.2 Generalizations

The definition of strong ellipticity is first generalized to second-order systems. They will be considered in Sections 11 and 12; see also Sections 16 and 17. Then it is generalized to higher-order systems with the Douglas–Nirenberg structure (see our Section 6.5). Take a set of positive (for simplicity) integers $m_1, \ldots, m_d$. In the notation of Section 6.5, we consider the case $t_j = m_j$. Let us write the system with leading part singled out:

$$Lu(x) = L_0 u(x) + \ldots = f(x). \tag{8.2.1}$$

The vector-valued functions $u(x)$ and $f(x)$ are columns of height d:

$$u(x) = (u_1(x), \ldots, u_d(x))', \quad f(x) = (f_1(x), \ldots, f_d(x))'.$$

The operators L and L_0 are $d \times d$ matrix operators. Let $L = (L^{r,s})$, and let $L_0 = (L_0^{r,s})$. Each of the scalar operators $L^{r,s}$ can be written in divergence form as

$$L^{r,s} = L^{r,s}(x, \partial) = \sum_{|\alpha| \leq m_r, |\beta| \leq m_s} (-1)^{|\alpha|} \partial^\alpha [a_{\alpha,\beta}^{r,s}(x) \partial^\beta \cdot]. \tag{8.2.2}$$

The matrix L_0 consists of the leading parts of these operators:

$$L_0^{r,s} = L_0^{r,s}(x, \partial) = \sum_{|\alpha| = m_r, |\beta| = m_s} (-1)^{m_r} \partial^\alpha [a_{\alpha,\beta}^{r,s}(x) \partial^\beta \cdot]. \tag{8.2.3}$$

The strong ellipticity condition has the form

$$\mathrm{Re} \sum_{r,s=1}^{d} \sum_{|\alpha|=m_r, |\beta|=m_s} a_{\alpha,\beta}^{r,s}(x) \xi^{\alpha+\beta} \zeta_s \bar{\zeta}_r \geq C_0 \sum_{j=1}^{d} |\xi|^{2m_j} |\zeta_j|^2, \tag{8.2.4}$$

where C_0 is a positive constant. The coordinates of the vector $\xi = (\xi_1, \ldots, \xi_n)$ are real, while the coordinates of $\zeta = (\zeta_1, \ldots, \zeta_d)'$ are complex. The definition of strongly elliptic systems in this generality is given by Nirenberg in [290]. So far we assume the functions $a_{\alpha,\beta}^{r,s}(x)$ to be infinitely differentiable on $\overline{\Omega}$.

First, consider the inner product $(Lu, v)_{\Omega}$ of compactly supported infinitely differentiable functions on Ω. Integration by parts yields

$$(f, v)_{\Omega} = \Phi_{\Omega}(u, v), \qquad (8.2.5)$$

where Φ_{Ω} is the form

$$\Phi_{\Omega}(u, v) = \int_{\Omega} \sum_{r,s=1}^{d} \sum_{|\alpha| \leq m_r, |\beta| \leq m_s} a_{\alpha,\beta}^{r,s}(x) \partial^{\beta} u_s(x) \overline{\partial^{\alpha} v_r(x)} \, dx \qquad (8.2.6)$$

with leading part

$$\int_{\Omega} \sum_{r,s=1}^{d} \sum_{|\alpha|=m_r, |\beta|=m_s} a_{\alpha,\beta}^{r,s}(x) \partial^{\beta} u_s(x) \overline{\partial^{\alpha} v_r(x)} \, dx. \qquad (8.2.7)$$

Of greatest importance and interest are the Dirichlet and Neumann problems in the weak setting with homogeneous boundary conditions. In the Dirichlet problem, we have

$$u_j, v_j \in \widetilde{H}^{m_j}(\Omega), \quad f_k \in H^{-m_k}(\Omega). \qquad (8.2.8)$$

By the *Neumann problem* we now mean the problem in which

$$u_j, v_j \in H^{m_j}(\Omega), \quad f_k \in \widetilde{H}^{-m_k}(\Omega). \qquad (8.2.9)$$

It is required that relation (8.2.5) hold for u, f, and any test functions v.

Let us introduce the space

$$H^m(\Omega) = H^{m_1}(\Omega) \times \ldots \times H^{m_d}(\Omega). \qquad (8.2.10)$$

The norm on this space is

$$\|u\|_{H^m(\Omega)} = \left(\sum_{1}^{d} \|u_j\|_{H^{m_j}(\Omega)}^2 \right)^{1/2}. \qquad (8.2.11)$$

Possibly, $m_1 = \ldots = m_d = m$. The form Φ_{Ω} is said to be *coercive* on the space (8.2.10) if all its elements satisfy the inequality

$$\|u\|_{H^m(\Omega)}^2 \leq C_1 \operatorname{Re} \Phi_{\Omega}(u, u) + C_2 \|u\|_{H^0(\Omega)}^2 \qquad (8.2.12)$$

with constants $C_1 > 0$ and $C_2 \geq 0$. This inequality with $C_2 = 0$ is called the *strong coercivity condition*. In a similar way, coercivity and strong coercivity on other solution spaces V are defined.

The most important of these spaces is

$$\widetilde{H}^m(\Omega) = \widetilde{H}^{m_1}(\Omega) \times \ldots \times \widetilde{H}^{m_d}(\Omega), \tag{8.2.13}$$

which is identified with

$$\overset{\circ}{H}{}^m(\Omega) = \overset{\circ}{H}{}^{m_1}(\Omega) \times \ldots \times \overset{\circ}{H}{}^{m_d}(\Omega). \tag{8.2.14}$$

It corresponds to the Dirichlet problem. More general solution spaces V are assumed to be subspaces of $H^m(\Omega)$ containing $\widetilde{H}^m(\Omega)$:

$$\widetilde{H}^m(\Omega) \subset V \subset H^m(\Omega). \tag{8.2.15}$$

In this case, we seek a solution u of the system $Lu = f$ in V and assume that the right-hand side f belongs to the space V' dual to V with respect to the extension of the inner product $(f, v)_\Omega$. Coercivity of the form Φ_Ω on $H^m(\Omega)$ implies coercivity on all subspaces V.

Theorem 8.2.1. *The strong ellipticity of the operator L implies coercivity on $\widetilde{H}^m(\Omega)$.*

This result is essentially due to Gårding [163], although in the generality considered here, this fact was mentioned in [290]. Unlike in the scalar case studied in Section 8.1, it needs to be proved, but the proof is simple and based on the method of freezing coefficients. For $m_1 = \ldots = m_d = 1$, i.e., in the case of a second-order system, it is given in Section 11; the general case is handled in a similar way.

The coercivity condition becomes the strong coercivity condition

$$\|u\|^2_{H^m(\Omega)} \leq C_1 \operatorname{Re} \Phi_\Omega(u, u) \tag{8.2.16}$$

if the real part of the form $(cu, u)_\Omega$ of the lower-order term is large enough.

Theorem 8.2.2. *If the form Φ_Ω is strongly coercive on the space $\widetilde{H}^m(\Omega)$, then the Dirichlet problem is uniquely solvable.*

This follows from the Lax–Milgram theorem.

If the form is only coercive (not strongly), i.e., the operator is only strongly elliptic, then, instead of unique solvability, the Fredholm property is obtained.

The situation with the Neumann and other problems is substantially more complicated. Sufficient conditions for coercivity on $H^m(\Omega)$ (and other subspaces V) have a very extensive literature. We give references in Section 19. The books most useful to read first are [8] and [286]. In Section 11 we give a convenient sufficient condition for the coercivity of second-order systems on $H^1(\Omega)$.

Theorem 8.2.3. *In the case of strong coercivity on $H^m(\Omega)$, the Neumann problem is uniquely solvable.*

This follows from the same Lax–Milgram theorem. In the case of nonstrong coercivity, we again obtain the Fredholm property.

The infinite differentiability of the coefficients is not required in these theorems. It suffices that the higher-order coefficients be continuous and the other coefficients be measurable and bounded. But if the coefficients, the boundary, and the function f are smooth, we can investigate the smoothness of the solution. As in Section 8.1, this can be done in two ways, by using Nirenberg's method of difference quotients [290] (see our Section 16.5) and by identifying the problem with the corresponding problem of the general theory of smooth elliptic problems. It should only be checked that the boundary operators of a variational problem satisfy the Lopatinskii condition.

Chapter 3
The Spaces H^s and Second-Order Strongly Elliptic Systems in Lipschitz Domains

9 Lipschitz Domains and Lipschitz Surfaces

9.1 Specifics of Lipschitz Domains and Surfaces

A surface Γ in $\mathbb{R}^n$ ($n \geq 2$; Γ is a curve for $n = 2$) is said to be *Lipschitz* if locally, in a neighborhood $U = U(x_0)$ of any point $x_0 \in \Gamma$, an appropriate rotation of the coordinate system transforms Γ into the graph of a *Lipschitz continuous function* $x_n = \varphi(x')$, or simply a *Lipschitz function*, i.e., a function satisfying the *Lipschitz condition*

$$|\varphi(y') - \varphi(x')| \leq C|y' - x'|. \tag{9.1.1}$$

Here $x' = (x_1, \ldots, x_{n-1}) \in O$ and $y' = (y_1, \ldots, y_{n-1}) \in O$, where O is, say, a ball in $\mathbb{R}^{n-1}$; of course, the constant C does not depend on x' and y'. A domain Ω in $\mathbb{R}^n$ is said to be *Lipschitz* if it has Lipschitz boundary and lies on one side of it, i.e., $\varphi(x) < 0$ (or $\varphi(x) > 0$) on all intersections $\Omega \cap U$. We usually assume the domain Ω to be bounded; in this case, the constants C in the local estimates (9.1.1) for a finite cover of the boundary by neighborhoods of its points have a finite maximum. The greatest lower bound for $\max C$ is called the *Lipschitz constant* of the surface Γ. We usually assume that the domain Ω is located under the graph of the function $x_n = \varphi(x')$, except in Section 10, where we do not want to deviate from the notation of [320].

A closed C^1 surface is "better" than a Lipschitz one, but it can also be considered as Lipschitz. In this case, the Lipschitz constant is zero, because the local Lipschitz constants are arbitrarily small. This is so because if the coordinate plane $x_n = 0$ is tangent to the graph of a C^1 function $y = \varphi(x')$ at the origin, then the first derivatives of $\varphi(x')$ at this point vanish.

Let us discuss examples. First, we show that all convex bounded domains (in particular, all convex polyhedra, cones, and cylinders in the three-dimensional case) are Lipschitz.

© Springer International Publishing Switzerland 2015
M.S. Agranovich, *Sobolev Spaces, Their Generalizations and Elliptic Problems in Smooth and Lipschitz Domains*, Springer Monographs in Mathematics, DOI 10.1007/978-3-319-14648-5_3

Let X be a linear manifold. As is known, a set in X is said to be *convex* if, together with any two points, it contains all points of the line segment joining them. Any intersection of convex sets is convex. The convex hull of given points in X is the minimal convex set containing these points.

In the next theorem, it is convenient to assume for a while that the function φ depends on n variables: $\varphi(x) = \varphi(x_1, \ldots, x_n)$.

A real function φ defined on an open set $U \subset \mathbb{R}^n$ is said to be *convex downward* if, given any two points x and y in U, the segment joining them is contained in U and

$$\varphi(\theta x + (1 - \theta)y) \le \theta\varphi(x) + (1 - \theta)\varphi(y)$$

for any $\theta \in (0, 1)$; if the reverse inequality holds, then φ is said to be *convex upward*. In the case of convexity downward, we omit the word "downward."

Theorem 9.1.1. *Any convex function is locally Lipschitz continuous.*

Proof [294, pp. 12–17]. 1. First, we check that the convex function φ on U is locally bounded from above.

If $x \in U$, then there exists a $\delta > 0$ and points $x^0, \ldots, x^n$ such that the convex hull of these points is contained in U and the ball $B_\delta(x)$ of radius δ centered at x is contained in this convex hull. Thus, for $y \in B_\delta(x)$, there exist numbers $t_k \in [0, 1]$ ($k = 0, \ldots, n$) such that

$$y = \sum t_k x^k \quad \text{and} \quad \sum t_k = 1.$$

The convexity of the function φ implies

$$\varphi(y) \le \sum t_k \varphi(x^k) \le \max \varphi(x^k) = c.$$

2. Now we show that φ is continuous.

Suppose that $\varphi(x) \le c$ on U and $x_0 \in U$. Without loss of generality, we assume that $x_0 = 0$ (i.e., x_0 coincides with the origin) and $\varphi(0) = 0$.

Suppose that a ball $B = B_r(0)$ is contained in U, $\varepsilon \in (0, c)$, and $V_\varepsilon = (\varepsilon/c)B$. This is a neighborhood of the origin. To verify the continuity of our function at 0, it suffices to prove that $|\varphi(x)| \le \varepsilon$ on V_ε. Let $x \in V_\varepsilon$.

We have $(c/\varepsilon)x \in B$ and

$$\varphi(x) \le \frac{\varepsilon}{c}\varphi\left(\frac{c}{\varepsilon}x\right) + \left(1 - \frac{\varepsilon}{c}\right)\varphi(0) \le \frac{\varepsilon}{c}c = \varepsilon.$$

Next, taking $y = -(c/\varepsilon)x$ and $\theta = (1 + \varepsilon/c)^{-1}$, we obtain

$$0 = \varphi(0) \le \frac{1}{1 + \frac{\varepsilon}{c}}\varphi(x) + \frac{\frac{\varepsilon}{c}}{1 + \frac{\varepsilon}{c}}\varphi\left(-\frac{c}{\varepsilon}x\right) \le \frac{1}{1 + \frac{\varepsilon}{c}}\varphi(x) + \frac{\frac{\varepsilon}{c}}{1 + \frac{\varepsilon}{c}},$$

so that $-\varepsilon \le \varphi(x)$. Thus, $|\varphi(x)| \le \varepsilon$.

3. Now we pass to the proof of the theorem. Suppose that $x_0 \in U$ and $\varphi(x_0) = 0$. Take positive numbers c, r, and δ for which $\overline{B_{r+\delta}(x_0)} \subset U$ and

$$|\varphi(y)| \le c \quad \text{for } y \in B_{r+\delta}(x_0).$$

We verify the Lipschitz condition on $B_r(x_0)$.

Take $x, y \in B_r(x_0)$, $x \ne y$. We set

$$z = y + \delta \frac{y - x}{|y - x|} \quad \text{and} \quad \theta = \frac{|y - x|}{|y - x| + \delta}.$$

We have $z \in B_{r+\delta}(x_0)$ and $y = \theta z + (1 - \theta)x$. Using the convexity of the function φ, we obtain

$$\varphi(y) \le \theta \varphi(z) + (1 - \theta)\varphi(x).$$

Therefore,

$$\varphi(y) - \varphi(x) \le \theta[\varphi(z) - \varphi(x)] \le \frac{|y - x|}{|y - x| + \delta} 2c \le \frac{2c}{\delta}|y - x|.$$

Since x and y are interchangeable, we have

$$|\varphi(y) - \varphi(x)| \le \frac{2c}{\delta}|y - x| \quad \text{for } x, y \in B_r(x_0). \qquad \square$$

Of course, a similar assertion holds for functions convex upward.

It can be shown that a bounded domain in $\mathbb{R}^n$ is convex if, near each boundary point, an appropriate rotation of the coordinate system transforms the boundary into the graph of a convex upward function $x_n = \varphi(x')$ and the adjacent part of the domain lies under this graph. We shall not dwell on this.

Corollary 9.1.2. *Any bounded convex domain is Lipschitz.*

In Maz'ya's book [250, Sec. 1.1.8], it was shown that if Ω is a bounded domain star-shaped with respect to some ball (i.e., star-shaped with respect to all points of this ball), then, in the spherical coordinate system (r, ω) with origin at the center of this ball, the boundary is given by the equation $r = r(\omega)$, where the right-hand side is a Lipschitz function. Obviously, any convex domain is star-shaped with respect to any of its points. It was also mentioned in [250] that, conversely, if the boundary of a bounded domain can be specified by $r = r(\omega)$ with Lipschitz right-hand side in a spherical coordinate system, then this domain is star-shaped with respect to some ball centered at the origin.

An example of a non-Lipschitz domain in $\mathbb{R}^3$ is two bricks laid across one another (see [258, p. 91] or [221, p. 5]). In [221, p. 5] an example of a (nonconvex) conical non-Lipschitz surface infinitely differentiable outside a point is also given. A plane domain bounded by a curve having a cusp point or cutting the domain near some of its points is not Lipschitz either.

Some authors use the term "Lipschitz" for more general domains; to our Lipschitz domains they often refer as strictly Lipschitz domains.

In particular, in the book [250, Sec. 1.1.9] Maz'ya used a more general definition of Lipschitz domains and referred to our Lipschitz domains as domains of class $C^{0,1}$. In the book [173, Sec. 1.2] Grisvard referred to our Lipschitz domains as domains with Lipschitz boundary and introduced more general Lipschitz submanifolds (in $\mathbb{R}^n$). He showed that the latter are invariant with respect to any bi-Lipschitz transformations $y = y(x)$ of coordinates, i.e., transformations with Lipschitz function $y = y(x)$ having Lipschitz inverse $x = x(y)$, but they do not always have Lipschitz boundary in our sense.

Note that our Lipschitz boundary Γ can be rectified by using the coordinate transformation

$$y' = x', \quad y_n = \varphi(x') - x_n.$$

This transformation is bi-Lipschitz, because the inverse transformation is

$$x' = y', \quad x_n = y_n - \varphi(y').$$

But the coefficients of the Jacobian matrices of these transformations are generally only bounded and measurable. Thus, if an operator whose coefficients possess some (possibly low) smoothness is given on the domain adjacent to Γ, then the smoothness may be violated under such a transformation.

Recall that a function φ defined in a neighborhood of a point $x' \in \mathbb{R}^{n-1}$ is said to be differentiable at this point if

$$\varphi(y') - \varphi(x') = \sum_{1}^{n-1} \partial_j \varphi(x')(y_j - x_j) + o(|y' - x'|) \quad \text{as } y' \to x'.$$

Theorem 9.1.3. *If a real-valued function $\varphi(x')$ is Lipschitz continuous on a domain U in $\mathbb{R}^{n-1}$, then it is differentiable almost everywhere on U.*

Proof. 1. For $n - 1 = 1$, this follows from well-known theorems (see, e.g., [214]). Indeed, a Lipschitz continuous function is of bounded variation; therefore, it equals the difference of two monotone functions, and a monotone function has a derivative almost everywhere, which is equivalent to differentiability at the same points in the one-dimensional case. Moreover, the absolute value of the derivative is bounded by the Lipschitz constant.

In the multidimensional case, we can assume that the domain U is a ball. Let v denote its Lebesgue measure (n-dimensional volume). The above considerations imply the existence and boundedness of all first partial derivatives on a set $U' \subset U$ of full measure. Each of them is the limit of the corresponding difference relation. By Egorov's theorem (see, e.g., [214]), for any $\delta > 0$, there exists a set $U'' \subset U'$ of Lebesgue measure at least $v - \delta$ on which the convergence is uniform. We can assume that this set is perfect, i.e., that it is closed and has no isolated points. On U'' all first partial derivatives are continuous and, hence, uniformly continuous. Moreover, they are uniformly bounded in absolute value by the Lipschitz constant C.

2. For simplicity, we first prove the theorem for $n - 1 = 2$. Here we follow [363]. Let $T \subset U''$ be the set of density points of U''. This means that, if $x \in T$ and $O_r(x)$

is a ball of radius r centered at x, then the ratio of the measure of the intersection $T \cap O_r(x)$ to the measure of this ball tends to 1 as $r \to 0$. The measures of the sets T and U'' are the same. We shall prove the differentiability of the function $\varphi(x)$ at each point x^0 of T.

Take any point x^1. Let $r = |x^1 - x^0|$. The assumption that x^1 belongs to U'' substantially simplifies the argument. Thus, suppose that x^1 does not belong to U''. The idea is to temporarily replace this point by a very close point $x^2 \in U''$. Take a small number $\varepsilon > 0$.

Since x^0 is a density point for U'', it follows that, for sufficiently small r, the intersection $O_r(x^0) \cap O_{\varepsilon r}(x^1)$ contains a point $x^2 \in U''$. We have

$$|x^1 - x^2| < \varepsilon |x^1 - x^0|. \tag{9.1.2}$$

We must show that

$$\varphi(x^1) - \varphi(x^0) = (x_1^1 - x_1^0)\partial_1\varphi(x^0) + (x_2^1 - x_2^0)\partial_2\varphi(x^0) + \alpha(x^1)|x^1 - x^0|, \tag{9.1.3}$$

where $\alpha(x^1) \to 0$ as $r \to 0$. Let us write the left-hand side in the form

$$\varphi(x^1) - \varphi(x^0) = [\varphi(x^1) - \varphi(x^2)] + [\varphi(x^2) - \varphi(x^0)]. \tag{9.1.4}$$

The first bracketed difference is estimated by using condition (9.1.1) and inequality (9.1.2) as

$$|\varphi(x^1) - \varphi(x^2)| \le C|x^1 - x^2| < C\varepsilon|x^1 - x^0|. \tag{9.1.5}$$

The second bracketed difference in (9.1.4) can be rewritten as

$$\varphi(x^2) - \varphi(x^0) = [\varphi(x_1^2, x_2^2) - \varphi(x_1^0, x_2^2)] + [\varphi(x_1^0, x_2^2) - \varphi(x_1^0, x_2^0)]. \tag{9.1.6}$$

We rewrite the first bracketed difference on the right-hand side in the form

$$(x_1^2 - x_1^0)\frac{\varphi(x_1^2, x_2^2) - \varphi(x_1^0, x_2^2)}{x_1^2 - x_1^0} = (x_1^2 - x_1^0)[\partial_1\varphi(x^2) + \beta_1]$$
$$= (x_1^2 - x_1^0)[\partial_1\varphi(x^0) + \beta_1 + \beta_2]. \tag{9.1.7}$$

As $r \to 0$, the quantity β_1 uniformly tends to zero by virtue of the uniform convergence of the difference quotients to partial derivatives. The quantity β_2 uniformly tends to zero by virtue of the uniform continuity of the partial derivatives on T. We mean uniformity not only in x^2 (we must change this point when letting $x^1 \to x^0$) but also in x^0.

The transformation of the second bracketed difference in (9.1.6) is somewhat simpler:

$$(x_2^2 - x_2^0)\frac{\varphi(x_1^0, x_2^2) - \varphi(x_1^0, x_2^0)}{x_2^2 - x_2^0} = (x_2^2 - x_2^0)[\partial_2\varphi(x^0) + \beta_3]. \tag{9.1.8}$$

Here $\beta_3 \to 0$ as $r \to 0$, again uniformly in x^2 and x^0. For sufficiently small r, all β_j are less than ε in absolute value.

Now we have

$$\varphi(x^1) - \varphi(x^0) = (x_1^2 - x_1^0)\partial_1\varphi(x^0) + (x_2^2 - x_2^0)\partial_2\varphi(x^0) + \gamma_1, \qquad (9.1.9)$$

where

$$|\gamma_1| \le |x^1 - x^0|C_1\varepsilon \qquad (9.1.10)$$

for sufficiently small $r = |x^1 - x^0|$. At the last step, we replace the linear multipliers:

$$\varphi(x^1) - \varphi(x^0) = (x_1^1 - x_1^0)\partial_1\varphi(x^0) + (x_2^1 - x_2^0)\partial_2\varphi(x^0) + \gamma_2, \qquad (9.1.11)$$

where

$$|\gamma_2| \le |x^1 - x^0|C_2\varepsilon;$$

thus, in (9.1.3), we have

$$|\alpha(x^1)| \le C_3\varepsilon, \qquad (9.1.12)$$

where the constant does not depend on x^1.

This proves the differentiability of φ at x^0, because ε is arbitrary. Note that the estimate for α in (9.1.3) is uniform in x^0, i.e., we have obtained a uniform estimate of the ratio of the difference between the increment of the function and its differential to the distance $|x^1 - x^0|$.

3. The proof can be completed by induction on the dimension n. We only outline it. Suppose that $n - 1 \ge 3$ and the required assertion with $n - 2$ instead of $n - 1$ is proved, as well as the corresponding uniform estimate.

We now write the points x of the ball U in the form (x_1, X_2), where $X_2 = (x_2, \ldots, x_{n-1})$. At all points of the set $U'' \subset U$ (which is perfect and has measure arbitrarily close to that of U), the function φ is differentiable with respect to X_2 and has a partial derivative with respect to x_1; moreover, all partial derivatives are continuous and uniformly bounded on U'', and the ratio of the difference between the increment of the function under variation of X_2 and its differential with respect to X_2 to the corresponding distance is uniformly small. Again, let T be the set of density points of U''. Total differentiability is again proved at each point $x^0 \in T$. If x^1 is another point at distance r from x^0, then, given small ε and sufficiently small r, we find a point $x^2 \in U''$ such that $|x^1 - x^2| \le \varepsilon|x^1 - x^0|$,

$$\varphi(x_1^2, X_2^2) - \varphi(x_1^0, X_2^2) = (x_1^2 - x_1^0)[\partial_1\varphi(x^0) + \beta],$$

and

$$\varphi(x_1^0, X_2^2) - \varphi(x_1^0, X_2^0) = \sum_{2}^{n-1}(x_j^2 - x_j^0)\partial_j\varphi(x^0) + \widetilde{\beta}|X_2^2 - X_2^0|,$$

where $|\beta| \le \varepsilon$ and $|\widetilde{\beta}| \le \varepsilon$ uniformly for sufficiently small r. The rest of the argument is similar to that used in the two-dimensional case, and the required uniform estimate is obtained again. $\square$

It follows from Theorem 9.1.3 that any Lipschitz surface has a tangent hyperplane and a normal at almost every point.

The Lebesgue measure on $\mathbb{R}^n$ induces a Lebesgue measure on Γ. Locally, in the same local coordinates as in (9.1.1), the Lebesgue integral of a function $f(x)$ on Γ over a fragment $\mathfrak{S}$ with projection O on the x'-plane is determined by

$$\int_{\mathfrak{S}} f(x)\,dS = \int_{O} f(x', \varphi(x')) \sqrt{1 + |\nabla\varphi(x')|^2}\,dx'. \qquad (9.1.13)$$

Here the gradient $\nabla\varphi(x')$ is bounded.

An important property of a Lipschitz surface Γ is that it satisfies the uniform cone condition, both interior and exterior. For the boundary Γ of a bounded Lipschitz domain Ω, this property can be formulated as follows. See [173].

Proposition 9.1.4. *Each point $x \in \Gamma$ is the common vertex of two closed circular cones $\Upsilon_+(x)$ and $\Upsilon_-(x)$ of finite height such that $\Upsilon_+(x) \setminus \{x\}$ lies inside Ω and $\Upsilon_-(x) \setminus \{x\}$ lies inside the complement of $\overline{\Omega}$. Moreover, all these cones are congruent to a fixed circular cone Υ and depend continuously on x.*

The cone Υ can be specified as

$$\Upsilon = \{z = (z', z_n): |z'|/\alpha \le z_n \le h\}, \qquad (9.1.14)$$

where h and α are sufficiently small positive numbers.

We say that, in this case, we have a *regular family of cones* $\{\Upsilon_\pm(x)\}$, $x \in \Gamma$.

Moreover, locally, the axes of these cones can be chosen parallel.

The converse is also true: the uniform cone condition implies the Lipschitz property of the boundary. (Also, the uniform interior cone condition implies the uniform exterior cone condition, and vice versa.) The proof is given in [173, pp. 10, 11]. The boundary point x is placed at the center of a small cylinder

$$S = \{y = (y', y_n): |y' - x'| < r, \ |y_n - x_n| < h\},$$

the function $y_n = \varphi(y')$ is defined for $|y' - x'| < r$ as the least upper bound of y_n for which $(y', y_n) \in S \cap \Omega$, and its Lipschitz continuity is verified.

Yet another important property of a Lipschitz boundary is that it can be approximated by an infinitely differentiable surface from inside and from outside. More precisely, the following assertion is valid, which was proved in [285] and [382].

Proposition 9.1.5. *Suppose that Ω^+ is a bounded Lipschitz domain, Ω^- is the complement to its closure, Γ is the common boundary of Ω^+ and Ω^-, and $\{\Upsilon_\pm(y)\}$, $y \in \Gamma$, is a regular family of cones for Γ.*

Then there exists a sequence of domains Ω_j^+ with infinitely differentiable boundaries Γ_j and closures contained inside Ω^+ and a sequence of Lipschitz diffeomorphisms $\Lambda_j: \Gamma \to \Gamma_j$ with the following properties.

1. *The sequence $\Lambda_j(y)$ converges uniformly with respect to $y \in \Gamma$ as $j \to \infty$, remaining in the cone $\Upsilon_+(y)$.*

2. *The Jacobians of the mappings Λ_j converge pointwise to 1 almost everywhere and in any space $L_p(\Gamma)$, $1 \le p < \infty$.*

3. *The unit outer normal vectors $\nu(\Lambda_j(y))$ to Γ_j converge to the unit outer normal vector $\nu(y)$ to Γ for almost all $y \in \Gamma$ and in any $L_p(\Gamma)$, $1 \le p < \infty$.*

4. *There exists a C^∞ vector field $h(x)$ on $\mathbb{R}^n$ such that, on Γ_j, the inner product $(h(\Lambda_j(y)), \nu(\Lambda_j(y))$ is bounded below by a positive constant depending only on the Lipschitz constant for Γ.*

There exists also a similar family of domains Ω_j^- with closures lying inside Ω^-.

The notations $\Omega_j^+ \uparrow \Omega$ and $\Omega_j^- \downarrow \Omega$ are often used.

The proof reduces to a local consideration of the boundary Γ in local coordinates. The corresponding function φ (see (9.1.1)) is smoothed, namely, replaced by its convolution with a function from a delta-shaped family of compactly supported C^∞ functions, and the graph of this convolution is shifted in an appropriate direction. We do not give a detailed proof, but in Section 9.3 the reader can find the most important details in the case $p = 2$.

We also need *special Lipschitz domains*. These are unbounded domains above or below the graph of a uniformly Lipschitz continuous function $x_n = \varphi(x')$ on $\mathbb{R}^{n-1}$ ("uniformly" means that the Lipschitz constant is fixed).

9.2 The Spaces H^s on Lipschitz Domains and Lipschitz Surfaces

Let Ω be a bounded Lipschitz domain with Lipschitz boundary Γ.

The definitions of the spaces $H^s(\Omega)$ given in Section 5, as well as those of $\widetilde{H}^s(\Omega)$ and $\mathring{H}^s(\Omega)$, remain the same for any s. But, in the general case, the space $H^s(\Gamma)$ can be defined invariantly (by using partitions of unity and norms on $H^s(\mathbb{R}^{n-1})$) only for $|s| \le 1$, because a Lipschitz surface is locally the graph of a Lipschitz function.

The trace theorem is stated as follows.

Theorem 9.2.1. *The operator of taking the trace on the boundary Γ of a bounded Lipschitz domain Ω acts boundedly from $H^s(\Omega)$ to $H^{s-1/2}(\Gamma)$ for $1/2 < s < 3/2$.*

As we shall see shortly, for $1/2 < s \le 1$, this assertion is obtained in almost the same way as in the case of a smooth boundary. For $1 < s < 3/2$, the result is due to Costabel [99].

Proof of Theorem 9.2.1. It is sufficient to consider the special Lipschitz domain below the graph of a uniformly Lipschitz continuous function $x_n = \varphi(x')$. In this case, x' stands for local coordinates on Γ. We assume that the function $u(x)$ is extended to a function in $H^s(\mathbb{R}^n)$. (The extension operator is constructed in the next section.) In this space the functions from $C_0^\infty(\mathbb{R}^n)$ are dense. Therefore, we can assume that $u \in C_0^\infty(\mathbb{R}^n)$.

Let

$$u_1(x) = u(x', \varphi(x') + x_n). \tag{9.2.1}$$

Then the boundary value $\gamma^+ u$ is $u_1(x', 0)$, and

$$\|\gamma^+ u\|_{H^{s-1/2}(\Gamma)} = \|u_1(x', 0)\|_{H^{s-1/2}(\mathbb{R}^{n-1})}. \tag{9.2.2}$$

Problem. Show that

$$\|u_1(x)\|_{H^s(\mathbb{R}^n)} \le C_1 \|u(x)\|_{H^s(\mathbb{R}^n)} \quad \text{for } 0 < s \le 1. \tag{9.2.3}$$

By virtue of the general trace theorem 1.10.1, we have

$$\|u_1(x', 0)\|_{H^{s-1/2}(\mathbb{R}^{n-1})} \le C_2 \|u_1(x)\|_{H^s(\mathbb{R}^n)} \quad \text{for } 1/2 < s \le 1;$$

therefore, the case $1/2 < s \le 1$ causes no difficulties. Suppose that $1 < s < 3/2$.

Following Costabel, we introduce the anisotropic space $E^s = E^s(\mathbb{R}^n)$ with norm

$$\|u\|_{E^s}^2 = \int a_s(\xi) |Fu(\xi)|^2 \, d\xi, \tag{9.2.4}$$

where

$$a_s(\xi) = |\xi_n|^{2s-2}(1 + |\xi|^2). \tag{9.2.5}$$

We claim that

$$\|u_1\|_{E^s} \le C_2 \|u\|_{H^s(\mathbb{R}^n)} \quad \text{and} \quad \|u_1(x', 0)\|_{H^{s-1/2}(\mathbb{R}^{n-1})} \le C_3 \|u_1\|_{E^s}. \tag{9.2.6}$$

This implies the required assertion. The first inequality in (9.2.6) is obvious; let us verify the second. Note that

$$\int a_s^{-1}(\xi) \, d\xi_n = C_s (1 + |\xi'|^2)^{-(s-1/2)}, \quad \text{where } C_s = \int \frac{dt}{t^{2s-2}(1 + t^2)}; \tag{9.2.7}$$

this integral converges for $1 < s < 3/2$. We have

$$\|u_1(x', 0)\|_{H^{s-1/2}(\mathbb{R}^{n-1})}^2 = \int_{\mathbb{R}^{n-1}} (1 + |\xi'|^2)^{s-1/2} \left| \int (Fu_1)(\xi', \xi_n) \, d\xi_n \right|^2 d\xi'.$$

Applying the Schwarz inequality, we see that the right-hand side does not exceed

$$\int_{\mathbb{R}^{n-1}} (1 + |\xi'|^2)^{s-1/2} \left(\int_{-\infty}^{\infty} a_s^{-1}(\xi) \, d\xi_n \right) \left(\int_{-\infty}^{\infty} a_s(\xi) |(Fu_1)(\xi)|^2 \, d\xi_n \right) d\xi',$$

and this expression equals $C_s \|u_1\|_{E^s}^2$. We have obtained the second inequality in (9.2.6), as required. $\qquad\square$

Already for $s = 3/2$, this theorem is no longer valid (see an example in [195]). Because of this limitation, the spaces $H^s(\Omega)$ with large $|s|$ are often inconvenient for considering boundary value problems in Lipschitz domains.

The construction of a right inverse operator from $H^{s-1/2}(\Gamma)$ to $H^s(\Omega)$ of the trace operator for $s \in (1/2, 1]$ causes no problem. For $s \in (1, 3/2)$, this construction is described in the book [197].

It is convenient to define $C^t(\overline{\Omega})$, where Ω is a bounded Lipschitz domain and $t > 0$, as the space consisting of the restrictions to $\overline{\Omega}$ of the functions from $C_b^t(\mathbb{R}^n)$ with the usual inf-norm. The linear manifold $C^\infty(\overline{\Omega})$ can also be understood as consisting of C^∞ functions on Ω admitting a continuous extension to $\overline{\Omega}$ together with all their derivatives. It is dense in all $H^s(\Omega)$.

The spaces $H^s(\Omega)$ and $\widetilde{H}^{-s}(\Omega)$ remain dual with respect to the extension of the inner product in $L_2(\Omega)$ to their direct product. The spaces $H^s(\Omega)$ and $\widetilde{H}^s(\Omega)$ with $|s| < 1/2$ can be identified, their norms are equivalent. The spaces $\widetilde{H}^s(\Omega)$ and $\mathring{H}^s(\Omega)$ can be identified for positive non-half-integer s, their norms are equivalent. For positive half-integer s, the space $\widetilde{H}^s(\Omega)$ is a nonclosed linear manifold in $\mathring{H}^s(\Omega)$ with norm of type (5.1.14); this norm is equivalent to the initial norm. The corresponding topology is stronger than that of $\mathring{H}^s(\Omega)$. Cf. [173].

The following remark is similar to Remarks 4.4.1 and 5.1.16.

Remark 9.2.2. For a non-half-integer $s > 1/2$, we can arrange the (nonstandard) strict inclusion

$$H^{-s}(\Omega) \subset \widetilde{H}^{-s}(\Omega), \tag{9.2.8}$$

as in (5.1.15), by agreeing that the functionals on the left-hand side are extended by zero to the orthogonal complement of $\widetilde{H}^s(\Omega)$ in $H^s(\Omega)$. The space on the right-hand side contains, in addition to that on the left-hand side, functionals on $H^s(\Omega)$ supported on the boundary. For $s \in (1/2, 3/2)$, they have the form

$$(h, v^+)_\Gamma, \tag{9.2.9}$$

where $h \in H^{-s+1/2}(\Gamma)$. This it is seen from (5.1.16); cf. [261]. For $0 \le s < 1/2$, the spaces in (9.2.8) can be identified.

As to multipliers, Theorem 5.1.6 and the remark after it remain valid.

The theorem on the embedding of the space $H^s(\Gamma)$ with $s > 1/2$ in the space $C(\overline{\Gamma})$ of continuous functions is valid on the boundary of a two-dimensional Lipschitz domain.

The spaces $H^s(\Gamma)$ and $H^{-s}(\Gamma)$, $|s| \le 1$, are dual with respect to the extension of the inner product in $L_2(\Gamma)$ to their direct product.

Suppose that a Lipschitz surface Γ is divided into two Lipschitz domains, Γ_1 and Γ_2, by their common boundary $\partial\Gamma_j$ (of dimension $n-2$), which is Lipschitz as well; $\Gamma = \Gamma_1 \cup \partial\Gamma_j \cup \Gamma_2$. Then, for s with $|s| \le 1$, we can consider the spaces $H^s(\Gamma_1)$, $\widetilde{H}^s(\Gamma_1)$, and $\mathring{H}^s(\Gamma_1)$ and the similar spaces on Γ_2. The space $\widetilde{H}^s(\Gamma_1)$ is a subspace of $H^s(\Gamma)$. There is an operator extending functions on Γ_1 to Γ, which is a bounded operator from $H^s(\Gamma_1)$ to $H^s(\Gamma)$ for $|s| \le 1$.

The spaces $H^s(\Gamma_1)$ and $\widetilde{H}^{-s}(\Gamma_1)$ are dual with respect to the extension of the inner product in $L_2(\Gamma_1)$ to their direct product. The spaces $H^s(\Gamma_1)$ and $\widetilde{H}^s(\Gamma_1)$ can be identified for $|s| < 1/2$. The spaces $\widetilde{H}^s(\Gamma_1)$ and $\mathring{H}^s(\Gamma_1)$ can be identified for $0 < s \le 1$, $s \ne 1/2$.

9.3 Integration by Parts

It is convenient to consider integration by parts in a separate subsection. Our main goal in this subsection is to explain Proposition 9.3.3. First, we must prove several assertions useful in considering Lipschitz functions.

1. Let $\varphi(x)$ be a Lipschitz function on a bounded domain $G \subset \mathbb{R}^n$. Then it is Lipschitz with respect to each variable x_j with the other variables fixed; therefore, it is absolutely continuous with respect to x_j and has a derivative with respect to x_j almost everywhere. Moreover, the Newton–Leibniz formula holds (see, e.g., [214, Chap. 6, Sec. 4]). Thus, for $j = n$ (for simplicity of notation), we have

$$\varphi(x',b) - \varphi(x',a) = \int_a^b \partial_n \varphi(x',t)\, dt.$$

Therefore, *if a function $\varphi(x)$ is compactly supported* (vanishes near the boundary ∂G), *then*

$$\int_G \partial_j \varphi(x)\, dx = 0 \tag{9.3.1}$$

for all j.

2. Let $\Delta_n \varphi(x)$ denote the increment of a function $\varphi(x)$ (arbitrary at the moment) with respect to x_n:

$$\Delta_n \varphi(x) = \varphi(x', x_n + h) - \varphi(x', x_n).$$

Here h is fixed and not indicated in the notation. Let φ_1 and φ_2 be any two functions. Then

$$\Delta_n[\varphi_1(x)\varphi_2(x)] = \Delta_n \varphi_1(x) \cdot \varphi_2(x', x_n + h) + \varphi_1(x)\Delta_n \varphi_2(x).$$

Now suppose that φ_1 and φ_2 are Lipschitz. Their product is, of course, Lipschitz as well. Dividing the last relation by h and passing to the limit, we obtain

$$\partial_n[\varphi_1(x)\varphi_2(x)] = \partial_n \varphi_1(x) \cdot \varphi_2(x) + \varphi_1(x)\partial_n \varphi_2(x) \tag{9.3.2}$$

on a set of full measure. Of course, a similar relation holds for ∂_j with any j.

Proposition 9.3.1. *If φ_1 and φ_2 are Lipschitz functions and one of them is compactly supported, then the following formula of integration by parts is valid:*

$$\int_G \partial_j \varphi_1(x) \cdot \varphi_2(x)\, dx = -\int_G \varphi_1(x) \partial_j \varphi_2(x)\, dx. \tag{9.3.3}$$

Indeed, integrating (9.3.2) and applying (9.3.1), we obtain (9.3.3).

3. Assuming for simplicity that the Lipschitz function $\varphi(x)$ is compactly supported, consider its convolution with the smoothing function $\omega_h(x)$ introduced in Section 1.11 for sufficiently small h:

$$(\varphi * \omega_h)(x) = \int_G \varphi(y)\omega_h(x-y)\, dy. \tag{9.3.4}$$

Proposition 9.3.2. 1. *The convolution (9.3.4) is infinitely differentiable and uniformly converges to φ as $h \to 0$.*
  2. *The relation*

$$\partial_j(\varphi * \omega_h) = (\partial_j \varphi) * \omega_h \tag{9.3.5}$$

holds, and this derivative converges to $\partial_j \varphi$ in $L_2(\Omega)$ as $h \to 0$.
  3. *There is a numerical sequence $\{h_k\}$ tending to 0 such that the derivatives (9.3.5) for these h_k converge to $\partial_j \varphi$ almost everywhere.*

Proof. The convolution (9.3.4) can be differentiated arbitrarily many times under the integral sign; its infinite differentiability causes no doubt.

Since $\int_G \omega_h(y)dy = 1$, it follows that

$$(\varphi * \omega_h)(x) - \varphi(x) = \int_G [\varphi(y) - \varphi(x)]\omega_h(x-y)\, dy.$$

Here the function $\omega_h(x-y)$ is nonzero only for small $|x-y|$, and the uniform convergence of the left-hand side to zero follows from the uniform continuity of φ.

Next, we have

$$\partial_j(\varphi * \omega_h)(x) = -\int_G \varphi(y)\partial_j \omega_h(x-y)\, dy$$

(the derivative on the right-hand side is with respect to y), so that (9.3.3) implies the relation

$$\partial_j(\varphi * \omega_h)(x) = \int_G [\partial_j \varphi(y)]\omega_h(x-y)\, dy,$$

which yields (9.3.5).

The bounded function $\partial_j \varphi$ belongs to $L_2(G)$; therefore, as in Section 1.11, this convolution converges to $\partial_j \varphi$ in $L_2(G)$.

This convergence implies convergence in measure: the measure of the set on which the absolute value of the difference $\partial_j(\varphi * \omega_h)(x) - \partial_j \varphi(x)$ is larger than an arbitrary fixed positive number tends to zero. It is known that any sequence of functions converging to zero in measure contains a subsequence converging to zero almost everywhere (see, e.g., [214, Chap. 5, Sec. 4]). This implies the last assertion of the proposition. $\qquad\square$

Relations of the form (9.3.5) are well known; see, e.g., [18] or [335]. Assertion 3 of Proposition 9.3.2 remains "jobless" in what follows, but it was mentioned in Proposition 9.1.5 in a close context.

Shifting the graphs of the convolutions downward, we can assume that they approximate the graph of the function φ from below.

4. Let Ω be a bounded domain with Lipschitz boundary Γ, and let ν be the unit outer normal vector to the boundary (it exists almost everywhere).

Proposition 9.3.3. *Let $u, v \in H^1(\Omega)$. Then*

$$(\partial_j u, v)_\Omega + (u, \partial_j v)_\Omega = (\nu_j u^+, v^+)_\Gamma. \tag{9.3.6}$$

Proof. We give only a part of the proof. It suffices to obtain the required formula for smooth functions u and v with supports adjacent to a small piece of the boundary. But we consider only the case where this piece lies above a small ball U in the coordinate hyperplane $x_n = 0$ and is the graph of a positive Lipschitz function $x_n = \varphi(x')$, the supports of the functions u and v lie above the same ball below this graph, and $j = 1, \dots, n-1$.

We approximate the function φ from below by C^∞ functions, also defined on U, as described above. We denote these functions by φ_m. Let $\theta_m(x)$ be the characteristic function of the part of Ω above the ball below the graph Γ of φ_m; then we have

$$(\theta_m \partial_j u, v)_\Omega + (\theta_m u, \partial_j v)_\Omega = (\nu_{m,j} u^+, v^+)_\Gamma, \tag{9.3.7}$$

where ν_m denotes the outer normal at the points of this graph. Obviously, as $m \to \infty$, the left-hand side of (9.3.7) converges to the left-hand side of (9.3.6). It remains to verify that the same is true for the right-hand side. It can be written in detail as

$$\int_U [\partial_j \varphi_m(x')] u(x', \varphi_m(x')) \overline{v(x', \varphi_m(x'))} \, dx'$$

after reduction by the square root of $1 + \sum_{k=1}^{n-1} (\partial_k \varphi_m)^2$, which is first written under $\partial_j \varphi_m(x')$ and in front of dx'. Since $\varphi_m \to \varphi$ uniformly and $\partial_j \varphi_m \to \partial_j \varphi$ in $L_2(U)$, the required result easily follows. $\qquad\square$

A more complete proof of (9.3.6) for functions in the more general Sobolev L_p-spaces W_p^1 (see Section 14) can be found in the book [286].

10 Discrete Norms, Discrete Representation of Functions, and a Universal Extension Operator

Our purpose in this section is threefold. First, we construct "discrete representations" of functions from $H^s(\mathbb{R}^n)$ and the corresponding "discrete norms," substantially more general than those constructed in Section 1.14. Secondly, we construct similar representations and norms in special Lipschitz domains Ω. Thirdly, on this basis, we construct a universal bounded operator extending functions in $H^s(\Omega)$ to functions in $H^s(\mathbb{R}^n)$, which is defined for all s. It is for the last purpose that the constructions of Section 1.14 are generalized. The universal extension operator is convenient, in particular, for applying theorems of interpolation theory.

The norms considered previously are more convenient for studying partial differential equations. This is seen from the extensive literature. But they do not fit to solve all problems of the theory of the H^s spaces. In particular, up to now we have no extension operator for negative s, and no norm on $H^s(\Omega)$ with negative s explicitly determined for functions or distributions on Ω, like norms (5.1.2) and (5.1.3). They will be defined in this section. In the construction of the extension operator it is essential that, in the discrete norms which we define, only numerical multipliers depend on s, as the reader will shortly see.

This entire section is a simplified version of the paper [320], in which much more general spaces were considered. But reading this section can hardly be easy.

10.1 Discrete Norms on $H^s(\mathbb{R}^n_x)$

We begin with a substantial generalization of the partition of unity used in Section 1.14. We shall shortly need this generalization.

Suppose that a function $\psi_0(\xi)$ belongs to the Schwartz space $S(\mathbb{R}^n_\xi)$ and is positive at $\xi = 0$; let, moreover,

$$|\psi_0(\xi) - \psi_0(0)| \le C_N|\xi|^N \quad (|\xi| \le 1) \tag{10.1.1}$$

for some integer $N \ge 1$ and constant C_N not depending on ξ. This condition is equivalent to the vanishing at 0 of all derivatives of ψ_0 of nonzero order lower than N. Of course, it does not impose any additional constraints for $N = 1$. But eventually we shall need $N = \infty$.

We set

$$\psi(\xi) = \psi_0(\xi) - \psi_0(2\xi) \tag{10.1.2}$$

and

$$\psi_j(\xi) = \psi(\xi/2^j) \quad (j = 1, 2, \ldots), \tag{10.1.3}$$

so that

$$\psi_j(\xi) = \psi_0(\xi/2^j) - \psi_0(\xi/2^{j-1}) \quad (j = 1, 2, \ldots). \tag{10.1.4}$$

We denote the Fourier preimages of the functions $\psi_0(\xi)$, $\psi(\xi)$, and $\psi_j(\xi)$ by $\varphi_0(x)$, $\varphi(x)$, and $\varphi_j(x)$, respectively. The conditions $\psi_0(0) \neq 0$ and (10.1.1) are equivalent to the constraints

$$\int \varphi_0(x)\,dx \neq 0 \quad \text{and} \quad \int x^\alpha \varphi_0(x)\,dx = 0 \quad \text{for } 0 < |\alpha| < N \qquad (10.1.5)$$

on the moments of the function φ_0 (if $N > 1$).

Since $\psi_0(\xi)$ belongs to $S(\mathbb{R}^n)$ and satisfies condition (10.1.1), it follows that

$$|\psi(\xi)| \leq C_{N,M} |\xi|^N (1 + |\xi|)^{-M} \qquad (10.1.6)$$

with arbitrarily large M and a constant not depending on ξ. Here $|\xi|^N$ can be replaced by 1 if $|\xi| \geq 1$. This implies, in particular, that for $j \geq 1$ (and $M = 0$) we have

$$2^{2js} |\psi_j(\xi)|^2 \leq C_{N,0}^2 |\xi|^{2N} 2^{2j(s-N)}. \qquad (10.1.7)$$

Note that

$$\sum_0^m \psi_j(\xi) = \psi_0(\xi/2^m) \to \psi_0(0) \quad \text{as } m \to \infty; \qquad (10.1.8)$$

therefore,

$$\sum_0^\infty \psi_j(\xi) = \psi_0(0) \qquad (10.1.9)$$

at each point ξ, and the convergence is uniform on any bounded domain.

The usual norm of a function $u(x)$ in $H^s(\mathbb{R}^n)$ with any $s \in \mathbb{R}$ was defined by

$$\|u(x)\|_{H^s(\mathbb{R}^n)} = \|(1 + |\xi|^2)^{s/2}(Fu)(\xi)\|_{L_2(\mathbb{R}^n)}, \qquad (10.1.10)$$

where F is the Fourier transform. Let us define a new norm on $H^s(\mathbb{R}^n)$ by

$$\|u(x)\|_{H^s(\mathbb{R}^n),\varphi_0} = \left(\sum_0^\infty 2^{2js} \|\psi_j(\xi)(Fu)(\xi)\|_{L_2(\mathbb{R}^n)}^2 \right)^{1/2}. \qquad (10.1.11)$$

This norm can be rewritten in the form

$$\|u\|_{H^s(\mathbb{R}^n),\varphi_0} = \left[\sum_0^\infty 2^{2js} \|(\varphi_j * u)(x)\|_{L_2(\mathbb{R}^n)}^2 \right]^{1/2}. \qquad (10.1.12)$$

Theorem 10.1.1. *For any function $\varphi_0 \in S$ satisfying conditions (10.1.5) and any $s < N$, the norm on $H^s(\mathbb{R}^n)$ defined by (10.1.12) is equivalent to the usual norm (10.1.10).*

This is a generalization of Proposition 1.14.2, where the function ψ_0 was compactly supported. The proof reduces to the verification of the following assertion.

Proposition 10.1.2. *Given $s < N$, the two-sided estimate*

$$C_1 \le \sum_0^\infty \frac{2^{2js}|\psi_j(\xi)|^2}{(1+|\xi|^2)^s} \le C_2 \quad \cdot \tag{10.1.13}$$

is valid, where C_1 and C_2 are positive constants not depending on ξ.

Note that, by virtue of (10.1.7), the series in (10.1.13) converges at any point ξ; moreover, for bounded ξ, it converges uniformly.

Proof of Proposition 10.1.2. Let us fix numbers s and N. We denote the sum of the series in (10.1.13) by $\Sigma(\xi)$. Let us derive the uniform upper bound for $\Sigma(\xi)$.

First, suppose that s is positive. Consider ξ with $2^k \le |\xi| \le 2^{k+1}$ for some nonnegative integer k. The sum of the terms with numbers $j \ge k$ in the series does not exceed

$$\sum_{j=k}^\infty 2^{2(j-k)(s-N)}$$

up to a constant multiplier; i.e., it is dominated by the sum of the infinite geometric progression with common ratio $2^{2(s-N)}$.

Let us estimate the sum of the other terms except the zeroth one. Taking into account the uniform boundedness of all $|\psi_j(\xi)|$, we see that this sum does not exceed

$$\sum_{j=1}^{k-1} 2^{2(j-k)s}$$

up to a constant factor; hence it is dominated by the sum of the infinite geometric progression with common ratio 2^{-2s}.

Of course, the zeroth term of the series is uniformly bounded as well. Thus, the function $\Sigma(\xi)$ is bounded above by a positive constant uniformly in ξ with $|\xi| \ge 1$. For $|\xi| \le 1$, a similar bound is obtained by using inequality (10.1.7).

Now let $s < 0$. In this case, we must estimate

$$\sum 2^{-2j|s|}|\psi_j(\xi)|^2(1+|\xi|^2)^{|s|}.$$

Again, let $2^k \le |\xi| \le 2^{k+1}$. As above, in the sum over $j \ge k$ we estimate $|\psi_j(\xi)|$ by a constant and conclude that this sum is not greater than

$$\sum_{j \ge k} 2^{2(k-j)|s|}$$

up to a constant multiplier; hence it is dominated by the sum of the geometric progression with common ratio $2^{-2|s|}$. The sum of the other terms, except the zeroth one, does not exceed

$$\sum_{j < k} 2^{2(k-j)|s|} 2^{2(j-k)M}$$

up to a constant multiplier (by virtue of the bound for $\psi_j(\xi)$ implied by (10.1.6)). Choosing $M > |s|$, we see that the last sum is bounded by the sum

$$\sum_0^\infty 2^{-2l(M-|s|)}$$

of the geometric progression. The zeroth term is uniformly bounded as well. For $|\xi| \leq 1$, there is again no problem: see inequality (10.1.7).

Finally, let $s = 0$. For $j \geq k$, we argue as in the case of positive s, and for $j < k$, as in the case of negative s: in the former case, we use the assumption $N > 0$, and in the latter, we take $M > 0$.

Now let us estimate the sum $\Sigma(\xi)$ from below. By virtue of (10.1.9), it is positive at each point ξ, and the uniform convergence implies that it is bounded below by a positive constant in any bounded domain. Let us find $r > 0$ such that $|\psi_0(\xi)| \leq \psi_0(0)/2$ for $|\xi| \geq r$. We denote the sum of the other terms in $\Sigma(\xi)$ (i.e., of all terms except the zeroth one) by $\Sigma_1(\xi)$. For $r \leq |\xi| \leq 2r$, this sum is uniformly bounded below. Let $s \geq 0$. Then $\Sigma_1(2\xi) \geq \Sigma_1(\xi)$. This it is seen from the relations $\psi_{j+1}(2\xi) = \psi_j(\xi)$ and

$$\frac{2^{2(j+1)s}|\psi_{j+1}(2\xi)|^2}{(1+|2\xi|^2)^s} \geq \frac{2^{2js}|\psi_j(\xi)|^2}{(1+|\xi|^2)^s}.$$

It follows that, for $2r \leq |\xi| \leq 4r$, $4r \leq |\xi| \leq 8r$, and so on, the function $\Sigma_1(\xi)$ is bounded below by the same constant as for $r \leq |\xi| \leq 2r$.

Now let $s < 0$. We rewrite $\Sigma_1(\xi)$ in the form

$$\frac{(1+|\xi|^2)^{|s|}}{|\xi|^{2|s|}}\Sigma_2(\xi), \quad \text{where } \Sigma_2(\xi) = \sum_1^\infty \left(\frac{|\xi|}{2^j}\right)^{2|s|}\left|\psi\left(\frac{\xi}{2^j}\right)\right|^2.$$

The boundedness of $\Sigma_2(\xi)$ from below is verified in the same way as in the preceding case, and the fraction in front of it is, obviously, bounded below.   □

10.2 Discrete Representation of Functions on $\mathbb{R}^n$

Now we assume that $\psi_0(0) = 1$, so that *the series* (10.1.9) *converges to* 1 *at each point uniformly in any bounded domain.* For the function $\varphi_0(x)$, we now have

$$\int \varphi_0(x)\,dx = 1 \quad \text{and} \quad \int x^\alpha \varphi_0(x)\,dx = 0 \quad \text{for} \quad 0 < |\alpha| < N \qquad (10.2.1)$$

(if $N > 1$).

Theorem 10.2.1. *In $S'(\mathbb{R}^n)$*

$$\sum_{j=0}^\infty \varphi_j(x) = \delta(x). \qquad (10.2.2)$$

Therefore, for $u \in S'(\mathbb{R}^n)$,

$$u = \sum_0^\infty \varphi_j * u \qquad (10.2.3)$$

in the sense of convergence in this space.

   Given any $s < N$, for any element $u \in H^s(\mathbb{R}^n)$, the same relation (10.2.3) holds in the sense of convergence in this space.

Proof. First we prove the last assertion. Let $v(\xi) = (Fu)(\xi)$. Consider the difference of the left-hand side of (10.2.3) and the mth partial sum on the right-hand side. The squared norm of this difference in $H^s(\mathbb{R}^n)$ equals

$$\int (1 + |\xi|^2)^s |v(\xi)|^2 \cdot |1 - \psi_0(\xi/2^m)|^2 \, d\xi. \tag{10.2.4}$$

We decompose this integral into the sum of the integrals over a ball of large radius R centered at the origin and over the complement to this ball. The second integral is uniformly (in m) small for large R, because the last factor under the integral sign in (10.2.4) is uniformly bounded. At fixed R, the first integral tends to 0 as $m \to \infty$, because so does the already mentioned factor. This proves the second assertion of the theorem.

   To prove (10.2.2), we verify the equality

$$\sum_{j=0}^{\infty} \psi_j(\xi) = 1 \tag{10.2.5}$$

in $S'(\mathbb{R}_\xi^n)$. For this we apply both sides of the equality to a test function $\chi(\xi) \in S(\mathbb{R}_\xi^n)$. It suffices to show that

$$\int |\chi(\xi)||\psi_0(\xi/2^m) - 1| \, d\xi \to 0$$

as $m \to \infty$. This convergence is proved in the same way as in the first part of the proof, by decomposing the integral into two integrals, over a ball of large radius and over its complement.   □

10.3 Discrete Representation of Functions and Norms on a Special Lipschitz Domain

In this subsection, by a *special Lipschitz domain* we mean an unbounded domain *above* the graph of a function $x_n = \varphi(x')$ defined on $\mathbb{R}^{n-1}$ and satisfying the uniform Lipschitz condition. Let Ω be such a domain, and let K be a convex cone with vertex at the origin and vertical axis such that $x + K \subset \Omega$ for any $x \in \Omega$. In what follows, we always *assume that the function $\varphi_0(x)$ belongs to S and its support lies in the cone* $-K = \{x : -x \in K\}$. Then the same is true for all $\varphi_j(x)$.

   We also assume that conditions (10.2.1) hold for some positive integer N.

The functions $\varphi_j(x)$ with compactly supported Fourier transforms $\psi_j(\xi)$ considered in Section 1.14 are now inconvenient. In Section 10.5 we shall separately consider the compatibility of the new assumption about φ_0 with (10.2.1).

Note that, under this assumption, if $u(x)$ is the restriction to Ω of an (at most) tempered continuous function on $\mathbb{R}^n$, then the convolution

$$(\varphi_j * u)(x) = \int_\Omega \varphi_j(x-y)u(y)\,dy$$

is defined at $x \in \Omega$, because, in this case, $x - y = z \in -K$ implies $y = x - z \in \Omega$.

Recall that all spaces $H^s(\mathbb{R}^n)$ are contained in $S'(\mathbb{R}^n)$. The elements v of the latter space are the derivatives $\partial^\alpha w(x)$ in the sense of distributions of tempered continuous functions on $\mathbb{R}^n$ (see, e.g., [18] or [335]). For a function $\varphi \in S(\mathbb{R}^n)$, the convolution $\varphi * v$ can be written in the form

$$(\varphi * v)(x) = \int \partial^\alpha \varphi(x-y)w(y)\,dy$$

by transferring the derivatives to φ. Obviously, this is a tempered smooth function, and all of its derivatives are tempered as well.

Let $S(\Omega)$ denote the linear manifold in $S(\mathbb{R}^n)$ consisting of all functions supported on $\overline{\Omega}$, and let $S'(\Omega)$ be the space of restrictions to $S(\Omega)$ of distributions in $S'(\mathbb{R}^n)$. We refer to such a restriction of a distribution in $S'(\mathbb{R}^n)$ as the restriction to Ω.

If $v \in S'(\mathbb{R}^n)$ has the form $\partial^\alpha w(x)$ specified above, then, considering such a restriction, we can replace $w(x)$ by zero outside Ω.

Now let $\varphi = \varphi_j$. Then the restriction $\varphi_j * v|_\Omega$ of the convolution $\varphi_j * v$ to Ω is completely determined by the restriction $v|_\Omega$. Indeed, if Ω' is the complement to Ω, then

$$\int_{\Omega'} \partial^\alpha \varphi_j(x-y)w(y)\,dy = 0 \quad \text{in } \Omega,$$

because the support of this convolution is contained in the arithmetic sum of the sets Ω' and $-K$ and, therefore, lies in Ω'.

It follows from what was said above that, for $v \in S'(\Omega)$, the convolution $\varphi_j * v$ in Ω is uniquely determined and does not depend on the extension of v to a distribution is $S'(\mathbb{R}^n)$. This is a smooth function on Ω (up to boundary) tempered at infinity together with all its derivatives.

The space $H^s(\Omega)$ is defined as the space of restrictions to Ω (in the sense of distributions) of elements of $H^s(\mathbb{R}^n)$ with the inf-norm; this norm can be defined, in particular, by

$$\|u\|_{H^s(\Omega),\varphi_0} = \inf \|v\|_{H^s(\mathbb{R}^n),\varphi_0}, \tag{10.3.1}$$

where the greatest lower bound is over all $v \in H^s(\mathbb{R}^n)$ with $v|_\Omega = u$.

Theorem 10.3.1. 1. *For distributions u in $S'(\Omega)$,*

$$u = \sum_{j=0}^{\infty} \varphi_j * u. \tag{10.3.2}$$

2. *For $s < N$,*

$$\|u\|'_{H^s(\Omega),\varphi_0} \leq C_s \|u\|_{H^s(\Omega)}, \tag{10.3.3}$$

where

$$\|u\|'_{H^s(\Omega),\varphi_0} = \left(\sum_{j=0}^{\infty} 2^{2js} \|\varphi_j * u\|^2_{L_2(\Omega)}\right)^{1/2}. \tag{10.3.4}$$

Proof. The first assertion follows from (10.2.3).

Let us verify the second assertion. Let $u = v|_\Omega$, where $v \in H^s(\mathbb{R}^n)$. Then

$$\varphi_j * u = (\varphi_j * v)|_\Omega;$$

therefore,

$$\|\varphi_j * u\|_{L_2(\Omega)} \leq \|\varphi_j * v\|_{L_2(\mathbb{R}^n)}$$

and

$$\|u\|'_{H^s(\Omega),\varphi_0} \leq \|v\|_{H^s(\mathbb{R}^n),\varphi_0}.$$

Passing to the greatest lower bound over the extensions v and taking into account Theorem 10.1.1, we obtain inequality (10.3.3). □

In the next subsection we shall show that $\|u\|'_{H^s(\Omega),\varphi_0}$ is a norm on $H^s(\Omega)$ equivalent to the usual norm.

Given a function w on Ω, we denote by $(w)_\Omega$ its extension to $\mathbb{R}^n$ by zero outside Ω.

To construct the extension operator, we need representations of functions and distributions on Ω in a somewhat more complicated form than (10.3.2), namely, in the form

$$u = \sum_{j=0}^{\infty} \Phi_j * \varphi_j * u \quad (u \in \mathcal{S}'(\Omega)) \tag{10.3.5}$$

with $\Phi_j * \varphi_j$ instead of φ_j. Anticipating the exposition, we mention that operator extending functions from Ω to $\mathbb{R}^n$ will be constructed in the form

$$\mathcal{E}u = \sum_{j=0}^{\infty} \Phi_j * (\varphi_j * u)_\Omega. \tag{10.3.6}$$

Here, in the Fourier images ($\Psi_j = F\Phi_j$),

$$\Psi_j(\xi) = \Psi(\xi/2^j) \quad (j = 1, 2, \ldots) \tag{10.3.7}$$

for some function $\Psi \in \mathcal{S}(\mathbb{R}^n)$, but instead of the relation $\Psi(\xi) = \Psi_0(\xi) - \Psi_0(2\xi)$, we shall have (10.3.16) below.

In [320] the following construction was proposed. Set

$$g_0(\xi) = \psi_0^2(\xi), \quad g(\xi) = g_0(\xi) - g_0(2\xi),$$
$$g_j(\xi) = g(\xi/2^j) \quad (j = 1, 2, \ldots), \tag{10.3.8}$$

where $\psi_0(\xi)$ is the function defined in Section 10.1 with $\psi_0(0) = 1$. Note that

$$1 = \sum_0^\infty g_j(\xi). \tag{10.3.9}$$

Thus, somewhat formal calculations yield

$$1 = \left[\sum_{j=0}^\infty g_j(\xi)\right]\left[\sum_{k=0}^\infty g_k(\xi)\right]$$

$$= g_0(\xi)\left[g_0(\xi) + 2\sum_{k=1}^\infty g_k(\xi)\right] + \sum_{j=1}^\infty g_j(\xi)\left[g_j(\xi) + 2\sum_{k=j+1}^\infty g_k(\xi)\right]$$

$$= g_0(\xi)[2 - g_0(\xi)] + \sum_{j=1}^\infty g_j(\xi)[g_j(\xi) + 2(1 - g_0(\xi/2^j))]$$

$$= \psi_0(\xi)[\psi_0(\xi)(2 - g_0(\xi))]$$

$$+ \sum_{j=1}^\infty [\psi_0^2(\xi/2^j) - \psi_0^2(\xi/2^{j-1})][g_j(\xi) + 2(1 - g_0(\xi/2^j))]. \tag{10.3.10}$$

We set

$$\Psi_0(\xi) = \psi_0(\xi)[2 - g_0(\xi)], \tag{10.3.11}$$

$$\Psi(\xi) = [\psi_0(\xi) + \psi_0(2\xi)][2 - g_0(\xi) - g_0(2\xi)], \tag{10.3.12}$$

$$\Psi_j(\xi) = \Psi(\xi/2^j) \quad (j = 1, 2, \ldots). \tag{10.3.13}$$

The equality of the left- and right-hand sides in (10.3.10) can be rewritten as (see (10.1.3))

$$1 = \sum_{j=0}^\infty \Psi_j(\xi)\psi_j(\xi). \tag{10.3.14}$$

We shall verify this relation more carefully a little further.

In the following proposition, $\varphi_0(x)$ is a given function, $\psi_0(\xi)$ is its Fourier image, $g_0(\xi) = \psi_0^2(\xi)$, the functions $\Psi_0(\xi)$, $\Psi(\xi)$, and $\Psi_j(\xi)$ are defined by (10.3.11)–(10.3.13), and $\Phi_0(x)$, $\Phi(x)$, and $\Phi_j(x)$ are their Fourier preimages.

Proposition 10.3.2. *Let $\varphi_0(x)$ be a function in S supported on the cone $-K$ and satisfying conditions (10.2.1) with some positive N. Then the following assertions are valid.*

1. The functions $\Phi_0(x)$ and $\Phi(x)$ belong to S, their supports lie in $-K$, and the functions $\Psi_0(\xi)$ and $\Psi(\xi)$ satisfy the conditions

$$|\Psi_0(\xi) - 1| \le C_N'|\xi|^N, \quad |\Psi(\xi)| \le C_N'|\xi|^N \tag{10.3.15}$$

with constants not depending on ξ.
2. Relation (10.3.5) holds.

In the second inequality, C'_N can be replaced by $C'_{N,M}(1+|\xi|)^{-M}$.

Proof. 1. The assertion that $\Phi_0(x)$ and $\Phi(x)$ belong to S is obvious, because the product of functions in S belongs to S. The assertion about their supports follows from that, as mentioned above, the support of the convolution of two functions is contained in the arithmetic sum of the supports of these functions, which lies in $-K$ in the case under consideration, and this argument can be applied repeatedly. Estimates (10.3.15) are obtained as follows. We have

$$1 - g_0(\xi) = [1 - \psi_0(\xi)][1 + \psi_0(\xi)] = O(|\xi|^N)$$

as $\xi \to 0$; therefore,

$$\Psi_0(\xi) - 1 = [\psi_0(\xi) - 1] + \psi_0(\xi)[1 - g_0(\xi)] = O(|\xi|^N).$$

Of course, $1 - g_0(2\xi) = O(|\xi|^N)$ as well, whence

$$\Psi(\xi) = [\psi_0(\xi) + \psi_0(2\xi)][(1 - g_0(\xi)) + (1 - g_0(2\xi))] = O(|\xi|^N).$$

2. Let us verify relation (10.3.14) more carefully. An elementary calculation yields (see (10.1.2))

$$\Psi(\xi)\psi(\xi) = \Psi_0(\xi)\psi_0(\xi) - \Psi_0(2\xi)\psi_0(2\xi). \qquad (10.3.16)$$

Therefore,

$$\sum_0^m \Psi_j(\xi)\psi_j(\xi) = \Psi_0(\xi/2^m)\psi_0(\xi/2^m). \qquad (10.3.17)$$

Since $\Psi_0(0) = 1$, it follows that *relation (10.3.14) is valid at each point ξ, and the convergence is uniform in any bounded domain.*

Using (10.3.14), we obtain the relation

$$\sum_{j=0}^\infty (\Phi_j * \varphi_j)(x) = \delta(x) \qquad (10.3.18)$$

in $S'(\mathbb{R}^n_x)$. This is a relation of the form (10.2.2) but with the functions $\Phi_j * \varphi_j$ instead of φ_j. In particular, this relation holds for the test functions in $S(\Omega)$. □

10.4 The Extension Operator

Let φ_0 be a function in S supported on the cone $-K$ and satisfying conditions (10.2.1). Using the system of functions $\Phi_j(x)$ constructed in Section 10.3, we set

$$v = \sum_0^\infty \Phi_j * (\varphi_j * u)_\Omega. \qquad (10.4.1)$$

We remind the reader that $(\varphi_j * u)_\Omega$ is the extension of the function in parentheses by zero outside Ω.

The following theorem is the main in this subsection.

Theorem 10.4.1. 1. *For $|s| < N$, this is a bounded operator which extends functions on $H^s(\Omega)$ to functions on $H^s(\mathbb{R}^n)$.*

2. *The norm on $H^s(\Omega)$ is equivalent to norm (10.3.4).*

The first assertion means that, for $u \in H^s(\Omega)$, series (10.4.1) converges in $H^s(\mathbb{R}^n)$, the restriction $v|_\Omega$ coincides with u in $H^s(\Omega)$, and

$$\|v\|_{H^s(\mathbb{R}^n)} \leq C\|u\|_{H^s(\Omega)}, \tag{10.4.2}$$

where the constant does not depend on u. The function $\varphi_j * u$ in brackets is smooth, because $\varphi_j \in S$. The second assertion has already been half verified in the preceding subsection.

The proof of this theorem uses two auxiliary assertions.

Proposition 10.4.2. *For any s,*

$$\sum_0^\infty (1 + |\xi|^2)^s 2^{-2js} |\Psi_j(\xi)|^2 \leq C, \tag{10.4.3}$$

where the constant does not depend on ξ.

This is an analogue of the already proved second inequality in (10.1.13) with s replaced by $-s$ and ψ_j by Ψ_j. The same proof applies.

Proposition 10.4.3. *Let $\{f_j(x)\}_0^\infty$ be a sequence of functions in $L_2(\mathbb{R}^n)$ such that*

$$\sum_0^\infty 2^{2js} \|f_j(x)\|_{L_2(\mathbb{R}^n)}^2 < \infty \tag{10.4.4}$$

for some s. Then series

$$v = \sum_{j=0}^\infty \Phi_j * f_j \tag{10.4.5}$$

converges in $H^s(\mathbb{R}^n)$, and

$$\|v\|_{H^s(\mathbb{R}^n)} \leq C' \Big(\sum_0^\infty 2^{2js} \|f_j\|_{L_2(\mathbb{R}^n)}^2 \Big)^{1/2}, \tag{10.4.6}$$

where the constant C' depends only on s.

Proof of Proposition 10.4.3. Let $h_j(\xi) = (Ff_j)(\xi)$. Using the preceding proposition and the Schwarz inequality, we obtain

$$(1+|\xi|^2)^s \left| \sum_0^\infty \Psi_j(\xi) h_j(\xi) \right|^2 \le C \sum_0^\infty 2^{2js} |h_j(\xi)|^2.$$

It remains to integrate this inequality. Summing from some m rather than from zero and integrating, we obtain the remainder of a convergent number series on the right-hand side.   □

Proof of Theorem 10.4.1. We already know that if $u \in H^s(\Omega)$, then

$$u = \sum \Phi_j * (\varphi_j * u)_\Omega$$

on Ω (see (10.3.18)) and that $\|u\|'_{H^s(\Omega)}$ does not exceed $C_s \|u\|_{H^s(\Omega)}$ (see Theorem 10.3.1).

Of course, the functions $(\varphi_j * u)_\Omega$ belong to $L_2(\mathbb{R}^n)$. By virtue of Proposition 10.4.3, series (10.4.1) converges in $H^s(\mathbb{R}^n)$, and the norm of its sum is bounded by $C' \|u\|'_{H^s(\Omega)}$.

It follows that the norm $\|u\|_{H^s(\Omega)}$ is bounded by $C' \|u\|'_{H^s(\Omega)}$, so that the norms under consideration are equivalent.   □

Remark 10.4.4. If u belongs to $H^s(\Omega)$, then the series in (10.3.2) converges to u in this space.

Indeed, if $u \in H^s(\Omega)$, then, as we saw above, the functions

$$\sum_0^\infty \varphi_j * u \quad \text{and} \quad \sum_0^m \varphi_j * u \qquad\qquad (10.4.7)$$

have the extensions

$$\sum_0^\infty \Phi_j * (\varphi_j * u)_\Omega \quad \text{and} \quad \sum_0^m \Phi_j * (\varphi_j * u)_\Omega$$

to $\mathbb{R}^n$, and the norm of their difference in $H^s(\mathbb{R}^n)$ tends to 0 as $m \to \infty$. Therefore, the usual norm of the difference of functions (10.4.7) in $H^s(\Omega)$ tends to zero as well.

Thus, we have almost completed the construction of the extension operator for functions on a special Lipschitz domain. It only remains to construct the function φ_0 supported on $-K$ with $N = \infty$.

After that, the extension operator for functions on a bounded Lipschitz domain is constructed in an obvious way by using a partition of unity.

10.5 Construction of φ_0 with $N=\infty$

As in [320], we first consider the single-valued analytic function

$$U(z) = \exp[-(z-1)^{1/8} - (z-1)^{-1/8}] \qquad (10.5.1)$$

of one complex variable on the complex plane cut along the ray $[1,\infty)$. This function rapidly tends to 0, together with all its derivatives, on the circles $|z-1| = R$ as $R \to 0$ and as $R \to \infty$. At $z = 0$, it is regular and different from zero, while, of course, all $z^k U(z)$ with positive integer k vanish. Let us apply the Cauchy integral theorem to the functions $(2\pi i)^{-1} z^{k-1} U(z)$ ($k = 0, 1, \ldots$) and two contours, (i) a small circle centered at the origin traversed in the positive direction and (ii) the contour γ going from infinity along the lower edge of the cut, then along a small circle around 1 in the clockwise direction, and finally to infinity along the upper edge of the cut (this contour is "closed by a circle of infinitely large radius"). We obtain

$$\int_\gamma \frac{U(z)}{z}\, dz \neq 0 \quad \text{and} \quad \int_\gamma \frac{z^k U(z)}{z}\, dz = 0 \quad (k = 1, 2, \ldots). \qquad (10.5.2)$$

Now we set

$$V(t) = \begin{cases} \dfrac{1}{t}[U(t+i0) - U(t-i0)], & t \geq 1, \\[2mm] 0, & t < 1. \end{cases} \qquad (10.5.3)$$

The support of this function is the ray $[1,\infty)$. At 1, all its right derivatives vanish, and this function belongs to $S(\mathbb{R})$. It follows from (10.5.2) that

$$\int V(t)\, dt \neq 0 \quad \text{and} \quad \int t^k V(t)\, dt = 0 \quad (k = 1, 2, \ldots). \qquad (10.5.4)$$

Separating the real and imaginary parts, we obtain a real function with the same properties.

Now we set

$$\varphi_0(x_1, \ldots, x_n) = CV(x_1) \ldots V(x_n), \quad C = \left(\int V(t)\, dt \right)^{-n}. \qquad (10.5.5)$$

The support of this function is contained in the cone consisting of all points with nonnegative coordinates, and

$$\int \varphi_0(x)\, dx = 1 \quad \text{and} \quad \int x^\alpha \varphi_0(x)\, dx = 0 \quad (|\alpha| > 0). \qquad (10.5.6)$$

Applying a homogeneous linear transformation and a rotation followed by normalization, we can transform this cone into the required cone $-K$ being the convex hull of n rays from the origin with preservation of properties (10.5.6). Thanks to these relations, the function φ_0 has the required properties (10.2.1) with $N = \infty$.

Then we determine functions φ, Φ_0, Φ, and Φ_j in the known way. We obtain the extension operator (10.4.1), which applies to the spaces H^s for all $s \in \mathbb{R}$.

11 Boundary Value Problems in Lipschitz Domains for Second-Order Strongly Elliptic Systems

As we saw in Section 9, boundaries of Lipschitz domains may be far from smooth, and such domains are often encountered in applications. Our purpose is to find out what general facts take place for the most important problems in all such special cases.

11.1 Basic Definitions and Results

In this section, unlike in Section 8, we consider a second-order system. Its leading part can also be written in the divergence form

$$Lu := -\sum_{j,k=1}^{n} \partial_j a_{j,k}(x)\partial_k u(x) + \sum_{j=1}^{n} b_j(x)\partial_j u(x) + c(x)u(x) = f(x). \qquad (11.1.1)$$

Here and in what follows u and f are (column) vector-valued functions of dimension d, so that the coefficients are $d \times d$ matrices. They consist of complex-valued functions, as well as u and f.

Writing the leading part in divergence form, we can minimize to some extent the smoothness assumptions on the coefficients. The simplest assumption is that the coefficients $a_{j,k}(x)$ and $b_j(x)$ (to be more precise, the elements of these matrices) are Lipschitz continuous on Ω, i.e., belong to $C^{0,1}(\overline{\Omega})$, and the coefficient $c(x)$ is bounded and measurable. *We shall assume this.* But in many cases, substantially weaker assumptions are sufficient, and we shall usually mention this. In this section, all coefficients *in problem statements* can be assumed to be bounded and measurable. But in Theorem 11.1.1 the continuity of the higher-order coefficients on $\overline{\Omega}$ is needed. In what follows, when the formal adjoint of L is used (see (11.1.22)), we have to differentiate the coefficients $b_j(x)$, and it is convenient to assume them to be Lipschitz.

Remark. Of interest are also systems with unbounded lower-order coefficients in suitable L_p classes; see, e.g., [226] and [269]. But we shall not consider such systems.

If desired, one can assume, as in Section 8, that

$$a_{j,k}(x) = a_{k,j}(x), \qquad (11.1.2)$$

but this is not necessary.

Let us write the corresponding form $\Phi_\Omega(u,v)$:

$$\int_{\Omega} \Big[\sum a_{j,k}(x) \partial_k u(x) \cdot \partial_j \overline{v(x)} + \sum b_j(x) \partial_j u(x) \cdot \overline{v(x)}$$

$$+ c(x) u(x) \cdot \overline{v(x)} \Big] dx; \quad (11.1.3)$$

it is defined and bounded on $H^1(\Omega)$, even if all coefficients belong only to $L_\infty(\Omega)$.

The first Green identity has the form

$$(Lu, v)_\Omega = \Phi_\Omega(u, v) - (T^+ u, v^+)_\Gamma \quad (11.1.4)$$

and is first derived for $u \in H^2(\Omega)$ and $v \in H^1(\Omega)$ by integration by parts under the assumption that the leading coefficients belong to $C^{0,1}(\overline{\Omega})$. Then this identity is extended by passing to the limit over $u \in H^s(\Omega)$, $s > 3/2$. Here the *smooth conormal derivative* $T^+ u$ is defined by

$$T^+ u(x) = \sum \nu_j(x) a_{j,k}(x) \gamma^+ \partial_k u(x). \quad (11.1.5)$$

But, as mentioned in Section 8, for general functions $u, v \in H^1(\Omega)$, identity (11.1.4) is *postulated* and considered as the definition of (i) the conormal derivative $T^+ u \in H^{-1/2}(\Gamma)$ for given u and $f = Lu$ (see Lemma 4.3 in [258]) and (ii) a solution of the Neumann problem for given function f and conormal derivative. In these definitions, it suffices to assume all coefficients to be bounded and measurable.

11.1.1

We explain and specify this as follows.

First, consider the Dirichlet problem with homogeneous boundary condition:

$$Lu = f_0 \text{ in } \Omega, \quad u^+ = 0. \quad (11.1.6)$$

In this case, the Green identity is written in the form

$$\Phi_\Omega(u, v) = (f_0, v)_\Omega \quad (u, v \in \widetilde{H}^1(\Omega)); \quad (11.1.7)$$

it does not contain the conormal derivative. If we take $u \in \widetilde{H}^1(\Omega)$, then the left-hand side of this identity is a continuous antilinear functional, and the right-hand side is its expression in terms of the uniquely determined element $f_0 \in H^{-1}(\Omega)$, which uses the duality between the spaces $\widetilde{H}^1(\Omega)$ and $H^{-1}(\Omega)$.

Note that f_0 treated as a functional on $\widetilde{H}^1(\Omega)$ is uniquely determined by u. (If the coefficients are infinitely differentiable, then f_0 is merely a distribution on $C_0^\infty(\Omega)$, and this functional can be extended to $\widetilde{H}^1(\Omega)$ by passing to the limit.)

Now consider the Neumann problem with homogeneous boundary condition:

$$Lu = f_1 \text{ in } \Omega, \quad T^+ u = 0. \quad (11.1.8)$$

The Green identity is written in the form

$$\Phi_\Omega(u,v) = (f_1,v)_\Omega \quad (u,v \in H^1(\Omega)), \tag{11.1.9}$$

and it does not contain the vanishing conormal derivative. If we take $u \in H^1(\Omega)$, then the left-hand side is a continuous antilinear functional on $H^1(\Omega)$, which is written in the form of the right-hand side with uniquely determined $f_1 \in \widetilde{H}^{-1}(\Omega)$ by using the duality between $H^1(\Omega)$ and $\widetilde{H}^{-1}(\Omega)$. The restriction f_0 of the functional f_1 to $\widetilde{H}^1(\Omega)$ is uniquely constructed inside the domain, as in the preceding case.

Can we replace f_1 by a different functional $f \in \widetilde{H}^{-1}(\Omega)$ without changing the left-hand side? Yes, we can, provided that its restriction to $\widetilde{H}^1(\Omega)$ coincides with f_0. But $(f_1 - f, v)_\Omega$ is then a functional on $H^1(\Omega)$ supported on the boundary, and we know that it can be represented in the form $(h,v)_\Gamma$ with $h \in H^{-1/2}(\Gamma)$ (see Section 9.2). This element h precisely is the conormal derivative T^+u, and the Green identity takes the form

$$\Phi_\Omega(u,v) = (f,v)_\Omega + (T^+u, v^+)_\Gamma \quad (u,v \in H^1(\Omega)). \tag{11.1.10}$$

This is the Green identity for the Neumann problem with inhomogeneous boundary condition. Specifying f, we uniquely determine $h = T^+u$.

There is an inconvenience, which consists in the mutual dependence of f and h. We shall return to it in what follows; but first, we shall consider the simpler Dirichlet and Neumann problems with homogeneous boundary conditions.

The notations f_0 and f_1 in these explanations were temporary; in what follows, we always denote the right-hand side of a system by f.

11.1.2

By $a(x,\xi)$ we denote the principal symbol $\sum a_{j,k}(x)\xi_j\xi_k$. The strong ellipticity condition [388] has the form

$$\operatorname{Re} a(x,\xi) := \frac{1}{2}[a(x,\xi) + a^*(x,\xi)] \geq C_0|\xi|^2 I \tag{11.1.11}$$

with $C_0 > 0$, where the inequality is in the sense of the corresponding quadratic forms. Condition (11.1.11) implies the *coercivity* of the form $\Phi_\Omega(u,u)$, or the *Gårding inequality*, on $\widetilde{H}^1(\Omega)$:

$$\|u\|^2_{\widetilde{H}^1(\Omega)} \leq C_1 \operatorname{Re}\Phi_\Omega(u,u) + C_2\|u\|^2_{H^0(\Omega)}. \tag{11.1.12}$$

More precisely, the following theorem is true.

Theorem 11.1.1. *Suppose that the leading coefficients $a_{j,k}(x)$ are continuous on $\overline{\Omega}$, the remaining coefficients are bounded and measurable, and the strong ellipticity condition (11.1.11) holds. Then there exist constants $C_1 > 0$ and $C_2 \geq 0$ such that, for functions in $\widetilde{H}^1(\Omega)$, inequality (11.1.12) is valid.*

Proof. In the scalar case, the proof is very simple; it was given in Section 8. In this case, it suffices to assume that $a_{j,k} \in L_\infty(\Omega)$. In the general case, the proof uses the method of freezing coefficients, as in Sections 6 and 7. First, suppose that the form $\Phi_\Omega(u, u)$ contains only the leading terms. Of course, we must estimate the L_2 norms of the first derivatives. If the coefficients are constant (frozen at a point x_0), then the inequality

$$\sum \int |\partial_j u|^2 \, dx \leq C_0^{-1} \operatorname{Re} \Phi_\Omega(u, u)$$

in $\mathbb{R}^n$ is obtained from (11.1.11) by using the Fourier transform and Parseval's identity:

$$\int \sum |\xi_j|^2 |(Fu)(\xi)|^2 \, d\xi \leq C_0^{-1} \operatorname{Re} \int \sum a_{j,k}(x_0)\xi_k\xi_j (Fu)(\xi) \cdot \overline{(Fu)(\xi)} d\xi.$$

It is essential here that, in fact, we must consider functions defined on $\mathbb{R}^n$. (We might state the theorem for functions in $\mathring{H}^1(\Omega)$; in this case, we would extend these functions by zero outside Ω.) If the coefficients are "almost constant" (i.e., close to their values at the point x_0), then we write

$$a_{j,k}(x) = a_{j,k}(x_0) + [a_{j,k}(x) - a_{j,k}(x_0)],$$

and if all differences in brackets are small enough, then we obtain

$$\|u\|_{\widetilde{H}^1(\Omega)}^2 \leq 2C_0^{-1} \operatorname{Re} \Phi_\Omega(u, u) + C_3 \|u\|_{H^0(\Omega)}^2.$$

In the general case, we use a partition of unity of the form $\sum \varphi_r^2 = 1$ consisting of C^∞ functions φ_r with sufficiently small supports and obtain

$$\|u\|_{\widetilde{H}^1(\Omega)}^2 = \sum \int_\Omega \varphi_r^2 (|\nabla u|^2 + |u|^2) \, dx$$

$$\leq C_4 \sum (\operatorname{Re} \Phi_\Omega(\varphi_r u, \varphi_r u)_\Omega + \|\varphi_r u\|_{L_2(\Omega)}^2) + \ldots \leq C_5 \operatorname{Re} \Phi_\Omega(u, u) + \ldots;$$

the terms denoted by dots are estimated by

$$C_7[\|u\|_{\widetilde{H}^1(\Omega)} \|u\|_{H^0(\Omega)} + \|u\|_{H^0(\Omega)}^2] \leq \varepsilon \|u\|_{\widetilde{H}^1(\Omega)}^2 + C_\varepsilon \|u\|_{H^0(\Omega)}^2$$

with arbitrarily small ε. This brings us to the goal.

If some of the b_j are nonzero, then we use estimate (5.1.9) for the intermediate norms. It remains valid in Lipschitz domains. $\qquad\square$

If the leading and the intermediate coefficients in L are fixed, then, for sufficiently large $\operatorname{Re} c$, the last term in (11.1.12) is not needed, and this inequality takes the form

$$\|u\|_{\widetilde{H}^1(\Omega)}^2 \leq C_1 \operatorname{Re} \Phi_\Omega(u, u). \tag{11.1.13}$$

This condition on c is not always necessary; see Remark 8.1.3.

We assume this *strong coercivity condition*, or *strong Gårding inequality*, to be satisfied. By the Lax–Milgram theorem (see Section 18.2), it implies the unique solvability of the Dirichlet problem with homogeneous boundary condition.

Inequality (11.1.12) implies that this problem is Fredholm with index zero by the same reasons as in Remark 8.1.7.

In the matrix case, the *coercivity* of the form $\Phi_\Omega(u,u)$ on $H^1(\Omega)$, i.e., the inequality

$$\|u\|_{H^1(\Omega)}^2 \leq C_1 \operatorname{Re} \Phi_\Omega(u,u) + C_2 \|u\|_{H^0(\Omega)}^2, \tag{11.1.14}$$

far from always holds for strongly elliptic systems (see Section 11.7). Sufficient conditions are known. Let us write the leading part of the integrand in $\Phi_\Omega(u,v)$ in the detailed form

$$\sum a_{j,k}^{r,s}(x)\partial_k u_s \partial_j \overline{v}_r. \tag{11.1.15}$$

The condition

$$\sum |\zeta_j^r|^2 \leq C_3 \operatorname{Re} \sum a_{j,k}^{r,s}(x)\zeta_k^s \overline{\zeta_j^r} \tag{11.1.16}$$

with arbitrary complex ζ_j^r, which is sufficient in the matrix case provided that $a_{j,k} = a_{k,j}$ (see the proof of Theorem 8.1.1 in the scalar case), is too burdensome (see, e.g., [258, p. 307]). In the author's view, in the case $d = n$, particularly convenient is the condition

$$\sum |\zeta_j^r + \zeta_r^j|^2 \leq C_4 \operatorname{Re} \sum a_{j,k}^{r,s}(x)\zeta_k^s \overline{\zeta_j^r} \tag{11.1.17}$$

for any complex ζ_j^r. The sufficiency of this condition follows from the remarkable (*second*) *Korn inequality*

$$\|u\|_{H^1(\Omega)}^2 \leq C_5 \left[\|u\|_{L_2(\Omega)}^2 + \sum_{j,r} \|\partial_j u_r + \partial_r u_j\|_{L_2(\Omega)}^2\right]. \tag{11.1.18}$$

Its proof and applications can be found, e.g., in [291] or [258]. Inequality (11.1.14) is obtained by substituting $\zeta_j^r = \partial_j u_r$ into (11.1.17) and integrating.

To generalize condition (11.1.17) to the case $d \neq n$, it suffices to set the "missing" ζ_j^r to zero.

Condition (11.1.17) can be applied to systems of elasticity theory. The most important of them is the *Lamé system* for isotropic homogeneous media:

$$-\mu \Delta u - (\lambda + \mu)\operatorname{grad}\operatorname{div} u = f. \tag{11.1.19}$$

Here u is the displacement vector and $d = n$. This is an important example, and we shall return to it in Section 11.7. This system and its closest generalizations in smooth three-dimensional domains were studied in detail, e.g., in the monograph [224], where many further references can be found. More general are systems of elasticity theory for anisotropic homogeneous and inhomogeneous media. These systems have real-valued coefficients $a_{j,k}(x)$ (not depending on x in the former case) satisfying the symmetry conditions

$$a_{j,k}^{r,s} = a_{k,j}^{r,s}, \quad a_{j,k}^{r,s} = a_{r,s}^{j,k}, \quad \text{and} \quad a_{j,k}^{r,s} = a_{r,k}^{j,s}. \tag{11.1.20}$$

The most important cases are $d = n = 2$ and 3. The three-dimensional system of anisotropic elasticity for homogeneous media was thoroughly studied in dissertation [281]; see also [282]. However, these systems have also been considered for larger $d = n$; see, e.g., [291].

Other conditions sufficient for coercivity (including higher-order equations and systems) can be found, e.g., in the books [8] and [286].

It should be mentioned that the presence or absence of coercivity on $H^1(\Omega)$ depends on the expression of the system in divergence form. The reader will find examples in Section 11.7. This expression may affect that of the form $\Phi_\Omega(u, u)$. In the case of the Dirichlet problem, it does not matter, as is seen from the proof of Theorem 11.1.1: in this case, only the principal symbol of the operator L is important, because when the coefficients are frozen, the Fourier transform and Parseval's identity are used.

In the case of coercivity on $H^1(\Omega)$, i.e., when inequality (11.1.14) holds, the last term in it again becomes redundant for sufficiently large $\operatorname{Re} c$:

$$\|u\|^2_{H^1(\Omega)} \le C_1 \operatorname{Re}\Phi_\Omega(u,u). \tag{11.1.21}$$

We say in this case that *the form Φ_Ω is strongly coercive on $H^1(\Omega)$*. If this inequality holds, then the Neumann problem with homogeneous boundary condition turns out to be uniquely solvable, again by virtue of the Lax–Milgram theorem (and if only inequality (11.1.14) holds, then this problem is Fredholm with index zero).

11.1.3

The operator formally adjoint to L has the form

$$\widetilde{L}v = -\sum_{j,k=1}^{n} \partial_j a^*_{k,j}(x)\partial_k v(x) - \sum_{j=1}^{n} \partial_j[b^*_j(x)v(x)] + c^*(x)v(x). \tag{11.1.22}$$

Here the structure of the intermediate terms differs from that in (11.1.1). But if $b_j \in C^{0,1}(\overline{\Omega})$, then this expression can be reduced to the following form similar to (11.1.1):

$$\widetilde{L}v = -\sum_{j,k=1}^{n} \partial_j a^*_{k,j}(x)\partial_k v(x) - \sum b^*_j(x)\partial_j v(x) \\ + \Big[c^*(x) - \sum \partial_j b^*_j(x)\Big]v(x). \tag{11.1.23}$$

In this case, we define the corresponding (smooth) conormal derivative for functions in $H^s(\Omega)$, $s > 3/2$, by

$$\widetilde{T}^+v(x) = \sum \nu_j(x)a^*_{k,j}(x)\gamma^+\partial_k v(x) + \sum \nu_j(x)b^*_j(x)v^+(x). \tag{11.1.24}$$

Under this definition, we have the first Green identity for $\widetilde{L}$ with the same form Φ_Ω, that is,

$$(u, \widetilde{L}v)_\Omega = \Phi_\Omega(u, v) - (u^+, \widetilde{T}^+v)_\Gamma, \tag{11.1.25}$$

and the second Green identity is

$$(Lu, v)_\Omega - (u, \widetilde{L}v)_\Omega = (u^+, \widetilde{T}^+v)_\Gamma - (T^+u, v^+)_\Gamma. \tag{11.1.26}$$

This identity holds if $u, v \in H^s(\Omega)$, $s > 3/2$. In the general case, where $u, v \in H^1(\Omega)$ and the coefficients belong to $L_\infty(\Omega)$, it follows from the postulated first Green identities for L and $\widetilde{L}$; in the first Green identity for $\widetilde{L}$, we take the same form Φ_Ω as for L.

Remark. Thanks to the coincidence of the forms for L and $\widetilde{L}$, the coercivity conditions on $\widetilde{H}^1(\Omega)$ and on $H^1(\Omega)$ hold for the operators L and $\widetilde{L}$ simultaneously.

For the formal self-adjointness of the operator L, the conditions

$$a_{j,k} = a_{k,j}^*, \quad b_j = 0, \quad c = c^*$$

are sufficient. More general conditions for the formal self-adjointness are

$$a_{j,k} = a_{k,j}^*, \quad b_j = -b_j^*, \quad c = c^* - \sum \partial_j b_j^*. \tag{11.1.27}$$

We dwell on this point, since we must consider problems for functions on the domain Ω rather than for compactly supported smooth functions on $\mathbb{R}^n$. Under these conditions on functions in $\widetilde{H}^1(\Omega)$, we have

$$(Lu, v)_\Omega = (u, Lv)_\Omega \tag{11.1.28}$$

and

$$\Phi_\Omega(u, v) = \overline{\Phi_\Omega(v, u)}. \tag{11.1.29}$$

If we want (11.1.29) to hold for all functions in $H^1(\Omega)$, then we must add the relation

$$\sum \nu_j(x) b_j(x) = 0 \text{ on } \Gamma, \tag{11.1.30}$$

which can be verified by integration by parts. In this case, the smooth conormal derivatives for L and $\widetilde{L}$ are the same, and we say that *the operator L is formally self-adjoint on $\overline{\Omega}$*. In considering the Dirichlet problem with homogeneous boundary condition, relation (11.1.30) is inessential. Of course, if all b_j vanish on Γ, then there is no problem at all.

As in Section 8, we denote the operators corresponding to the Dirichlet and Neumann problems with homogeneous boundary conditions by $\mathcal{L}_D$ and $\mathcal{L}_N$, respectively. Here we treat them as bounded operators from $\widetilde{H}^1(\Omega)$ to $H^{-1}(\Omega)$ and from $H^1(\Omega)$ to $\widetilde{H}^{-1}(\Omega)$, respectively. Their adjoints with respect to the corresponding extensions of the inner product in $L_2(\Omega)$ are the operators $\widetilde{\mathcal{L}}_D$ and $\widetilde{\mathcal{L}}_N$. This is adjointness in the sense of the theory of bounded operators on Banach spaces (see Section 18.1).

Let us formulate the main results.

Theorem 11.1.2. *If the form Φ_Ω is strongly coercive on $\widetilde{H}^1(\Omega)$, then the operator $\mathcal{L}_D$ is invertible. If Φ_Ω is only coercive (strongly elliptic), then $\mathcal{L}_D$ is Fredholm with index zero. Moreover, the equation $\mathcal{L}_D u = f$ is solvable if and only if f is orthogonal to all elements v of the kernel of the operator $\widetilde{\mathcal{L}}_D$ with respect to the extension of the form $(f,v)_\Omega$.*

If the form Φ_Ω is strongly coercive on $H^1(\Omega)$, then the operator $\mathcal{L}_N$ is invertible. If Φ_Ω is only coercive, then $\mathcal{L}_N$ is Fredholm with index zero. Moreover, the equation $\mathcal{L}_N u = f$ is solvable if and only if f is orthogonal to all solutions of the equation $\widetilde{\mathcal{L}}_N v = 0$ with respect to the extension of the form $(f,v)_\Omega$.

Similar assertions concerning the solvability of the equations with operators $\widetilde{\mathcal{L}}_D$ and $\widetilde{\mathcal{L}}_N$ are also valid.

Remark. The Dirichlet problem with homogeneous boundary condition for the equation $Lu = f$ is uniquely solvable if and only if any continuous antilinear functional on $\widetilde{H}^1(\Omega)$ can be uniquely represented as $\Phi_\Omega(u,v)$ with some $u \in \widetilde{H}^1(\Omega)$. Similarly, the Neumann problem with homogeneous boundary condition is uniquely solvable if and only if any continuous antilinear functional on $H^1(\Omega)$ can be uniquely represented as $\Phi_\Omega(u,v)$ with some $u \in H^1(\Omega)$.

11.1.4

Now consider the Dirichlet problem with inhomogeneous boundary condition:

$$Lu = f \text{ in } \Omega, \quad u^+ = g. \tag{11.1.31}$$

This requires additional conventions. In (11.1.31) $u \in H^1(\Omega)$ and $g \in H^{1/2}(\Gamma)$. As to f, *we assume that $f \in H^{-1}(\Omega)$.*

Let u_0 be *any* function in $H^1(\Omega)$ with boundary value g, i.e., such that $u_0^+ = g$. We define $f_0 = Lu_0 \in H^{-1}(\Omega)$ by the formula

$$(f_0, v)_\Omega = \Phi_\Omega(u_0, v), \quad v \in \widetilde{H}^1(\Omega). \tag{11.1.32}$$

We say that a *function u is a solution of problem* (11.1.31) if $u = u_0 + u_1$, where u_1 is a solution of the problem

$$Lu_1 = f - f_0 \text{ in } \Omega, \quad u_1^+ = 0, \tag{11.1.33}$$

i.e.,

$$(f - f_0, v)_\Omega = \Phi_\Omega(u_1, v), \quad v \in \widetilde{H}^1(\Omega). \tag{11.1.34}$$

Here $u_1 \in \widetilde{H}^1(\Omega)$.

We assume that the form Φ_Ω is strongly coercive on $\widetilde{H}^1(\Omega)$, so that $u_1 \in \widetilde{H}^1(\Omega)$ is determined uniquely.

Let us show that u *does not depend on the choice of* u_0. Suppose that $\widehat{u}_0$ is another function in $H^1(\Omega)$ with boundary value g. Let $\widehat{f}_0$ be determined from $\widehat{u}_0$

and $\widehat{u}_1$ by using relations of types (11.1.32) and (11.1.34) with $\widetilde{f}_0$ instead of f_0. Then

$$(\widetilde{f}_0 - f_0, v)_\Omega = \Phi_\Omega(u_1 - \widehat{u}_1, v), \quad v \in \widetilde{H}^1(\Omega),$$

i.e.,

$$\Phi_\Omega(\widehat{u}_0 - u_0, v) = \Phi_\Omega(u_1 - \widehat{u}_1, v), \quad v \in \widetilde{H}^1(\Omega).$$

It follows that

$$\widehat{u}_0 - u_0 = u_1 - \widehat{u}_1$$

in $\widetilde{H}^1(\Omega)$, so that $u_0 + u_1 = \widehat{u}_0 + \widehat{u}_1$.

We have shown that problem (11.1.31) is uniquely solvable.

The inhomogeneous Neumann problem

$$Lu = f \text{ in } \Omega, \quad Tu^+ = h \tag{11.1.35}$$

is simpler in that, as mentioned in Section 8, we can include the boundary term of the Green identity in the form $(f, v)_\Omega$ by changing f.

Thus, the following theorem is valid.

Theorem 11.1.3. *Under the assumptions of Theorem 11.1.2, the assertions of this theorem are valid for complete Dirichlet problems (with $f \in H^{-1}(\Omega)$ and $g \in H^{1/2}(\Gamma)$) and complete Neumann problems (with $f \in \widetilde{H}^{-1}(\Omega)$ and $h \in H^{-1/2}(\Gamma)$).*

Remark. For the complete Dirichlet problem, we can also give a variational setting, namely, write the Green identity

$$\Phi_\Omega(u, v) = (f, v)_\Omega, \quad v \in \widetilde{H}^1(\Omega).$$

But here we must point out that $u \in H^1(\Omega)$ and $f \in H^{-1}(\Omega)$. Moreover, this setting does not take into account the condition $u^+ = g$, which must be specified separately (as in [258]).

How will this identity change if we extend it to $v \in H^1(\Omega)$? In this case, we must extend f to a functional in $\widetilde{H}^{-1}(\Omega)$, and u will become a solution of the new equation $Lu = f$, while, generally, a term with a conormal derivative will arise on the right-hand side of the identity.

In the classical setting, the function u always satisfies the equation inside the domain, but in the weak setting with right-hand side in $\widetilde{H}^{-1}(\Omega)$ and test functions in $H^1(\Omega)$, this is no longer the case. As mentioned in [261], including a term supported on the boundary on the right-hand side of the equation, we obtain an equation in $\overline{\Omega}$.

For solutions of the Dirichlet problem (11.1.31), there is a *two-sided a priori estimate*. If the form Φ_Ω is strongly coercive on $\widetilde{H}^1(\Omega)$, this estimate has the form

$$\|u\|_{H^1(\Omega)} \le C_3[\|f\|_{H^{-1}(\Omega)} + \|g\|_{H^{1/2}(\Gamma)}] \le C_4\|u\|_{H^1(\Omega)}. \tag{11.1.36}$$

In this subsection we verify the first inequality and prove the second one only for $f = 0$ or $g = 0$. The proof of the second inequality will be completed in Section 11.2.

If $g = 0$, then, using the Green identity and the strong coercivity of Φ_Ω on $\widetilde{H}^1(\Omega)$ and applying the generalized Schwarz inequality, we obtain

$$\|u\|^2_{\widetilde{H}^1(\Omega)} \le C_1 \operatorname{Re}\Phi_\Omega(u, u) = C_1 \operatorname{Re}(f, u)_\Omega \le C_2 \|f\|_{H^{-1}(\Omega)} \|u\|_{\widetilde{H}^1(\Omega)}.$$

Therefore,

$$\|u\|_{\widetilde{H}^1(\Omega)} \le C_2 \|f\|_{H^{-1}(\Omega)}. \tag{11.1.37}$$

On the other hand, since the form $\Phi_\Omega(u, v)$ is bounded on the direct product of two copies of $\widetilde{H}^1(\Omega)$, it follows that

$$\|f\|_{H^{-1}(\Omega)} \le C_3 \sup_{v \ne 0} \frac{|(f, v)_\Omega|}{\|v\|_{\widetilde{H}^1(\Omega)}} = C_3 \sup_{v \ne 0} \frac{|\Phi_\Omega(u, v)|}{\|v\|_{\widetilde{H}^1(\Omega)}} \le C_4 \|u\|_{\widetilde{H}^1(\Omega)}. \tag{11.1.38}$$

Next, if $f = 0$, then we assume that $\|u_0\|_{H^1(\Omega)} \le C_5 \|g\|_{H^{1/2}(\Gamma)}$ (using the boundedness of the right inverse for the trace operator) and obtain, similarly to (11.1.38),

$$\|f_0\|_{H^{-1}(\Omega)} \le C_4 \|u_0\|_{H^1(\Omega)} \le C_6 \|g\|_{H^{1/2}(\Gamma)},$$

so that

$$\|u_1\|_{\widetilde{H}^1(\Omega)} \le C_7 \|g\|_{H^{1/2}(\Gamma)} \quad \text{and} \quad \|u\|_{H^1(\Omega)} \le (C_5 + C_7) \|g\|_{H^{1/2}(\Gamma)}.$$

On the other hand,

$$\|g\|_{H^{1/2}(\Gamma)} \le C_8 \|u\|_{H^1(\Omega)}$$

by the trace theorem.

The complete Dirichlet problem reduces to two problems, with $g = 0$ and with $f = 0$, which implies the first inequality in (11.1.36).

Below we state the final result on the Dirichlet problem, assuming the two-sided estimate to be verified.

Theorem 11.1.4. *If the form Φ_Ω is strongly coercive on $\widetilde{H}^1(\Omega)$, then the operator $u \mapsto (f, g)$ corresponding to the Dirichlet problem (11.1.31) maps bijectively and continuously the space $H^1(\Omega)$ to the space $H^{-1}(\Omega) \times H^{1/2}(\Gamma)$ and is invertible; moreover, the two-sided estimate (11.1.36) holds. In the case of only strong ellipticity, this operator is Fredholm with index zero; in this case, to the middle expression in the estimate, the term $\|u\|_{H^0(\Omega)}$ is added.*

The a priori estimate for the uniquely solvable Neumann problem has the form

$$\|u\|_{H^1(\Omega)} \le C_5 [\|f\|_{\widetilde{H}^{-1}(\Omega)} + \|h\|_{H^{-1/2}(\Gamma)}]. \tag{11.1.39}$$

If $f = 0$ or $h = 0$, then the two-sided estimate is also valid. We postpone its verification to Section 11.2. In the general case, we do not have a two-sided estimate, because f and h are not independent.

The Dirichlet, Neumann, and other problems in more general spaces will be discussed in Sections 16 and 17.

11.1.5

We return to the multivalence in the definition of the right-hand side of the equation and of the conormal derivative in the Neumann problem.

In Section 8, we mentioned that this multivalence can sometimes be removed by introducing a special convention. For example, if the right-hand side f of the system $Lu = f$ belongs to $L_2(\Omega)$ (e.g., vanishes), then a functional supported on the boundary is not added to it.

Below we outline Costabel's reasoning [99]. Suppose for simplicity that the coefficients in L are infinitely differentiable. Let $H_L^1(\Omega)$ denote the subspace in $H^1(\Omega)$ consisting of those u for which Lu on Ω (in the sense of distributions) belongs to $L_2(\Omega)$. This space is endowed with the norm

$$\|u\|^2_{H_L^1(\Omega)} = \|u\|^2_{H^1(\Omega)} + \|Lu\|^2_{L_2(\Omega)}. \tag{11.1.40}$$

It was shown in [173] that if L is the Laplacian, then the linear manifold $C^\infty(\overline{\Omega})$ is dense in $H_L^1(\Omega)$. Costabel noticed that precisely the same proof applies to any L, which makes it possible to approximate functions $u \in H_L^1(\Omega)$ by functions $u_j \in C^\infty(\overline{\Omega})$ and define the conormal derivative of any such function as

$$(T^+ u, v)_\Gamma = \lim_{j \to \infty} [\Phi_\Omega(u_j, v) - (Lu_j, v)] \quad (v \in H^1(\Omega)), \tag{11.1.41}$$

i.e., as the limit of smooth conormal derivatives. For smooth solutions, these are smooth conormal derivatives.

In [261, 262] Mikhailov proposed a construction of a "canonical" conormal derivative for solutions such that $f = Lu$ belongs to $\widetilde{H}^{-1/2}(\Omega)$. In this case, f contains no term supported on the boundary, and the conormal derivative is determined uniquely.

In Chapter 5 of the book [186] by Hsiao and Wendland, the right-hand sides of the system $Lu = f$ are assumed to belong to the orthogonal complement in $\widetilde{H}^{-1}(\Omega)$ of the subspace of functionals supported on Γ. Thus, these functionals are not included in the right-hand sides. What is said on this question in the next section (in Subsection 11.2.5) can be regarded as another realization of this idea.

The point is that, for $v \in H^1(\Omega)$, the functionals $\Phi_\Omega(u, v)$ admit different representations by pairs (f, h) in the form

$$(f, v)_\Omega + (h, v^+)_\Gamma.$$

There is no isomorphism between the space of solutions u and the space of such pairs. To each pair (f, h) there corresponds precisely one solution u, but to each u there correspond infinitely many pairs (f, h). This is not a pathology, but is inconvenient. But we can construct an isomorphism between the space of solutions u and the space of some particular pairs (f, h) uniquely representing all (continuous antilinear) functionals on $H^1(\Omega)$, and we shall do this in the next section.

The results explained in this section are a prerequisite for the considerations of Section 11.2.

11.2 The Weyl Decomposition of the Space $H^1(\Omega)$ and the Choice of f and h

11.2.1

The idea to decompose the space $H^1(\Omega)$, applied here, goes back to Weyl, who used it in his paper [395].

The space $H^1(\Omega)$ is Hilbert, and $\widetilde{H}^1(\Omega) = \mathring{H}^1(\Omega)$ is a (closed) subspace in it. Therefore, $\widetilde{H}^1(\Omega)$ is a complemented subspace, and we have already used this fact (see Remark 9.2.2). Now we can conveniently specify another direct complement of this subspace.

Theorem 11.2.1. *If the form Φ_Ω is strongly coercive on $\widetilde{H}^1(\Omega)$, then the space $H^1(\Omega)$ is the direct sum $H_1 + H_2$ of the subspace $H_1 = \widetilde{H}^1(\Omega) = \mathring{H}^1(\Omega)$ of functions with zero boundary values and the subspace H_2 of solutions to the homogeneous system $Lu = 0$ in Ω.*

Here by "in Ω" we mean "inside Ω" in the sense that test functions are taken in $\widetilde{H}^1(\Omega)$.

Proof of Theorem 11.2.1. Obviously, H_2 is a (closed) subspace in $H^1(\Omega)$. By virtue of uniqueness for the Dirichlet problem, $\widetilde{H}^1(\Omega)$ contains no nonzero solutions of the equation $Lu = 0$ in Ω. Therefore, the intersection $H_1 \cap H_2$ comprises only zero.

The projections P_1 and P_2 of $H^1(\Omega)$ onto H_1 and H_2, respectively, are constructed by using the unique solvability of the Dirichlet problem as follows. We take $u \in H^1(\Omega)$, denote by u_2 the solution of the Dirichlet problem

$$Lu_2 = 0 \text{ in } \Omega, \quad u_2^+ = u^+ \tag{11.2.1}$$

and set $u_1 = u - u_2$. We have $u_2 \in H_2$ and $u_1 \in H_1$. Thus, $P_1u = u_1$ and $P_2u = u_2$ are the required projections.

It is seen from the estimate (11.1.36) of the solution of the Dirichlet problem with $f = 0$ that P_2 is a continuous projection. Therefore, P_1 is continuous as well. $\square$

We refer to the relation $H^1(\Omega) = H_1 + H_2$ as the *Weyl decomposition* of the space $H^1(\Omega)$ corresponding to the operator L and to the relation $u = u_1 + u_2$, where $u_j \in H_j$, as the Weyl decomposition of the function u corresponding to this operator.

Remarks and Corollaries

11.2.2

The space $H_1 = \widetilde{H}^1(\Omega)$ does not depend on L (although P_1, of course, depends on L). The space H_2 depends on L, and we write $H_2 = H_2(L)$. But the solutions of the homogeneous system $Lu = 0$ are parametrized by the Dirichlet data, which form the space $H^{1/2}(\Gamma)$, and this space does not depend on L.

11.2.3

If the form $\Phi_\Omega(u, u)$ is strongly coercive on $H^1(\Omega)$, then $\Phi_\Omega(u, v)$ provides a general representation of any continuous antilinear functional of $v \in H^1(\Omega)$. If, in addition, the operator L is formally self-adjoint, then $\Phi_\Omega(u, v)$ can be taken for an inner product in $H^1(\Omega)$. It is seen from the Green identity that the form $\Phi_\Omega(u, v)$ vanishes when $u \in H_2$ and $v \in H_1$, so that H_1 and H_2 are orthogonal with respect to this inner product, and P_1 and P_2 are orthogonal projections. But even in the general case, the spaces H_1 and H_2 in the decomposition corresponding to the operator L are "orthogonal" in the sense that the first of the two forms $\Phi_\Omega(u, v)$ and $\Phi_\Omega(v, u)$ vanishes when $v \in H_1$ and $u \in H_2$.

11.2.4

Now we see that the a priori estimate (11.1.36) is two-sided. Indeed, for $u = u_1 + u_2$, where $u_j \in H_j$, the estimates for the norms of the functions u_j are two-sided, and the norm of u is equivalent to the sum of the norms of u_j. This completes the proof of Theorem 11.1.4.

11.2.5

Let us substitute the Weyl decomposition $u = u_1 + u_2$ corresponding to the operator L and the Weyl decomposition $v = v_1 + v_2$ corresponding to $\widetilde{L}$ into the form $\Phi_\Omega(u, v)$. We now assume that the operators L and $\widetilde{L}$ are defined on all functions in $H^1(\Omega)$. Here $u_1^+ = v_1^+ = 0$. However, *we assume the zero functional Lu_2 on H_1 to be extended by zero to the complementary space $H_2(\widetilde{L})$ and hence to $H^1(\Omega)$*:

$$(Lu_2, v)_\Omega = 0 \quad (v \in H^1(\Omega)). \tag{11.2.2}$$

This agrees with the tradition; see the beginning of Subsection 11.1.5. In particular, *we propose not to include a term supported on the boundary in Lu_2*. The operator $\widetilde{L}$ is treated similarly: we assume that $(u, \widetilde{L}v_2)_\Omega = 0$.

We obtain

$$\Phi_\Omega(u,v) = \sum_{i,j=1}^{2} \Phi_\Omega(u_i, v_j),$$

where

$$\Phi_\Omega(u_1, v_1) = (Lu_1, v_1)_\Omega,$$
$$\Phi_\Omega(u_2, v_1) = (Lu_2, v_1)_\Omega + (T^+ u_2, v_1^+)_\Gamma = 0,$$
$$\Phi_\Omega(u_1, v_2) = (Lu_1, v_2)_\Omega + (T^+ u_1, v_2^+)_\Gamma$$
$$= (u_1, \widetilde{L} v_2)_\Omega + (u_1^+, \widetilde{T}^+ v_2)_\Gamma = 0,$$
$$\Phi_\Omega(u_2, v_2) = (T^+ u_2, v_2^+)_\Gamma.$$

On the first line Lu_1 is the restriction of a functional on $H^1(\Omega)$ to $\widetilde{H}^1(\Omega)$. Thus,

$$\Phi_\Omega(u,v) = (Lu_1, v_1)_\Omega + (T^+ u_2, v_2^+)_\Gamma. \tag{11.2.3}$$

This expression is a new representation for the form on the left-hand side, in which $f = Lu_1$ *and* $h = T^+ u_2$ *are determined uniquely.*

Expression (11.2.3) can be interpreted as follows. In the first term on the right-hand side, $f = Lu_1 = Lu$ is a functional given on $H_1 = \widetilde{H}^1(\Omega)$. To $H_2(\widetilde{L})$ (not to $H_2(L)$ if $L \neq \widetilde{L}$) it is actually extended by zero. We denote the subspace of such extensions in $\widetilde{H}^{-1}(\Omega)$ by $\widetilde{H}_1^{-1}(\Omega)$. The second term on the right-hand side is a functional from $\widetilde{H}^{-1}(\Omega)$ supported on the boundary. We denote the subspace of such functionals by $\widetilde{H}_2^{-1}(\Omega)$.

Proposition 11.2.2. *The space* $\widetilde{H}^{-1}(\Omega)$ *admits the decomposition*

$$\widetilde{H}^{-1}(\Omega) = \widetilde{H}_1^{-1}(\Omega) + \widetilde{H}_2^{-1}(\Omega). \tag{11.2.4}$$

The first subspace on the right-hand side is isomorphic to $H^{-1}(\Omega)$, *and its elements do not contain terms supported on* Γ. *The second subspace consists of functionals supported on* Γ *and is isomorphic to* $H^{-1/2}(\Gamma)$.

Cf. Remarks 4.4.1, 5.1.16, and 9.2.2.

If u is not specified but the complete Neumann problem (11.1.35) is posed, then this problem decomposes now into two problems: the Dirichlet problem in Ω for the inhomogeneous system with right-hand side in $H^{-1}(\Omega)$ and homogeneous boundary condition

$$Lu_1 = f \text{ in } \Omega, \quad u_1^+ = 0 \tag{11.2.5}$$

and the Neumann problem in $\overline{\Omega}$ for the homogeneous system with inhomogeneous boundary condition

$$Lu_2 = 0 \text{ in } \Omega, \quad T^+ u_2 = h_2,$$
$$(h_2, v^+)_\Gamma = (h, v^+)_\Gamma + (f, v)_\Omega - \Phi_\Omega(u_1, v), \quad v \in H^1(\Omega). \tag{11.2.6}$$

The right-hand sides are here determined uniquely and specified independently; moreover, naturally, f contains no terms supported on the boundary and is uniquely determined by u inside the domain. In the first problem, the functional f is assumed to be extended by zero to v_2, but this is essentially the simplest Dirichlet problem, with which we began our consideration in the preceding section.

In solving the complete Neumann problem, from its solution a particular solution of the inhomogeneous equation is often subtracted. But the Weyl decomposition suggests to *subtract the solution of the Dirichlet problem* (11.2.5) from the complete solution in $H^1(\Omega)$. This is the essence of the "endeavor" which we propose.

Let us see what we have in the case of the usual Neumann problem. Suppose that the functions u and v belong to $H^2(\Omega)$. In the usual setting, which we are now considering, the function $f = Lu$ belongs to $L_2(\Omega)$ and contains no component supported on the boundary. It determines a functional on $H^1(\Omega)$ by means of the usual L_2 duality $(f,v)_\Omega$, and this functional is not extended by zero to the functions v_2. The functions u and u_1 have usual conormal derivatives T^+u and T^+u_1, which are included in the first Green identity (see (11.1.4)). Under the subtraction of u_1 from u, the conormal derivative T^+u_1 is subtracted from T^+u. If the function u_1 is found, then this conormal derivative T^+u_1 can be considered as known. Thus, $h = T^+u$ and $h_2 = T^+u_2$ are different.

But we have seen that

$$(Lu_1,v_2)_\Omega + (T^+u_1,v_2^+)_\Gamma = (u_1,\widetilde{L}v_2)_\Omega + (u_1^+,\widetilde{T}^+v_2)_\Gamma = 0. \qquad (11.2.7)$$

The first relation can also be obtained from the second Green identity (11.1.26). The terms on the left-hand side do not affect the form $\Phi_\Omega(u,v)$ and, therefore, the solution u of the complete Neumann problem. These terms can be replaced by zeros.

In the general case, it is more convenient to extend the form $(Lu_1,v)_\Omega$ by zero. It may happen that even the smooth function u_1 has no usual conormal derivative. Indeed, possibly $f = Lu_1$ is a usual function which grows near boundary so that the integrals $(f,v_1)_\Omega$ converge, while the integrals $(f,v_2)_\Omega$ may diverge.

This can be compared with the impossibility to specify arbitrarily the Neumann boundary condition for functions in $\widetilde{H}^1(\Omega)$, because they already obey the homogeneous Dirichlet condition. The Cauchy problem for elliptic equations (with Dirichlet and Neumann conditions) is ill-posed.

11.2.6

Now we can derive a two-sided estimate for the solutions $u = u_1 + u_2$ of the complete Neumann problem represented in the new form (11.2.5)–(11.2.6).

We assume the form Φ_Ω to be strongly coercive on $H^1(\Omega)$. The estimate has the form

$$\|u\|_{H^1(\Omega)} \le C_1[\|f\|_{H^{-1}(\Omega)} + \|h_2\|_{H^{-1/2}(\Gamma)}] \le C_2\|u\|_{H^1(\Omega)}. \qquad (11.2.8)$$

Proof of inequalities (11.2.8). We write a solution u in the form $u = u_1 + u_2$, where u_1 is a solution of the equation $Lu_1 = f$ in $H_1 = \widetilde{H}^1(\Omega)$. This is a solution of the Dirichlet problem with homogeneous boundary condition.

For u_1, we already have the estimates

$$\|u_1\|_{\widetilde{H}^1(\Omega)} \le C_3 \|f\|_{H^{-1}(\Omega)} \le C_4 \|u_1\|_{\widetilde{H}^1(\Omega)}. \tag{11.2.9}$$

We also have

$$Lu_2 = 0 \quad \text{and} \quad T^+ u_2 = h_2$$

in the sense that

$$\Phi_\Omega(u_2, v) = (h_2, v^+)_\Gamma.$$

Setting $v = u_2$ and using the strong coercivity of Φ_Ω on $H^1(\Omega)$ together with the generalized Schwarz inequality, we obtain

$$\|u_2\|^2_{H^1(\Omega)} \le C_5 \|h_2\|_{H^{-1/2}(\Gamma)} \|u_2^+\|_{H^{1/2}(\Gamma)},$$

whence

$$\|u_2\|_{H^1(\Omega)} \le C_6 \|h_2\|_{H^{-1/2}(\Gamma)}. \tag{11.2.10}$$

On the other hand,

$$\|h_2\|_{H^{-1/2}(\Gamma)} \le C_7 \sup_{v^+ \ne 0} \frac{|(h_2, v^+)_\Gamma|}{\|v^+\|_{H^{1/2}(\Gamma)}}$$

$$= C_7 \sup_{v^+ \ne 0} \frac{|\Phi_\Omega(u_2, v)|}{\|v^+\|_{H^{1/2}(\Gamma)}} \le C_8 \|u_2\|_{H^1(\Omega)}. \tag{11.2.11}$$

Here we have used the expression for v in terms of v^+ and the right inverse of the trace operator.

Relations (11.2.9)–(11.2.11) imply (11.2.8). □

Let us summarize the results on the Neumann problem.

Theorem 11.2.3. *If the form Φ_Ω is strongly coercive on $H^1(\Omega)$, then the operator $u \mapsto (f, h_2)$ corresponding to the Neumann problem (11.2.5)–(11.2.6) implements a one-to-one continuous mapping of the space $H^1(\Omega)$ onto the direct product of $H^{-1}(\Omega)$ and $H^{-1/2}(\Gamma)$. It is invertible, and the two-sided a priori estimate (11.2.8) holds. If the form is only coercive, then this operator is Fredholm with index zero, and the same estimate with $\|u\|_{L_2(\Omega)}$ added to the middle expression holds.*

11.2.7

We do *not exclude* the possibility of combining the right-hand side of the system and the conormal derivative into a single functional in $\widetilde{H}^{-1}(\Omega)$ when this is useful, e.g., for applying the Lax–Milgram theorem.

11.3 The Poincaré–Steklov Operators

Consider the solutions of the homogeneous system $Lu = 0$. We know what this means from the preceding section: $u \in H_2(L)$ and the functional Lu is extended by zero to $H_2(\widetilde{L})$.

If the form Φ_Ω is strongly coercive on $\widetilde{H}^1(\Omega)$, there naturally arises an operator acting on these solutions, which maps Dirichlet data to Neumann data:

$$D: u^+ \mapsto T^+ u. \tag{11.3.1}$$

This is a bounded operator from $H^{1/2}(\Gamma)$ to $H^{-1/2}(\Gamma)$. Under the above assumption we cannot assert that it is invertible (see Example 2 in Section 11.7). But if the form Φ_Ω is strongly coercive on $H^1(\Omega)$, then on the solutions of the homogeneous system the following operator also acts in a natural way:

$$N: T^+ u \mapsto u^+; \tag{11.3.2}$$

this is a bounded operator from $H^{-1/2}(\Gamma)$ to $H^{1/2}(\Gamma)$, and it is inverse to the operator D, so that, in this case, both operators D and N are invertible. They are called the *Dirichlet-to-Neumann operator* and the *Neumann-to-Dirichlet operator*, respectively. These operators are also known as the *Poincaré–Steklov operators*. We touched on these operators in Section 7.4 in a different context.

Theorem 11.3.1. *If the form Φ_Ω is strongly coercive on $H^1(\Omega)$, then*

$$\|\varphi\|^2_{H^{1/2}(\Gamma)} \leq C_1 \operatorname{Re}(D\varphi, \varphi)_\Gamma \quad and \quad \|\psi\|^2_{H^{-1/2}(\Gamma)} \leq C_2 \operatorname{Re}(N\psi, \psi)_\Gamma \tag{11.3.3}$$

for $\varphi \in H^{1/2}(\Gamma)$ and $\psi \in H^{-1/2}(\Gamma)$, respectively.

Proof. Let us derive the first inequality. We take a solution u of the system $Lu = 0$ with Dirichlet condition $u^+ = \varphi$ and write the Green identity in the form

$$\Phi_\Omega(u, u) = (D\varphi, \varphi)_\Gamma. \tag{11.3.4}$$

Using the trace theorem and strong coercivity on $H^1(\Omega)$, we obtain

$$\|\varphi\|^2_{H^{1/2}(\Gamma)} \leq C' \|u\|^2_{H^1(\Omega)} \leq C'' \operatorname{Re} \Phi_\Omega(u, u) = C'' \operatorname{Re}(D\varphi, \varphi)_\Gamma.$$

We have verified the first inequality. The simplest way to verify the second is to set $D\varphi = \psi$. We then have $\varphi = N\psi$ and

$$\|\psi\|^2_{H^{-1/2}(\Gamma)} \leq C''' \|\varphi\|^2_{H^{1/2}(\Gamma)} \leq \widetilde{C} \operatorname{Re}(D\varphi, \varphi)_\Gamma = \widetilde{C} \operatorname{Re}(N\psi, \psi)_\Gamma. \qquad \square$$

We refer to (11.3.3) as the *strong coercivity inequalities for the forms of the operators D and N.*

In the case where the surface Γ and the coefficients in L are infinitely differentiable, the operators D and N are strongly elliptic pseudodifferential operators on Γ of orders 1 and -1, respectively.

11.4 The Mixed Problem

The simplest mixed problem is as follows. Suppose that the boundary Γ consists of two domains Γ_1 and Γ_2 and their common boundary $\partial\Gamma_j$, which is a closed Lipschitz surface if $n > 2$ of dimension $n - 2$ (less by 1 than that of Γ). Consider the problem

$$Lu = f \text{ in } \Omega, \quad u^+ = g \text{ on } \Gamma_1, \quad T^+u = h \text{ on } \Gamma_2. \qquad (11.4.1)$$

Here

$$g \in H^{1/2}(\Gamma_1) \quad \text{and} \quad h \in H^{-1/2}(\Gamma_2). \qquad (11.4.2)$$

Let us introduce the *space* $H^1(\Omega; \Gamma_1)$ as the (closed) subspace in $H^1(\Omega)$ consisting of all functions vanishing on Γ_1. We assume that the test functions v belong to this space. Thus, we must also assume that f belongs to the dual space

$$\widetilde{H}^{-1}(\Omega; \Gamma_1) := [H^1(\Omega; \Gamma_1)]^* \qquad (11.4.3)$$

with respect to the form $(f, v)_\Omega$.

Arguing as in the case of the Dirichlet problem, we first suppose that $g = 0$. Then the solution u and the test function v must belong to $H^1(\Omega; \Gamma_1)$. The Green identity has the form

$$\Phi_\Omega(u, v) = (f, v)_\Omega + (h, v^+)_{\Gamma_2}. \qquad (11.4.4)$$

Suppose that the form Φ_Ω is strongly coercive on $H^1(\Omega; \Gamma_1)$ (this follows from strong coercivity on $H^1(\Omega)$). Then the problem is uniquely solvable by virtue of the Lax–Milgram theorem.

Now suppose that $g \neq 0$. We proceed as in the case of the Dirichlet problem with inhomogeneous boundary condition. First, we extend g to a function in $H^{1/2}(\Gamma)$, then we take a function $u_0 \in H^1(\Omega)$ with boundary value g, and finally we define $f_0 = Lu_0 \in \widetilde{H}^{-1}(\Omega; \Gamma_1)$ by

$$(f_0, v)_\Omega = \Phi_\Omega(u_0, v), \quad v \in H^1(\Omega; \Gamma_1), \qquad (11.4.5)$$

assuming thereby that $T^+u_0 = 0$ on Γ_2. We *say* that the solution of our problem is the function $u = u_0 + u_1 \in H^1(\Omega)$, where $u_1 \in H^1(\Omega; \Gamma_1)$ is a solution of the problem

$$Lu_1 = f - f_0 \text{ in } \Omega, \quad u_1^+ = 0 \text{ on } \Gamma_1, \quad T^+u_1 = h \text{ on } \Gamma_2, \qquad (11.4.6)$$

i.e.,

$$\Phi_\Omega(u_1, v) = (f - f_0, v)_\Omega + (h, v^+)_{\Gamma_2}, \quad v \in H^1(\Omega; \Gamma_1). \qquad (11.4.7)$$

The strong coercivity of the form Φ_Ω on $H^1(\Omega; \Gamma_1)$, which we assume, implies the existence and uniqueness of the solution u_1. The verification of the independence of u of the choice of u_0 and the method for extending the function g are the same as in the case of the Dirichlet problem.

Thus, we have proved the following theorem.

Theorem 11.4.1. *If the form $\Phi_\Omega(u,v)$ is strongly coercive on $H^1(\Omega;\Gamma_1)$, then the mixed problem stated above is uniquely solvable.*

The Dirichlet and Neumann problems can be regarded as special cases of the mixed problem; in these cases, all considerations agree with those in the preceding sections.

Now let us show that the mixed problem can be reduced to both the Dirichlet and the Neumann problem by solving certain equations on a part of the boundary. For this purpose, we must introduce analogues of the operators N and D on a part of the boundary.

We introduce the operator

$$D_1\varphi = (D\varphi)|_{\Gamma_1} \tag{11.4.8}$$

from $\widetilde{H}^{1/2}(\Gamma_1)$ to $H^{-1/2}(\Gamma_1)$. It is obtained from the operator D by narrowing its domain to the subspace $\widetilde{H}^{1/2}(\Gamma_1)$ of $H^{1/2}(\Gamma)$ and restricting the functions $D\varphi$ thus obtained to Γ_1. Obviously, this operator is well defined and bounded.

Theorem 11.4.2. *If the form Φ_Ω is strongly coercive on $H^1(\Omega)$, then the form of the operator D_1 satisfies the strong coercivity inequality*

$$\|\varphi\|^2_{\widetilde{H}^{1/2}(\Gamma_1)} \leq C\operatorname{Re}(D_1\varphi,\varphi)_{\Gamma_1} \quad (\varphi \in \widetilde{H}^{1/2}(\Gamma_1)), \tag{11.4.9}$$

and this operator is invertible.

Proof. Inequality (11.4.9) follows from the first inequality in (11.3.3): the operator D_1 inherits it. The functions $D_1\varphi$ and φ in (11.4.9) belong to dual spaces. The invertibility of D_1 follows by applying the Lax–Milgram theorem to the form $(D_1\varphi_1,\varphi_1)_{\Gamma_1}$ and to the equation $D_1\varphi_1 = \psi_1$. $\qquad\Box$

The operator D_2 is defined in the same way and has similar properties.

Now we introduce the operator

$$N_1\psi = (N\psi)|_{\Gamma_1} \tag{11.4.10}$$

from $\widetilde{H}^{-1/2}(\Gamma_1)$ to $H^{1/2}(\Gamma_1)$. It is obtained from the operator N by narrowing its domain to the subspace $\widetilde{H}^{-1/2}(\Gamma_1)$ of $H^{-1/2}(\Gamma)$ and restricting the functions $N\psi$ thus obtained to Γ_1. Obviously, this operator is also well defined and bounded.

Theorem 11.4.3. *If the form Φ_Ω is strongly coercive on $H^1(\Omega)$, then the form of the operator N_1 satisfies the strong coercivity inequality*

$$\|\psi\|^2_{\widetilde{H}^{-1/2}(\Gamma_1)} \leq C\operatorname{Re}(N_1\psi,\psi)_{\Gamma_1} \quad (\psi \in \widetilde{H}^{-1/2}(\Gamma_1)), \tag{11.4.11}$$

and this operator is invertible.

Proof. The proof is similar to that of Theorem 11.4.2. Inequality (11.4.11) follows from the second inequality in (11.3.3). $\qquad\Box$

The operator N_2 is defined similarly and has similar properties.

Note that the operators N_1 and D_1^{-1} do not coincide: they have different domains and different ranges. Similarly, the operators N_2 and D_2^{-1} do not coincide.

Now we proceed to the reduction of the mixed problem to Dirichlet and Neumann problems.

Suppose that $f = 0$. (We assume that we can construct a solution of the equation $Lu = f$; cf. Section 12.1 below.)

Let us extend the right-hand side g of the boundary condition on Γ_1 to a function in $H^{1/2}(\Gamma)$, which we denote by the same letter g. Then the solution u of the problem under consideration must satisfy the condition $u^+ = g + g_1$, where $g_1 \in \widetilde{H}^{1/2}(\Gamma_2)$.

On the other hand,

$$h = (D(g + g_1))|_{\Gamma_2}.$$

We obtain the following equation for g_1:

$$D_2 g_1 = h_1, \tag{11.4.12}$$

where $h_1 = h - (Dg)|_{\Gamma_2}$ is a known function. But we know that the operator D_2 is invertible. If we find g_1, then the mixed problem will be reduced to the Dirichlet problem.

Now let us show how the mixed problem is reduced to the Neumann problem. The reduction is similar. We extend h to an element of $H^{-1/2}(\Gamma)$ and obtain $T^+ u = h + h_2$, where $h_2 \in \widetilde{H}^{-1/2}(\Gamma_1)$. For h_2, we have the equation

$$N_1 h_2 = g_2, \tag{11.4.13}$$

where $g_2 = g - (Nh)|_{\Gamma_1}$ is a known function. The operator N_1 is invertible, and solving this equation, we reduce the mixed problem to the Neumann problem.

In the next section we explain how to reduce the Dirichlet and Neumann problems to equations on the boundary Γ by using potential-type operators. Thereby, we show that the mixed problem reduces to equations on the boundary as well.

Now we give an analogue of the Weyl decomposition for the space $H^1(\Omega; \Gamma_1)$.

Proposition 11.4.4. *If the Dirichlet problem is uniquely solvable, then the space $H^1(\Omega; \Gamma_1)$ is the direct sum of the subspace $\widetilde{H}^1(\Omega)$ and the solution space of the Dirichlet problem for a homogeneous system $Lu = 0$ with Dirichlet data vanishing on Γ_1. The latter space is isomorphic to $\widetilde{H}^{1/2}(\Gamma_2)$.*

We leave the proof to the reader.

11.5 Duality Relations on Γ

The second Green identity (11.1.26) implies the relations

$$(D\varphi_1, \varphi_2)_\Gamma = (\varphi_1, \widetilde{D}\varphi_2)_\Gamma \quad \text{and} \quad (N\psi_1, \psi_2)_\Gamma = (\psi_1, \widetilde{N}\psi_2)_\Gamma. \tag{11.5.1}$$

Here and in what follows, we mark the operators on the boundary corresponding to $\widetilde{L}$ by a tilde. The functions φ_j and ψ_j belong to $H^{1/2}(\Gamma)$ and $H^{-1/2}(\Gamma)$, respectively.

As a consequence, we have

$$(D_1\varphi_1, \varphi_2)_{\Gamma_1} = (\varphi_1, \widetilde{D}_1\varphi_2)_{\Gamma_1} \quad \text{and} \quad (N_1\psi_1, \psi_2)_{\Gamma_1} = (\psi_1, \widetilde{N}_1\psi_2)_{\Gamma_1}. \qquad (11.5.2)$$

Here $\varphi_j \in \widetilde{H}^{1/2}(\Gamma_1)$ in the first relation and $\psi_j \in \widetilde{H}^{-1/2}(\Gamma_1)$ in the second.

Relations (11.5.1) and (11.5.2) mean that, in each of the four cases under consideration, the operators corresponding to formally adjoint systems are mutually adjoint with respect to the extension of the inner product in $L_2(\Gamma)$ or $L_2(\Gamma_1)$.

11.6 Spectral Problems

In this subsection we briefly consider the following spectral problems.

1° The Dirichlet spectral problem

$$Lu = \lambda u \text{ in } \Omega, \quad u^+ = 0 \text{ on } \Gamma. \qquad (11.6.1)$$

2° The Neumann spectral problem

$$Lu = \lambda u \text{ in } \Omega, \quad T^+u = 0 \text{ on } \Gamma. \qquad (11.6.2)$$

3° The mixed spectral problem

$$Lu = \lambda u \text{ in } \Omega, \quad u^+ = 0 \text{ on } \Gamma_1, \quad T^+u = 0 \text{ on } \Gamma_2. \qquad (11.6.3)$$

These are problems with spectral parameter in the system. The fourth problem contains a spectral parameter in the boundary condition:

4° The Poincaré–Steklov problem

$$Lu = 0 \text{ in } \Omega, \quad T^+u = \lambda u^+ \text{ on } \Gamma. \qquad (11.6.4)$$

The fifth and sixth problems are also mixed; these are analogues of the Poincaré–Steklov problem with a spectral parameter on a part of the boundary:

5° $Lu = 0$ in Ω, $T^+u = \lambda u^+$ on Γ_1, $u^+ = 0$ on Γ_2. $\qquad (11.6.5)$

6° $Lu = 0$ in Ω, $T^+u = \lambda u^+$ on Γ_1, $T^+u = 0$ on Γ_2. $\qquad (11.6.6)$

Four more spectral problems will be considered in Section 12.9.

It should be mentioned that if the boundary Γ and the coefficients of the operator L are smooth, then at least problems 1°, 2°, and 4° (provided that they are elliptic) can be considered in the framework of the general theory developed in Section 7. In this case, stronger results are obtained. In particular, the generalized eigenfunctions are smooth (on $\overline{\Omega}$ or on Γ), and sharper estimates for the residual terms in asymptotic expressions for eigenvalues are known. Here we dwell on what can be found out for a Lipschitz boundary and coefficients of low smoothness. For simplicity we assume the $a_{j,k}$ and b_j to be Lipschitz continuous. All facts are given without proof.

Throughout this subsection we assume that the strong coercivity condition (11.1.21) holds. As in Section 18.3, we number the eigenvalues of unbounded self-adjoint positive operators with compact resolvent by positive integers, counting multiplicities, in nondecreasing order; the eigenvalues of non-self-adjoint operators with compact resolvent are also numbered with multiplicities taken into account but in the order of nondecreasing absolute value. The material of Section 18.3 can be used for references.

We begin with the first problem. Denote the corresponding operator by $\mathcal{L}_D$. In the general theory, under additional smoothness assumptions, $\mathcal{L}_D$ is usually regarded as an unbounded operator in $L_2(\Omega)$ with domain $H^2(\Omega) \cap \mathring{H}^1(\Omega)$. We discussed this in Section 7.4 and shall touch upon in Section 11.9. But now we do not have an exact description of the domain of this operator and shall treat $\mathcal{L}_D$ as an invertible operator from $\widetilde{H}^1(\Omega)$ to $H^{-1}(\Omega)$. We distinguish between three cases.

1. First, suppose that L is a formally self-adjoint operator. Then it follows from the second Green identity that

$$(\mathcal{L}_D u, v)_\Omega = (u, \mathcal{L}_D v)_\Omega \quad (u, v \in \widetilde{H}^1(\Omega)) \tag{11.6.7}$$

and, therefore,

$$(\mathcal{L}_D^{-1} f, g)_\Omega = (f, \mathcal{L}_D^{-1} g)_\Omega \quad (f, g \in H^{-1}(\Omega)). \tag{11.6.8}$$

This makes it possible to endow $H^{-1}(\Omega)$ with the new inner product

$$\langle f, g \rangle_{H^{-1}(\Omega)} = (\mathcal{L}_D^{-1} f, g)_\Omega. \tag{11.6.9}$$

Here we use the duality between the spaces $\widetilde{H}^1(\Omega)$ and $H^{-1}(\Omega)$. The operator $\mathcal{L}_D^{-1}$ turns out to be a self-adjoint compact operator in $H^{-1}(\Omega)$ with inner product (11.6.9), and $\mathcal{L}_D$ is a self-adjoint operator with compact resolvent in this space. Recall that $\widetilde{H}^1(\Omega)$ is identified with $\mathring{H}^1(\Omega)$ and is contained in $H^{-1}(\Omega)$; therefore, the spectral problem makes sense. The spectrum of $\mathcal{L}_D$ is discrete and consists of positive eigenvalues, so that this operator is positive self-adjoint, and its eigenfunctions belong to $\widetilde{H}^1(\Omega)$; in this space, as well as in $H^{-1}(\Omega)$, they form an orthogonal basis, namely, with respect to the inner product

$$\langle u, v \rangle_{\widetilde{H}^1(\Omega)} = (\mathcal{L}_D u, v)_\Omega, \tag{11.6.10}$$

which *coincides* with $\Phi_\Omega(u, v)$ by virtue of the first Green identity. The eigenfunctions belong also to intermediate spaces, in which they form orthogonal bases with respect to inner products introduced by means of powers of the operator $\mathcal{L}_D$ using its positivity. As we shall see in Section 16.4, the scale of these spaces can be extended to the left and right beyond the interval $[-1, 1]$ of index values, namely, to the interval $(-3/2, 3/2)$, where the eigenfunctions again form an orthogonal basis.

The eigenvalues have the asymptotics

$$\lambda_j(\mathcal{L}_D) = c_L j^{2/n} + O(j^{3/2n}), \tag{11.6.11}$$

where c_L is a positive constant. The estimate of the remainder term is derived from results of Métivier [260].

We add that Métivier considered variational spectral problems for the operator A determined by the following coercive form $\Phi(u, v)$ on $L_2(\Omega)$:

$$\Phi(u, v) = (Au, v)_{L_2(\Omega)} \quad (u \in D(A), v \in V),$$

where the domain V of the form can be a subspace of $H^1(\Omega)$ containing $\mathring{H}^1(\Omega)$; cf. Subsection 3 in Section 11.8 below. In the notation of Section 11.9 this is the operator A_2. Its eigenvalues are the same (see Section 11.9).

In more special cases of problems in Lipschitz domains, a sharper estimate of the remainder can be derived; see [287]. But already estimate (11.6.11) is useful. For general smooth scalar problems, the estimate is $O(j^{1/n})$.

2. Now suppose that the operator L is not formally self-adjoint but has formally self-adjoint leading part. Consider $\frac{1}{2}(L + \widetilde{L})$. We denote the operator corresponding to the Dirichlet problem for $\frac{1}{2}(L + \widetilde{L})$ by $\mathcal{L}_{D,0}$. We can assume that $\mathcal{L}_{D,0}$ is invertible as well. We endow $H^{-1}(\Omega)$ with an inner product of the form (11.6.9) but with the operator $\mathcal{L}_{D,0}$ instead of $\mathcal{L}_D$. The operator $\mathcal{L}_D$ corresponding to the Dirichlet problem for L turns out to be a weak perturbation of the self-adjoint positive operator $\mathcal{L}_{D,0}$:

$$\mathcal{L}_D = \mathcal{L}_{D,0} + \mathcal{L}_{D,1}, \quad \text{where } \|\mathcal{L}_{D,1}\mathcal{L}_{D,0}^{-q}\| < \infty. \tag{11.6.12}$$

Here $q = 1/2$ in the general case and $q = 0$ if $\mathcal{L}_{D,1}$ corresponds to an operator of order zero. The following assertions are valid.

The spectrum of $\mathcal{L}_D$ is discrete, and its eigenvalues belong to a domain of the form

$$\{\lambda = \sigma + i\tau : \sigma > 0, |\tau| < C\sigma^q\} \tag{11.6.13}$$

with $C > 0$, at least beginning with some number; there are no eigenvalues in the closed left half-plane. The eigenvalues have the asymptotics

$$\lambda_j = c_L j^{2/n} + o(j^{2/n}) \tag{11.6.14}$$

with the same coefficient c_L. The generalized eigenfunctions of the operator $\mathcal{L}_D$ on $H^{-1}(\Omega)$ belong to $\widetilde{H}^1(\Omega)$ and to all intermediate spaces, and they form a complete set in each of them.

3. Finally, consider the entirely non-self-adjoint case. The operator $\mathcal{L}_D$ remains an operator on $H^{-1}(\Omega)$ with domain $\widetilde{H}^1(\Omega)$ and compact resolvent. Suppose that the values of the quadratic form $\Phi_\Omega(u, u)$ are contained in the sector

$$\Theta_\theta = \{\lambda : |\arg \lambda| \le \theta\}. \tag{11.6.15}$$

Here θ is always less than $\pi/2$, because, by virtue of the strong coercivity condition,

$$|\operatorname{Im}\varPhi_\Omega(u,u)| \le C\operatorname{Re}\varPhi_\Omega(u,u). \tag{11.6.16}$$

All eigenvalues of $\mathcal{L}_D$ are contained in this sector. Outside the sector $\Theta_{\theta+\varepsilon}$ with arbitrarily small $\varepsilon > 0$ the optimal estimate for the resolvent holds:

$$\|(\mathcal{L}_D - \lambda I)^{-1}\| \le C(1 + |\lambda|)^{-1}. \tag{11.6.17}$$

(In the preceding case, such an estimate is valid outside Θ_ε for any ε if $|\lambda|$ is large enough.)

The completeness of the system of generalized eigenfunctions in the extreme and the intermediate spaces can be ensured if

$$\theta < \pi/n. \tag{11.6.18}$$

For the Neumann problem $2°$, all main results remain valid, but for different spaces: the operator $\mathcal{L}_N$ corresponding to this problem acts boundedly from $H^1(\Omega)$ to $\widetilde{H}^{-1}(\Omega)$, and, e.g., in the case of formal self-adjointness (on $\overline{\Omega}$; see Section 11.1.3), we endow the latter space with the inner product

$$\langle f,g\rangle_{\widetilde{H}^{-1}(\Omega)} = (\mathcal{L}_N^{-1}f,g)_\Omega. \tag{11.6.19}$$

The constant c_L in the asymptotics is the same as previously, and Métivier's result remains valid.

The convention of Section 11.2 is here inconvenient. But the generalized eigenfunctions belong to $H^1(\Omega)$.

We proceed to the mixed problem $3°$. Here the picture is similar, but the operator corresponding to the problem acts from $H^1(\Omega; \Gamma_1)$ to $\widetilde{H}^{-1}(\Omega; \Gamma_1)$. We denote it by $\mathcal{L}_m$ (m is for "mixed"). Assuming this operator to be invertible, we again consider three cases. In the first case, if L is formally self-adjoint, we introduce the inner product

$$\langle f,g\rangle_{\widetilde{H}^{-1}(\Omega;\Gamma_1)} = (\mathcal{L}_m^{-1}f,g)_\Omega \tag{11.6.20}$$

in $\widetilde{H}^{-1}(\Omega; \Gamma_1)$. The results are of the same character as previously; in particular, the asymptotics for the eigenvalues retains the constant c_L.

Now consider the Poincaré–Steklov problem $4°$. Clearly, at the *eigenfunctions*, it is equivalent to the equation

$$D\varphi = \lambda\varphi, \quad \varphi = u^+. \tag{11.6.21}$$

Suppose that the operator L is formally self-adjoint on $\overline{\Omega}$. Then (see Section 11.5)

$$(D\varphi_1,\varphi_2)_\Gamma = (\varphi_1, D\varphi_2)_\Gamma \quad (\varphi_1,\varphi_2 \in H^{1/2}(\Gamma)) \tag{11.6.22}$$

and

$$(N\psi_1,\psi_2)_\Gamma = (\psi_1, N\psi_2)_\Gamma \quad (\psi_1,\psi_2 \in H^{-1/2}(\Gamma)). \tag{11.6.23}$$

The latter relation makes it possible to endow $H^{-1/2}(\Gamma)$ with the new inner product

$$\langle\psi_1,\psi_2\rangle_{H^{-1/2}(\Gamma)} = (N\psi_1,\psi_2)_\Gamma. \qquad (11.6.24)$$

Now the Neumann-to-Dirichlet operator N is a compact self-adjoint operator on $H^{-1/2}(\Gamma)$, and the Dirichlet-to-Neumann operator D inverse to it is a self-adjoint unbounded operator with compact resolvent and discrete spectrum, whose domain is the space $H^{1/2}(\Gamma)$. The eigenfunctions belong to $H^{1/2}(\Gamma)$ and to the intermediate spaces. The eigenvalues of both operators are positive, and the eigenvalues of D tend to $+\infty$.

In the case where the surface Γ and the coefficients of L are infinitely differentiable, general results of the spectral theory of elliptic pseudodifferential operators imply the following asymptotics for the eigenvalues:

$$\lambda_j(D) = c'j^{1/(n-1)} + O(1). \qquad (11.6.25)$$

Definition [27]. A Lipschitz surface Γ is said to be *almost smooth* if it is smooth outside a closed subset of measure zero.

Clearly, the class of almost smooth surfaces contains an extensive set of, e.g., two-dimensional surfaces in $\mathbb{R}^3$ with singularities, such as convex curvilinear polyhedra, cones, cylinders, etc., which are smooth outside their singularities of the type of edge, vertex, and so on.

In [23] it was shown that if the surface Γ is almost smooth, then

$$\lambda_j(D) = c'j^{1/(n-1)} + o(j^{1/(n-1)}). \qquad (11.6.26)$$

Now suppose that only the principal part of the operator L is formally self-adjoint. Then both N and D can be regarded as weak perturbations of self-adjoint operators on $H^{-1/2}(\Gamma)$; we denote them by N_0 and D_0, respectively. These operators correspond to a formally self-adjoint operator on Ω with the same principal part. An inner product is introduced by using N_0 as

$$\langle\psi_1,\psi_2\rangle_{H^{-1/2}(\Gamma)} = (N_0\psi_1,\psi_2)_\Gamma.$$

The corresponding exponent q (cf. (11.6.12)) is less than 1, but we shall not determine its exact value. The results are as follows. The generalized eigenfunctions belong to $H^{1/2}(\Gamma)$ and to the intermediate spaces, in which they form complete systems. The eigenvalues of the operator D lie in an arbitrarily narrow angle Θ_ε starting with some number depending on ε and have the same asymptotics. There are no eigenvalues in the left half-plane $\mathrm{Re}\,\lambda < 0$. Outside the angle Θ_ε, for large $|\lambda|$, the optimal estimate for the resolvent of D holds:

$$\|(D - \lambda I)^{-1}\| \leq C|\lambda|^{-1}. \qquad (11.6.27)$$

In the entirely non-self-adjoint case, suppose that Θ_θ is the angle containing all values of the quadratic form $\Phi_\Omega(u,u)$ for $u \in H^1(\Omega)$. All eigenvalues of the operator N (and of D) are contained in Θ_θ. Outside the angle $\Theta_{\theta+\varepsilon}$ with arbitrarily small $\varepsilon > 0$, the same optimal estimate (11.6.27) for the resolvent holds. The generalized eigenfunctions again belong to $H^{1/2}(\Gamma)$ and to the intermediate spaces. They form a complete system in these spaces if $2(n-1)\theta < \pi$.

We proceed to problems 5° and 6°.

In problem 5°, the functions $\varphi = u^+$ vanish on Γ_2, and at the *eigenfunctions* φ, the problem is equivalent to the equation $D_1\varphi = \lambda\varphi$. The basic Hilbert space is $H^{-1/2}(\Gamma_1)$, and if the operator L is formally self-adjoint, then D_1 is a self-adjoint operator in this space with inner product

$$\langle\psi_1,\psi_2\rangle_{H^{-1/2}(\Gamma_1)} = (D_1^{-1}\psi_1,\psi_2)_{\Gamma_1} \tag{11.6.28}$$

and has domain $\widetilde{H}^{1/2}(\Gamma_1)$.

In problem 6°, the functions $\psi = T^+u$ vanish on Γ_2. At the *eigenfunctions* the problem is equivalent to the equation $\psi = \lambda N_1\psi$. The basic Hilbert space for it is $\widetilde{H}^{-1/2}(\Gamma_1)$, and N_1 is a compact operator in this space with range $H^{1/2}(\Gamma_1)$. If the operator L is formally self-adjoint, then N_1 is a self-adjoint operator in the space $\widetilde{H}^{-1/2}(\Gamma_1)$ with inner product

$$\langle\psi_1,\psi_2\rangle_{\widetilde{H}^{-1/2}(\Gamma_1)} = (N_1\psi_1,\psi_2)_{\Gamma_1}. \tag{11.6.29}$$

In other respects, the spectral properties of the operators N_1 and D_1^{-1} are similar to those of N. Note that the coefficients in the spectral asymptotics for N_1 and D_1^{-1} are the same, although the eigenfunctions and the eigenvalues are different.

Example. Consider the Laplace equation in the square

$$\{(x,y)\colon 0 < x < \pi, 0 < y < \pi\}.$$

Let Γ_1 be the left edge of this square, and let Γ_2 be formed by the three other edges. Problem 5° has the solutions $\sinh j(\pi - x)\sin jy$ $(j = 1,2,\ldots)$, and problem 6° has the solutions $\cosh j(\pi - x)\cos jy$ $(j = 0,1,\ldots)$. The eigenfunctions of the operator D_1^{-1} are $\sin jy$, and the eigenfunctions of N_1 are $\cos jy$. The eigenvalues equal $j^{-1}\tanh(j\pi)$ and $j^{-1}\coth(j\pi)$, respectively. They are different in these two problems, but their asymptotics coincide.

11.7 Examples

1. First, we consider the well-known strongly elliptic system with two different Neumann problems, which arise for different representations of this system in divergence form, or, equivalently, under different choices of the quadratic form. This is the *Lamé system*, already mentioned in Section 11.1:

$$Lu := -\mu\Delta u - (\lambda + \mu)\operatorname{grad}\operatorname{div} u = f. \tag{11.7.1}$$

Here $d = n$, so that this is an $n \times n$ system; λ and μ are real numbers, called the *Lamé parameters* of the elastic medium. The symbol of this homogeneous operator equals

$$\mu |\xi|^2 I + (\lambda + \mu)(\xi_1, \ldots, \xi_n)'(\xi_1, \ldots, \xi_n), \qquad (11.7.2)$$

and the corresponding quadratic form is

$$\mu |\xi|^2 |\zeta|^2 + (\lambda + \mu)|\xi \cdot \zeta|^2. \qquad (11.7.3)$$

We see that the system is strongly elliptic for

$$\mu > 0 \quad \text{and} \quad \lambda > -2\mu. \qquad (11.7.4)$$

These conditions are necessary and sufficient (to verify necessity, it suffices to write out the upper left element of the matrix (11.7.2); it must be positive).

The first possible expression of the operator L in divergence form is suggested by (11.7.1):

$$(Lu)_k = -\mu \operatorname{div}\operatorname{grad} u_k - (\lambda + \mu)\partial_k \operatorname{div} u \quad (k = 1, \ldots, n). \qquad (11.7.5)$$

We have the Green identity

$$(Lu, v)_\Omega = \Phi_\Omega(u, v) - (T^+ u, v^+)_\Gamma \qquad (11.7.6)$$

with (smooth) conormal derivative

$$T^+ u = T_1^+ u = \mu \partial_\nu u + (\lambda + \mu)(\operatorname{div} u)\nu \qquad (11.7.7)$$

and form

$$\Phi_\Omega(u, v) = \Phi_{\Omega,1}(u, v)$$
$$= \int_\Omega \left[\sum_1^n \mu \operatorname{grad} u_k \cdot \operatorname{grad} \bar{v}_k + (\lambda + \mu)\operatorname{div} u \cdot \operatorname{div} \bar{v} \right] dx. \qquad (11.7.8)$$

Clearly, the form is coercive on $H^1(\Omega)$ if $\mu > 0$ and $\lambda \geq -\mu$. But since

$$|\operatorname{div} u|^2 \leq n \sum_1^n |\partial_j u_j|^2 \leq n|\operatorname{grad} u|^2,$$

we can write the coercivity conditions in the form

$$\mu > 0, \quad \lambda > -\mu(n+1)/n. \qquad (11.7.9)$$

In physics, of special importance is the traction problem. This is the Neumann problem for the for the system $Lu = 0$, where

$$(Lu)_k = -\mu \sum_j \partial_j (\partial_k u_j + \partial_j u_k) - \lambda \partial_k \operatorname{div} u \quad (k = 1, \dots, n). \qquad (11.7.10)$$

In the Green identity, the (smooth) conormal derivative is

$$T^+ u = T_2^+ u = \mu (\partial_j u_k + \partial_k u_j)\nu + \lambda (\operatorname{div} u)\nu \qquad (11.7.11)$$

(the first parenthesized expression is a matrix), and the form equals

$$\Phi_{\Omega,2}(u,v) = \int_\Omega \left[\frac{\mu}{2} \sum_{j,k=1}^n (\partial_j u_k + \partial_k u_j)(\partial_j \overline{v}_k + \partial_k \overline{v}_j) + \lambda \operatorname{div} u \cdot \operatorname{div} \overline{v} \right] dx. \quad (11.7.12)$$

Obviously, by virtue of the second Korn inequality (11.1.18), the form is coercive on $H^1(\Omega)$ for $\mu > 0$ and $\lambda \geq 0$. But actually, coercivity takes place for

$$\mu > 0 \quad \text{and} \quad \lambda > -2\mu/n. \qquad (11.7.13)$$

This follows from the inequalities [115]

$$|\operatorname{div} u|^2 \leq n \sum_1^n |\partial_j u_j|^2 \leq \frac{n}{4} \sum_{j,k=1}^n |\partial_j u_k + \partial_k u_j|^2.$$

2. Now we construct examples of second-order strongly elliptic systems for which the corresponding forms are not coercive on $H^1(\Omega)$. These examples are suggested by Agmon's considerations of higher-order scalar strongly elliptic equations in [8, Sec. 11].

Let $P(D)$ be a first-order matrix homogeneous elliptic operator with constant real-valued coefficients, and let $d \geq 2$. For example,

$$P(D) = \begin{pmatrix} D_1 & D_2 \\ -D_2 & D_1 \end{pmatrix}. \qquad (11.7.14)$$

Here $d = n = 2$. We set (in the general case)

$$L(D) = P'(D) \cdot P(D) + \tau I, \qquad (11.7.15)$$

where the prime denotes transposition, τ is an arbitrarily large positive constant, and $D_j = -i\partial_j$. The operator $L(D)$ is strongly elliptic; indeed, at real ξ, its leading part satisfies the following relation by virtue of the homogeneity and nonsingularity of the matrix $P(\xi)$:

$$L_0(\xi)\zeta \cdot \overline{\zeta} = |P(\xi)\zeta|^2.$$

The domain Ω can be arbitrary, in particular, smooth. We define the form Φ_Ω by

$$\Phi_\Omega(u,v) = \int_\Omega [P(D)u(x) \cdot \overline{P(D)v(x)} + \tau u(x) \cdot \overline{v(x)}] \, dx, \qquad (11.7.16)$$

$\tau > 0$. The form $\Phi_\Omega(u, u)$ is strongly coercive on $\widetilde{H}^1(\Omega)$.

Suppose that the determinant of the matrix $P(\zeta)$ vanishes at some (nonreal) ζ, and let u_0 be the corresponding eigenvector. Then the leading part of the form $\Phi_\Omega(u, u)$ vanishes at the functions

$$u_\lambda(x) = e^{i\lambda\zeta \cdot x} u_0. \tag{11.7.17}$$

These functions span an infinite-dimensional linear space. Assuming that

$$\|u\|^2_{H^1(\Omega)} \leq C\Phi_\Omega(u, u)$$

for some τ, we obtain a contradiction. Indeed, for the functions specified above and their linear combinations, this inequality takes the form

$$\|u\|^2_{H^1(\Omega)} \leq C\tau\|u\|^2_{L_2(\Omega)}$$

and extends to the limits of these linear combinations in the space $H^1(\Omega)$. But such an inequality implies the finite-dimensionality of this space: its unit ball in the L_2-norm turns out to be compact.

Moreover, for $\tau = 0$, the conormal derivatives $P'(\nu)[P(D)u]^+$ of all of these functions turn out to be zero. Therefore, the Neumann problem for the systems considered above is non-Fredholm for any τ.

Note that in the case of operator (11.7.14), the operator (11.7.15) decomposes into two scalar operators $-\Delta + \tau$. If the corresponding form is written as

$$\int_\Omega (\nabla u \cdot \nabla \overline{v} + \tau u \cdot \overline{v}) \, dx,$$

then it is, of course, strongly coercive on $H^1(\Omega)$. This example supplements the preceding one and again shows that the presence or absence of coercivity on $H^1(\Omega)$ depends not only on the principal symbol of the operator L but also on the choice of the expression of this operator in divergence form. The smooth conormal derivative depends on this expression.

11.8 Other Problems

1. We begin with elementary considerations concerning the third boundary value problem, or the *Robin problem* (cf., e.g., [116] and [227])

$$Lu = f \text{ in } \Omega, \quad T^+u + \beta u^+ = h. \tag{11.8.1}$$

Here $\beta = \beta(x)$ is a $d \times d$ matrix-valued function defined on the boundary; for simplicity, we assume it to have nonnegative real part:

$$\text{Re}\,\beta(x) \ge 0. \tag{11.8.2}$$

It suffices to assume that $\beta(x)$ is measurable and bounded. The Green identity for $u, v \in H^1(\Omega)$ is

$$(f,v)_\Omega + (h,v^+)_\Gamma = \Phi_\Omega(u,v) + (\beta u^+, v^+)_\Gamma. \tag{11.8.3}$$

If the form $\Phi_\Omega(u,u)$ is strongly coercive on $H^1(\Omega)$, then so is the form $\Phi_\Omega(u,u) + (\beta u^+, u^+)_\Gamma$, and the Lax–Milgram theorem implies the unique solvability of the problem in $H^1(\Omega)$ for $f \in \widetilde{H}^{-1}(\Omega)$ and $h \in H^{-1/2}(\Gamma)$. In essence, this is a slight generalization of what we know about the Neumann problem.

This topic might be developed further, but we stop here.

2. We also mention the *oblique derivative problem*. In the case of the Laplace equation, it has the form

$$\Delta u = 0 \text{ in } \Omega, \quad \nabla u \cdot W = h \text{ on } \Gamma. \tag{11.8.4}$$

In [211], W is a vector field forming an acute angle with the normal vectors. In the literature, this problem has also been considered for more general equations in Lipschitz domains [272].

3. There is a fairly large class of problems which are easy to formulate in the case of homogeneous boundary conditions. Let V be a subspace in $H^1(\Omega)$ containing $\overset{\circ}{H}{}^1(\Omega) = \widetilde{H}^1(\Omega)$:

$$\overset{\circ}{H}{}^1(\Omega) \subset V \subset H^1(\Omega). \tag{11.8.5}$$

By V^* we denote the space dual to V with respect to the extension of the inner product in $L_2(\Omega)$. Consider the problem of finding a solution $u \in V$ of the equation $Lu = f$ with $f \in V^*$. Its weak setting is obvious:

$$(f,v)_\Omega = \Phi_\Omega(u,v) \quad (v \in V). \tag{11.8.6}$$

The strong coercivity of the form Φ_Ω on $H^1(\Omega)$ implies its strong coercivity on V and, therefore, the unique solvability of the problem. Obviously, the mixed problems with homogeneous boundary conditions belong to this class.

11.9 Two Classical Operator Approaches to Variational Problems

First, we consider the following abstract situation.[1] Suppose given three Hilbert spaces, H_{-1}, H_0, and H_1, with compact dense embeddings

$$H_{-1} \supset H_0 \supset H_1. \tag{11.9.1}$$

---

[1] We ask the reader to bear in mind that the subscripts have a different meaning in other subsections of this section.

We denote the inner products and norms on these spaces by $(\cdot,\cdot)_k$ and $\|\cdot\|_k$, $k = -1,0,1$. Suppose also that the spaces H_{-1} and H_1 are dual with respect to the extension of the form $(u,v)_0$ to the direct product of these spaces (with the generalized Schwarz inequality). Finally, suppose that on $H_1 \times H_1$ a bounded sesquilinear form $\Phi(u,v)$ is defined, which satisfies the (strong) coercivity inequality

$$\|u\|_1^2 \leq C \operatorname{Re} \Phi(u,u). \tag{11.9.2}$$

Then the relation

$$\Phi(u,v) = (A_1 u, v)_0 \quad (u,v \in H_1) \tag{11.9.3}$$

defines a bounded linear operator $A_1 \colon H_1 \to H_{-1}$, and this operator is invertible by the Lax–Milgram theorem (see Section 18.2). This is a model of the *first* approach to variational problems with homogeneous boundary conditions, which has already been considered in this section. In the case of a second-order strongly elliptic system, the space H_0 is, of course, $L_2(\Omega)$. In the Dirichlet, Neumann, and mixed problems, we have, respectively,

$$
\begin{aligned}
H_1 &= \widetilde{H}^1(\Omega), & H_{-1} &= H^{-1}(\Omega), & A_1 &= \mathcal{L}_D, \\
H_1 &= H^1(\Omega), & H_{-1} &= \widetilde{H}^{-1}(\Omega), & A_1 &= \mathcal{L}_N, \\
H_1 &= H^1(\Omega;\Gamma_1), & H_{-1} &= \widetilde{H}^{-1}(\Omega;\Gamma_1), & A_1 &= \mathcal{L}_m.
\end{aligned}
\tag{11.9.4}
$$

This will remain our main approach in the further exposition of the theory in Lipschitz domains (in Sections 12, 16, and 17).

The same form Φ determines the operator A_1^* "adjoint" to A_1 in the sense that

$$\Phi(u,v) = (u, A_1^* v)_0 \quad (u,v \in H_1), \tag{11.9.5}$$

or, equivalently,

$$\overline{\Phi(v,u)} = (A_1^* u, v)_0 \quad (u,v \in H_1). \tag{11.9.6}$$

Let Λ be the least closed angle (sector) with bisector $\mathbb{R}_+$ containing all values of the form $\Phi(u,u)$. It follows from (11.9.2) that the imaginary part of this form is bounded by the real one; therefore, the opening of this sector is less than π. The operator A_1 on H_{-1} has domain H_1 and discrete spectrum, which is contained in Λ; moreover, some neighborhood of the origin contains no eigenvalues. The same is true for the operator A_1^*. The generalized eigenvectors lie in H_1.

Now suppose that the form $\Phi(u,v)$ is Hermitian:

$$\Phi(u,v) = \overline{\Phi(v,u)} \quad (u,v \in H_1). \tag{11.9.7}$$

In this case, the sector Λ is merely the ray $\mathbb{R}_+$. We redenote the form by $\Psi(u,v)$ and the corresponding operator by S_1. It is invertible. We have $(S_1 u, v)_0 = (u, S_1 v)_0$. The Hermitian form

$$\langle u,v \rangle_1 = \Psi(u,v) = (S_1 u, v)_0 = (u, S_1 v)_0 \tag{11.9.8}$$

satisfies the coercivity condition

$$\|u\|_1^2 \le C\langle u, u \rangle_1 \qquad (11.9.9)$$

on H_1, and we can take it for an inner product in H_1. The corresponding norm is equivalent to the initial norm on H_1. Replacing u by $S_1^{-1}u$ and v by $S_1^{-1}v$, we obtain a new inner product in H_{-1}:

$$\langle u, v \rangle_{-1} = (S_1^{-1}u, v)_0 = (u, S_1^{-1}v)_0. \qquad (11.9.10)$$

The corresponding norm is equivalent to the initial norm on H_{-1}, and the operator S_1 (with domain H_1) is self-adjoint in H_{-1} with respect to this inner product. We already used this in Section 11.6. The compact operator S_1^{-1} in H_{-1} is self-adjoint as well with respect to this inner product.

In the general case, since the operators A_1 and A_1^* in H_{-1} have common domain H_1, the real part S_1 of these operators can be easily introduced as their half-sum. The corresponding form $\Psi(u, v)$ is the half-sum of the forms $\Phi(u, v)$ and $\overline{\Phi(v, u)}$; for $u = v$, it is positive and satisfies inequality (11.9.9):

$$\|u\|_1^2 \le C\Psi(u, u). \qquad (11.9.11)$$

By the operator S_1 the inner products (11.9.8) and (11.9.10) are determined. In fact, fact, this operator is constructed from the form Φ.

We also want to obtain an estimate for the resolvent of the operator A_1 similar to (8.1.33).

Proposition 11.9.1. *Let Λ' be an open angle with vertex at the origin containing $\Lambda \setminus \{0\}$. Then, for $\lambda \in \mathbb{C} \setminus \Lambda'$, the solution $v(\lambda)$ of the equation $(A_1 - \lambda I)v(\lambda) = f$, $f \in H_{-1}$, satisfies the inequality*

$$\|v(\lambda)\|_1 + |\lambda| \|v(\lambda)\|_{-1} \le C_1 \|f\|_{-1} \qquad (11.9.12)$$

with a constant C_1 not depending on f and λ.

The verification of this assertion is similar to that of estimate (8.1.33).

The *second* approach to the problems under consideration consists in considering the relation

$$\Phi(u, v) = (A_2 u, v)_0 \quad (v \in H_1), \qquad (11.9.13)$$

which essentially coincides with (11.9.3), as the definition of an unbounded linear operator A_2 in H_0 with domain $H_2 = D(A_2)$ contained in H_1:

$$H_1 \supset H_2. \qquad (11.9.14)$$

In the abstract framework, the domain $D(A_2)$ of the operator A_2 is defined as follows. It consists of those $u \in H_1$ for which the antilinear functional $\Phi(u, v)$ on H_1 is continuous in the norm $\|v\|_0$:

$$|\Phi(u,v)| \le C_u\|v\|_0. \tag{11.9.15}$$

Hence it extends to a continuous antilinear functional on H_0 and can be written in the form (11.9.13). On H_2 we introduce the graph norm

$$\|u\|_2 = (\|u\|_0^2 + \|A_2u\|_0^2)^{1/2}, \tag{11.9.16}$$

which turns H_2 into a Hilbert space and A_2 into a bounded operator from H_2 to H_0.

· Note that if $A_1u = f \in H_0$, then the form $\Phi(u,v) = (f,v)_0$ is, of course, continuous in v on H_0; moreover, $A_2u = A_1u$ and $A_2^{-1}f = A_1^{-1}f$. Therefore, A_2 is the restriction of A_1 to H_2. In our particular problems these are realizations of the corresponding operator L in $L_2(\Omega)$. The spectra and the generalized eigenvectors of the operators A_1 and A_2 are the same. This follows from the fact that the eigenvectors of A_1 belong to H_0 and, hence, to H_2, because the eigenvalues are nonzero; cf. Section 11.6. Similar assertions hold for the operators A_2^* and A_1^*, as well as for the operators S_2 and S_1, the first of which is determined by the form $\Psi(u,v)$ on H_0 and is self-adjoint on this space:

$$\Psi(u,v) = (S_2u,v)_0 = (u,S_2v)_0 \quad (u,v \in D(S_2)). \tag{11.9.17}$$

The operator A_2^* is adjoint to A_2.

Proposition 11.9.2. *The operator A_2 is invertible. The same is true for A_2^* and S_2.*

Proof [235, p. 11]. It is seen from (11.9.3) and (11.9.13) that $A_2u = 0$ implies $u = 0$. Let us show that the equation $A_2u = f$ is solvable in H_2 for any $f \in H_0$ (cf. Section 18.2).

The forms $\Phi(u,v)$ and $(u,v)_1$ are two general representations of a continuous antilinear functional on H_1. Hence there exists a bounded invertible operator B in H_1 for which

$$\Phi(u,v) = (Bu,v)_1 \quad (u,v \in H_1). \tag{11.9.18}$$

If $f \in H_0$, then $(f,v)_0$ is a continuous antilinear functional on H_0 and, therefore, a continuous functional on H_1. Hence there exists a continuous operator J from H_0 to H_1 such that

$$(f,v)_0 = (Jf,v)_1 \quad (v \in H_1). \tag{11.9.19}$$

If $A_2u = f$, then

$$(Bu,v)_1 = \Phi(u,v) = (f,v)_0 = (Jf,v)_1 \quad (v \in H_1).$$

We obtain $Bu = Jf$, whence $u = B^{-1}Jf$. It is easy to verify that this relation defines the sought solution in H_1, but this solution belongs to H_2, because $f \in H_0$. □

As in the Lax–Milgram theorem, this assertion remains true under the replacement of the real part of the form on the right-hand side of (11.9.2) by its absolute value.

Corollary 11.9.3. *The space $H_2 = D(A_2)$ is dense and compactly embedded in H_1. The same is true for $D(A_2^*)$ and $D(S_2)$.*

Proof. The required embedding is implemented by the operator $\mathcal{E}_{2,1} = S_1^{-1}\mathcal{E}_{0,-1}S_2$, where $\mathcal{E}_{0,-1}$ is the operator of embedding of H_0 into H_{-1}. The latter operator is compact, as well as S_1. Therefore, $\mathcal{E}_{2,1}$ is compact.

Next, if $u_0 \in H_1$, then $v_0 = S_1u_0$ can be approximated in H_{-1} by elements $v_k \in H_0$. The $w_k = S_2^{-1}v_k$ belong to H_2 and converge to u_0 in H_1. □

See also [202, Chap. VI, Sec. 2].

For the operator A_2, the following analogue of Proposition 11.9.1 is also valid.

Proposition 11.9.4. *For the solutions* $v(\lambda) \in H_2$ *of the equation*

$$(A_2 - \lambda I)v(\lambda) = f, \quad f \in H_0,$$

with $\lambda \in \mathbb{C} \setminus \Lambda'$, *the estimate*

$$\|v(\lambda)\|_2 + |\lambda|\|v(\lambda)\|_0 \le C\|f\|_0 \qquad (11.9.20)$$

with a constant not depending on f *and* λ *holds. The same is true for* A_2^* *and* S_2 *instead of* A_2.

Proof. First, suppose that λ lies in an angle of opening less than π with bisector $\mathbb{R}_-$. Then

$$\operatorname{Re}\Phi(v(\lambda), v(\lambda)) - \operatorname{Re}\lambda(v(\lambda), v(\lambda)) = \operatorname{Re}(f, v(\lambda)),$$

and the first term on the left-hand side is nonnegative. We obtain

$$|\operatorname{Re}\lambda|\|v(\lambda)\|_0 \le \|f\|_0,$$

which gives the required estimate of the second term on the left-hand side in (11.9.20), since $|\operatorname{Im}\lambda|$ is estimated by $|\operatorname{Re}\lambda|$. This estimate is extended to $\lambda \in \mathbb{C} \setminus \Lambda'$ by passing from Φ to the forms $e^{i\theta}\Phi$ with small $|\theta|$. To estimate the first term on the left-hand side in (11.9.20), we must estimate $\|v(\lambda)\|_0$ and $\|A_2v(\lambda)\|_0$. The former norm is in fact already estimated, and for the latter, we have $A_2v(\lambda) = f + \lambda v(\lambda)$; the second summand on the right-hand side has just been estimated. □

The second approach to problems for strongly elliptic systems is generally known as well; it was presented, e.g., in the books [235] and [202, Chap. VI, Sec. 2] and considered by many authors. Of course, of special importance is a maximally complete description of the space H_2. For second-order equations with smooth coefficients in a smooth domain, such a description is obtained, under the ellipticity conditions, from results of the general theory which we presented in Section 7. This is the subspace of $H^2(\Omega)$ determined by homogeneous boundary conditions. Clearly, in the case of, e.g., the Neumann problem, the domains of the operators A_2 and A_2^* may be different.

Both approaches can be generalized to the situation where the problem is initially Fredholm and reduces to a uniquely solvable one by adding the term μu with sufficiently large positive μ to L. Both approaches can also be extended to problems for higher-order systems.

The first approach is convenient in that the spaces are given in advance and "adequate" to the problem under examination, as in "smooth" problems. For the second approach, generally, this is so only in the smooth case. But this approach is convenient in that the operator A_2 always acts in $L_2(\Omega)$.

The operators A_1 and A_2 have discrete spectrum, because their domains are compactly embedded in H_{-1} and H_0, respectively.

In Section 16.6 we shall discuss important facts related to fractional powers of the operators A_1 and A_2.

12 Potential Operators and Transmission Problems

12.1 A System on the Torus

It is convenient to expose the material of this section, which is very important for the theory of strongly elliptic systems in Lipschitz domains, by considering a system in two adjacent domains, inner and outer. But in the unbounded outer domain, we must consider conditions at infinity. To simplify exposition, we assume that the inner domain, $\Omega = \Omega^+$, is contained in the standard torus $\mathbb{T} = \mathbb{T}^n$ with 2π-periodic coordinates $x = (x_1, \ldots, x_n)$ and separated by a connected Lipschitz boundary Γ from the outer domain Ω^-, which complements the closure $\overline{\Omega^+}$ to the torus, so that

$$\mathbb{T} = \Omega^+ \cup \Gamma \cup \Omega^-.$$

We assume that the normal ν (at those points of Γ where it exists) is directed to Ω^-, so that it is inner for Ω^-. This should be taken into account in relations for Ω^-. We shall denote the conormal derivative of a function u on Γ on the side of Ω^- by T^-u. If the conormal derivative is smooth, then it is formed by using the inner normal vector with respect to Ω^-. In the case of the unique solvability of the Dirichlet and Neumann problems in $\Omega^\pm$, we now have four Poincaré–Steklov operators:

$$N^\pm T^\pm u = \pm u^\pm \quad \text{and} \quad D^\pm u^\pm = \pm T^\pm u. \tag{12.1.1}$$

We shall obtain, in particular, a series of formulas whose analogues for classical equations are well known.

Note that $\widetilde{H}^s(\Omega^\pm)$ are now subspaces in $H^s(\mathbb{T})$ (consisting of elements supported on $\overline{\Omega^\pm}$).

Here we assume system (11.1.1) to be given and strongly elliptic on the torus. In the general case, the coefficients $a_{j,k}$ and b_j can be assumed to be Lipschitz, but this assumption can often be weakened. *We assume that the form $\Phi_{\mathbb{T}}(u,v)$ is strongly coercive on $H^1(\mathbb{T})$.* This implies the strong coercivity of the forms $\Phi_{\Omega^\pm}(u,v)$ on $\widetilde{H}^1(\Omega^\pm)$ (as on subspaces) and, therefore, the unique solvability of the Dirichlet problems in $\Omega^\pm$. When needed, we assume the strong coercivity of these forms on $H^1(\Omega^\pm)$, to ensure also the unique solvability of the Neumann problems. Our assumptions hold automatically for the formally adjoint system, because the corresponding forms are the same. As a rule, if a system is given only in Ω^+ and satisfies the required conditions there, it can be extended to the torus so that the required conditions are satisfied on the torus and on Ω^-. We do not comment on what "as a rule" means.

An equation (system) $Lu = f$ on the torus is an equation on a smooth manifold, although its coefficients may be not very smooth. There arise no "Lipschitz difficulties" with this equation. Our smoothness assumptions on the coefficients make it possible to regard L as an operator from $H^{1+\sigma}(\mathbb{T})$ to $H^{-1+\sigma}(\mathbb{T})$ with $|\sigma| \leq 1$. Moreover, the variational (weak) definition

$$(Lu, v)_{\mathbb{T}} = \Phi_{\mathbb{T}}(u, v) \quad (u \in H^{1+\sigma}(\mathbb{T}), \ v \in H^{1-\sigma}(\mathbb{T})) \tag{12.1.2}$$

of a solution is equivalent to the usual definition

$$Lu = f, \tag{12.1.3}$$

at least in the sense of distribution theory. Indeed, when both sides of (12.1.3) act on a test function $v \in C^\infty(\mathbb{T})$, the equality $(Lu, v)_{\mathbb{T}} = \Phi_{\mathbb{T}}(u, v)$ is obtained by transferring the derivatives ∂_j to v, after which it can be extended to $v \in H^{1-\sigma}(\Omega)$ by passing to the limit. Conversely, we can pass from (12.1.2) to (12.1.3) by taking C^∞ functions for v.

If the coefficients of L are sufficiently smooth, then a generalization of the assertion about the invertibility of the operator L can be obtained by using results of Section 6. However, low smoothness of the coefficients is sufficient.

Theorem 12.1.1. *The above assumptions on the coefficients and strong coercivity on the torus imply the invertibility of the operator* $L: H^{1+\sigma}(\mathbb{T}) \to H^{-1+\sigma}(\mathbb{T})$ *for* $|\sigma| \leq 1$.

In this section we prove this theorem for $\sigma = 1$ by employing a left parametrix construction (cf. Section 6). Interestingly, this construction uses Fourier series instead of Fourier transforms. The proof will be completed in Section 17.2 by means of interpolation (although the consideration of all σ at once is not much more difficult); in the same section, a generalization of this theorem to other spaces will also be given.

Let us make some preliminary remarks.

As we know, there are two equivalent systems of norms on the spaces $H^s(\mathbb{T})$. The first system consists of norms defined by using a partition of unity on the torus and norms on $\mathbb{R}^n$; the corresponding local coordinates can be assumed to be unique (periodic). The second system consists of norms determined by using Fourier series; these norms were mentioned in Section 2.2:

$$\|u\|^2_{H^s(\mathbb{T})} = \sum |c_\alpha(u)|^2 (1 + |\alpha|^2)^s, \tag{12.1.4}$$

where $c_\alpha(u)$ are the Fourier coefficients of the function u. Accordingly, we shall refer to these norms as "the first norms" and "the second norms."

Consider the operator Φ of multiplication by a function $\varphi(x)$ in $C^{0,1}(\mathbb{T})$. Using the first norms, we obtain

$$\|\varphi u\|_{H^0(\mathbb{T})} \le \sup|\varphi(x)|\|u\|_{H^0(\mathbb{T})}. \tag{12.1.5}$$

Here the norm of Φ is small if so is the least upper bound of the absolute value of the function. But

$$\|\varphi u\|_{H^1(\mathbb{T})} \le \sup|\varphi(x)|\|u\|_{H^1(\mathbb{T})} + K\|u\|_{H^0(\mathbb{T})}. \tag{12.1.6}$$

Here the constant K is generally not small. Instead, we can apply a trick used in Section 1.9 and represent this operator as the sum of an operator with small norm on $H^1(\mathbb{T})$ and a bounded operator from $H^0(\mathbb{T})$ to $H^1(\mathbb{T})$. We shall need this in what follows.

For this purpose, we introduce the operator

$$\Theta_N u(x) = \sum_{|\alpha| \ge N} c_\alpha(u)e^{i\alpha \cdot x}. \tag{12.1.7}$$

The second norms of the operators Θ_N and $I - \Theta_N$ equal 1 on any $H^s(\mathbb{T})$. The operator $\Phi' = \Phi(I - \Theta_N)$ acts boundedly from $H^0(\mathbb{T})$ to $H^1(\mathbb{T})$. For the operator $\Phi'' = \Phi\Theta_N$, we obtain the estimate

$$\|\Phi'' u\|_{H^1(\mathbb{T})} \le [\sup|\varphi(x)| + KN^{-1}]\|u\|_{H^1(\mathbb{T})}; \tag{12.1.8}$$

here the constant on the right-hand side is small for large N if $\sup|\varphi(x)|$ is small.

Proof of Theorem 12.1.1 for $\sigma = 1$. Let us write the operator L in the form

$$L = a(x, D) + L_1, \quad \text{where } a(x, D) = \sum D_j a_{j,k}(x)D_k \tag{12.1.9}$$

and L_1 is a first-order operator, so it acts boundedly from $H^1(\mathbb{T})$ to $H^0(\mathbb{T})$.

Take a small number $\eta > 0$ (later on we shall see how small it must be) and consider a C^∞ partition of unity

$$\sum_1^M \varphi_l(x) \equiv 1 \tag{12.1.10}$$

on the torus, fine enough for all coefficients and all l to satisfy the inequality

$$\|a_{j,k}(x) - a_{j,k}(y)\| \le \eta \quad \text{for } x, y \in \text{supp}\,\varphi_l \tag{12.1.11}$$

(as the norm of a matrix we use, say, the square root of the sum of the squared moduli of its elements). Fix a point x_l in the support of each function φ_l. Consider the operator $a(x_l, D)$ obtained from $a(x, D)$ by freezing the coefficients at x_l:

$$a(x_l, D) = \sum D_j a_{j,k}(x_l)D_k. \tag{12.1.12}$$

Here it is not necessary to write derivatives on both sides of the coefficients. This operator can be expressed as the Fourier-type series

$$a(x_l, D)u(x) = \sum_{0 \neq \alpha \in \mathbb{Z}^n} a(x_l, \alpha)c_\alpha(u)e^{i\alpha \cdot x}, \qquad (12.1.13)$$

where the $c_\alpha(u)$ are the Fourier coefficients of $u \in H^2(\mathbb{T})$. By virtue of ellipticity, all coefficients $a(x_l, \alpha)$ are nonzero except the zeroth one. To construct a left parametrix, we introduce the operators

$$b(x_l, D)f(x) = \sum_{0 \neq \alpha \in \mathbb{Z}^n} a^{-1}(x_l, \alpha)c_\alpha(f)e^{i\alpha \cdot x} \qquad (12.1.14)$$

($f \in H^0(\mathbb{T})$) and

$$b(x, D) = \sum_{1}^{M} \varphi_l(x)b(x_l, D). \qquad (12.1.15)$$

Note that these are bounded operators from $H^0(\mathbb{T})$ to $H^2(\mathbb{T})$, and the norm of the operator (12.1.14) is bounded by a constant not depending on the choice of x_l.

We have

$$b(x, D)a(x, D) = \sum_{1}^{M} \varphi_l(x)b(x_l, D)a(x, D)$$

$$= \sum_{1}^{M} \varphi_l(x)b(x_l, D)a(x_l, D) + \sum_{1}^{M} \varphi_l(x)b(x_l, D)[a(x, D) - a(x_l, D)].$$

The first sum on the right-hand side differs from the identity operator by a one-dimensional operator. Consider the second sum. In each term of the sum over l, j, and k (see (12.1.12)), we transpose the operator of multiplication by the smooth function φ_l and $b(x_l, D)D_j$, which is a pseudodifferential operator of order -1 on the torus with symbol not depending on x. Note that the commutator of these operators is an operator of order -2 (because φ_l is a scalar function; see Section 18.4). Thus,

$$b(x, D)a(x, D) = I + \sum_{l,j,k} b(x_l, D)D_j a_{j,k,l}(x)D_k + T_1, \qquad (12.1.16)$$

where

$$a_{j,k,l}(x) = \varphi_l(x)[a_{j,k}(x) - a_{j,k}(x_l)] \qquad (12.1.17)$$

and T_1 is a bounded operator from $H^1(\mathbb{T})$ to $H^2(\mathbb{T})$. Let $A_{j,k,l}$ denote the operator of multiplication by the function (12.1.17). This operator has small norm $\sup |a_{j,k,l}(x)|$ in $H^0(\mathbb{T})$. Let us replace it by $A_{j,k,l}\Theta_N + A_{j,k,l}[I - \Theta_N]$ with large N.

It is easy to see that we obtain

$$b(x, D)a(x, D) = I + T_2 + T_3,$$

where T_2 is an operator with small norm in $H^1(\mathbb{T})$ and $H^2(\mathbb{T})$ (we can assume that this norm is smaller than 1) and T_3 is a bounded operator from $H^1(\mathbb{T})$ to $H^2(\mathbb{T})$.

Now suppose that $f \in H^0(\mathbb{T})$. A solution u of the equation $Lu = f$ surely exists and is unique in $H^1(\mathbb{T})$. It has the form

$$u = (I + T_2)^{-1}[b(x, D)(f - L_1 u) - T_3 u].$$

We conclude that $u \in H^2(\mathbb{T})$. This smoothness result proves the existence of the required solution. Uniqueness follows from uniqueness in $H^1(\mathbb{T})$. $\square$

The operator inverse to L is integral:

$$(L^{-1}f)(x) = \int_{\mathbb{T}} E(x, y)f(y)\, dy. \tag{12.1.18}$$

The kernel $E(x, y)$ of this integral operator is a fundamental solution for L:

$$L_x E(x, y) = \delta(y).$$

If the function f is given only on $\Omega^\pm$, then this is a *Newton*, or *volume*, potential in $\Omega^\pm$.

12.2 Definition of Single- and Double-Layer Potentials

As in the scalar case, a *single-layer potential* for functions in, say, $H^0(\Gamma)$ is usually defined by

$$u = \mathcal{A}\psi(x) = \int_{\Gamma} E(x, y)\psi(y)\, dS_y \quad (x \in \mathbb{T}). \tag{12.2.1}$$

This is an integral with a weak (integrable) singularity in the kernel. But we largely use it for $\psi \in H^{-1/2}(\Gamma)$. In what follows, we give a more general definition of this operator.

A *double-layer potential* on a sufficiently smooth boundary is defined, in the general case, by

$$u = \mathcal{B}\varphi(x) = \int_{\Gamma} [\widetilde{T}_y^+ E^*(x, y)]^* \varphi(y) dS_y \quad (x \notin \Gamma) \tag{12.2.2}$$

(cf. [258, p. 202]). The asterisk denotes the passage to the Hermitian conjugate matrix. In particular,

$$\widetilde{L}_y E^*(x, y) = \delta(x). \tag{12.2.3}$$

In the cases of $L = -\Delta$ or $-\Delta + I$, the asterisks and the tilde are not needed. The function φ is assumed to belong to, at least, $H^0(\Gamma)$. But under the passage to the boundary ($x \in \Gamma$), in the general case, an integral with a critical singularity arises, which must be treated as a singular integral operator (see (18.4.8)). This is also possible in the case of a Lipschitz surface but involves analytical difficulties. The same difficulties arise in considering the first derivatives of the single-layer potential. A brief discussion of the corresponding very important approach is contained in Section 19. Moreover, as already mentioned, we need more general functions ψ.

These difficulties can be obviated by using a different approach, which we present and follow here. In particular, we give a generalization of the definition of the operator (12.2.2) convenient in the case $\varphi \in H^{1/2}(\Gamma)$.

This approach is based on the assumption that the Dirichlet problem is uniquely solvable. It was proposed by Costabel [99], who relied on Nečhas' result on the regularity of solutions to the Dirichlet problem [286], and systematically exposed by McLean in the book [258]. A development of this method based on the unique solvability of the Dirichlet and Neumann problems was undertaken by the author. Here we reproduce and complete it in the framework needed now.

Let γ denote the operator of taking the trace on Γ of a function on the torus. The adjoint operator γ^* is naturally defined as

$$(\gamma^*\psi, v)_{\mathbb{T}} = (\psi, \gamma v)_\Gamma. \tag{12.2.4}$$

Since γ is a bounded operator from $H^{1+\sigma}(\mathbb{T})$ to $H^{1/2+\sigma}(\Gamma)$ for $|\sigma| < 1/2$, it follows that γ^* is a bounded operator from $H^{-1/2+s}(\Gamma)$ to $H^{-1+s}(\mathbb{T})$ for $|s| < 1/2$. Following [99], we define the operator $\mathcal{A}$ by the formula

$$\mathcal{A}\psi = L^{-1}\gamma^*\psi. \tag{12.2.5}$$

We automatically obtain the following result.

Proposition 12.2.1. *The operator $\mathcal{A}$ acts boundedly from $H^{-1/2+s}(\Gamma)$ to $H^{1+s}(\mathbb{T})$ and, therefore, to $H^{1+s}(\Omega^\pm)$ for $|s| < 1/2$.*

In particular, this is a bounded operator from $H^{-1/2}(\Gamma)$ to $H^1(\mathbb{T})$. On the other hand, a comparison of (12.2.1) and (12.2.5) shows that these relations define one and the same operator for $\psi \in H^0(\Gamma)$. Indeed, let $g(x)$ be a smooth test function with support lying inside Ω^+ or Ω^-. Then

$$\int_{\mathbb{T}}\int_{\mathbb{T}} E(x,y)(\gamma^*\psi)(y)\,dy \cdot \overline{g(x)}\,dx = \int_{\mathbb{T}}(\gamma^*\psi)(y) \cdot \int_{\mathbb{T}}\overline{E^*(x,y)g(x)}\,dx\,dy$$

$$= \int_\Gamma \psi(y) \cdot \gamma_y \int_{\mathbb{T}}\overline{E^*(x,y)g(x)}\,dx\,dS_y = \int_{\mathbb{T}}\int_\Gamma (\gamma_y E^*(x,y))^*\psi(y)dS_y \cdot \overline{g(x)}\,dx.$$

Here the two stars "cancel" each other.

It is seen from definition (12.2.5) that the function $u = \mathcal{A}\psi$ with $\psi \in H^{-1/2+s}(\Gamma)$, $|s| < 1/2$, satisfies the homogeneous equation $Lu = 0$ outside Γ. On the torus it satisfies the equation $Lu = \gamma^*\psi$ with right-hand side supported on Γ. This is also seen from (12.2.5).

Since the function $\mathcal{A}\psi$ belongs to $H^{1+s}(\mathbb{T})$, $|s| < 1/2$, it has the same boundary values $\gamma^{\pm}\mathcal{A}\psi$. We define the operator

$$A\psi = \gamma^{\pm}\mathcal{A}\psi. \tag{12.2.6}$$

It acts boundedly from $H^{-1/2+s}(\Gamma)$ *to* $H^{1/2+s}(\Gamma)$. It is also clear that $T^{\pm}\mathcal{A}$ are *bounded operators in* $H^{-1/2+s}(\Gamma)$, $|s| < 1/2$.

Now we want to define the operator $\mathcal{B}$ in similar terms. This is a little more complicated.

Note that the expression

$$Tv = T^{\pm}v = \sum \nu_j(x)a_{j,k}(x)\gamma\partial_k v(x) \tag{12.2.7}$$

well defines a two-sided smooth conormal derivative for functions $v \in H^{2-s}(\mathbb{T})$, $0 < s < 1/2$. Indeed, differentiation and passage to the boundary decrease the index $2-s$ by $3/2$, the traces $\gamma^{\pm}\partial_k v(x)$ coincide, and T is a bounded operator from $H^{2-s}(\mathbb{T})$ to $H^{1/2-s}(\Gamma)$.

We define the adjoint operator by

$$(T^*\varphi, v)_{\mathbb{T}} = (\varphi, Tv)_{\Gamma}. \tag{12.2.8}$$

Here it can be assumed that $\varphi \in H^{s-1/2}(\Gamma)$, $v \in H^{2-s}(\mathbb{T})$, and $T^*\varphi \in H^{s-2}(\mathbb{T})$, $0 < s < 1/2$.

We need a similar operator $(\widetilde{T})^*$ for the formally adjoint system. It is introduced in a similar way.

Now we define the operator $\mathcal{B}$ by

$$\mathcal{B} = L^{-1}(\widetilde{T})^*. \tag{12.2.9}$$

Let us outline the connection of this formula with (12.2.2). We have, with the same $g(x)$,

$$\int_{\mathbb{T}}\int_{\mathbb{T}} E(x,y)(\widetilde{T}^*\varphi)(y)\,dy \cdot \overline{g(x)}dx = \int_{\mathbb{T}}(\widetilde{T}^*\varphi)(y) \cdot \int_{\mathbb{T}}\overline{E^*(x,y)g(x)}\,dx\,dy$$

$$= \int_{\Gamma}\varphi(y) \cdot \overline{\widetilde{T}_y}\int_{\mathbb{T}}\overline{E^*(x,y)g(x)}\,dx\,dS_y = \int_{\mathbb{T}}\int_{\Gamma}(\widetilde{T}_y E^*(x,y))^*\varphi(y)dS_y \cdot \overline{g(x)}dx.$$

At the moment, we know that (12.2.9) is a bounded operator from $H^{s-1/2}(\Gamma)$ to $H^s(\mathbb{T})$ for $0 < s < 1/2$. First, we must extend it to a bounded operator from $H^{1/2}(\Gamma)$ to $H^1(\Omega^{\pm})$.

12.3 Solution Representation and Its Consequences.
The Transmission Problem

To achieve the goal set forth at the end of the preceding section, we first derive a *representation of a solution* in terms of the jumps of this solution and its conormal derivative (see expression (12.3.4) below), following [99] and [258]. This representation is important by itself, and we repeatedly use it in what follows.

We introduce the following notation for jumps:

$$[u]_\Gamma = u^- - u^+, \quad [Tu]_\Gamma = T^-u - T^+u. \tag{12.3.1}$$

Theorem 12.3.1. *Let u be a function on the torus for which*

$$u|_{\Omega^\pm} \in H^1(\Omega^\pm) \quad and \quad Lu|_{\Omega^\pm} = f^\pm \in \widetilde{H}^{-1}(\Omega^\pm),$$

so that $f = f^+ + f^-$ belongs to $H^{-1}(\mathbb{T})$. Then u treated as an element of $H^s(\mathbb{T})$, $0 < s < 1/2$, and $v \in H^{2-s}(\mathbb{T})$ satisfy the relation

$$(Lu, v)_\mathbb{T} = (f, v)_\mathbb{T} + ([u]_\Gamma, \widetilde{T}v)_\Gamma - ([Tu]_\Gamma, \gamma v)_\Gamma. \tag{12.3.2}$$

We prove this theorem a little later. We emphasize that u, v, and $f^\pm$ are assumed to be given, so that the conormal derivatives $T^\pm u$ are uniquely determined by the Green identities. The conormal derivatives of v are smooth. In particular, Theorem 12.3.1 applies when $f = 0$.

Corollary 12.3.2. *Under the assumptions of Theorem 12.3.1,*

$$Lu = f + (\widetilde{T})^*[u]_\Gamma - \gamma^*[Tu]_\Gamma \tag{12.3.3}$$

and

$$u = L^{-1}f + \mathcal{B}[u]_\Gamma - \mathcal{A}[Tu]_\Gamma. \tag{12.3.4}$$

Proof of Theorem 12.3.1. Let us write the four Green identities for the operators L and $\widetilde{L}$ and the domains Ω^+ and Ω^-:

$$(Lu, v)_{\Omega^\pm} = \Phi_{\Omega^\pm}(u, v) \mp (T^\pm u, v^\pm)_\Gamma, \tag{12.3.5}$$

$$(u, \widetilde{L}v)_{\Omega^\pm} = \Phi_{\Omega^\pm}(u, v) \mp (u^\pm, \widetilde{T}^\pm v)_\Gamma. \tag{12.3.6}$$

We have

$$(Lu, v)_\mathbb{T} = (u, \widetilde{L}v)_\mathbb{T} = (u, \widetilde{L}v)_{\Omega^+} + (u, \widetilde{L}v)_{\Omega^-}. \tag{12.3.7}$$

Into (12.3.7) we substitute the expressions for the terms on the right-hand side taken from (12.3.6) and the expression for $\Phi_{\Omega^\pm}(u, v)$ taken from (12.3.5). As is easy to see, we obtain (12.3.2). □

Representation (12.3.4) resembles a representation of, say, harmonic functions from textbooks on mathematical physics. But this is a fairly delicate representation.

We first consider u as a distribution on the torus and v as a test function. Relation (12.3.3) resembles the formula for the derivative in the sense of distributions for a piecewise smooth function on the line (see, e.g., [18] or [165]). Then we use the assumption $u \in H^1(\Omega^\pm)$, thanks to which $[u]_\Gamma \in H^{1/2}(\Gamma)$ and $[Tu]_\Gamma \in H^{-1/2}(\Gamma)$.

Representation (12.3.4) implies the following assertion for solutions of the homogeneous system in $\Omega^\pm$ extended by zero to the complementary domain $\Omega^\mp$.

Corollary 12.3.3. *The solutions of the homogeneous system $Lu = 0$ in $\Omega^\pm$ belonging to $H^1(\Omega^\pm)$ satisfy the relations*

$$-\mathcal{B}u^+ + \mathcal{A}T^+u = \begin{cases} u \ in \ \Omega^+, \\ 0 \ in \ \Omega^-; \end{cases} \tag{12.3.8}$$

$$\mathcal{B}u^- - \mathcal{A}T^-u = \begin{cases} u \ in \ \Omega^-, \\ 0 \ in \ \Omega^+. \end{cases} \tag{12.3.9}$$

Of course, the inhomogeneous system in Ω^+ or Ω^- can also be considered; in this case, the term $L^{-1}f$ should be added to the left-hand side.

We remind the reader that the form $\Phi_{\mathbb{T}}$ is assumed to be strongly coercive on $H^1(\mathbb{T})$, so that the form $\Phi_{\Omega^\pm}$ is strongly coercive on $\widetilde{H}^1(\Omega^\pm)$.

Proposition 12.3.4. *The operator $\mathcal{B}$ can be extended to a bounded operator from $H^{1/2}(\Gamma)$ to $H^1(\Omega^\pm)$.*

Proof. Consider, e.g., Ω^+. Since the Dirichlet problem in Ω^+ is uniquely solvable, for u^+ in (12.3.8) we can take any function in $H^{1/2}(\Gamma)$. The remaining terms in this formula belong to $H^1(\Omega^+)$ (see, in particular, Proposition 12.2.1). Thus, we can define the operator $\mathcal{B}$ on $\Omega^\pm$ by (12.3.8)–(12.3.9). $\qquad\square$

Now we can treat representation (12.3.4) as an equality in $H^1(\Omega^\pm)$.

Remark 12.3.5. Since we plan to discuss a generalization of the results presented below to more general spaces in Sections 16 and 17, we point out that, up to the end of Section 12.7, we in fact use only the unique solvability of the Dirichlet and, later, Dirichlet and Neumann problems rather than strong coercivity. The only exceptions are Theorems 12.5.1 and 12.5.2.

For $\varphi \in H^{1/2}(\Gamma)$, the function $\mathcal{B}\varphi$ satisfies the homogeneous equation $Lu = 0$ in $\Omega^\pm$. On the torus it satisfies an equation with right-hand side supported on Γ.

Since $\mathcal{B}$ is a bounded operator from $H^{1/2}(\Gamma)$ to $H^1(\Omega^\pm)$, it follows that $\gamma^\pm \mathcal{B}$ are bounded operators in $H^{1/2}(\Gamma)$ and $T^\pm \mathcal{B}$ are bounded operators from $H^{1/2}(\Gamma)$ to $H^{-1/2}(\Gamma)$.

Proposition 12.3.6. *If $\varphi \in H^{1/2}(\Gamma)$ and $\psi \in H^{-1/2}(\Gamma)$, then the following relations for jumps hold:*

$$[\mathcal{A}\psi]_\Gamma = 0, \quad [T\mathcal{A}\psi]_\Gamma = -\psi, \quad [\mathcal{B}\varphi]_\Gamma = \varphi, \quad [T\mathcal{B}\varphi]_\Gamma = 0. \tag{12.3.10}$$

Proof. We already know the first relation in (12.3.10): see (12.2.6). Applying it, we obtain the third relation from (12.3.4) with $f = 0$ and $[u]_\Gamma = \varphi$. Here we use arbitrariness in the choice of $[u]_\Gamma$.

Now let $u = \mathcal{A}\psi$. Then $Lu = \gamma^*\psi$ on the torus by the definition of the operator $\mathcal{A}$. On the other hand, the relations (12.3.3) and $[u]_\Gamma = 0$ imply $Lu = -\gamma^*[T\mathcal{A}\psi]_\Gamma$. It is easy to show that the kernel of γ^* (see (12.2.4)) is trivial. This gives the second relation in (12.3.10).

Now, applying this relation and (12.3.4) with $f = 0$, we obtain

$$[Tu]_\Gamma = [T\mathcal{B}\varphi]_\Gamma - [T\mathcal{A}[Tu]_\Gamma]_\Gamma = [T\mathcal{B}\varphi]_\Gamma + [Tu]_\Gamma,$$

which implies the fourth relation in (12.3.10).   □

Now we introduce the operator

$$H = -T^\pm\mathcal{B}. \tag{12.3.11}$$

This is the so-called *hypersingular operator*. It acts boundedly from $H^{1/2}(\Gamma)$ to $H^{-1/2}(\Gamma)$.

Now consider the following *transmission problem*:

$$Lu = 0 \text{ in } \Omega^\pm, \quad [u]_\Gamma = \varphi, \quad [Tu]_\Gamma = \psi. \tag{12.3.12}$$

Theorem 12.3.7. *For $\varphi \in H^{1/2}(\Gamma)$ and $\psi \in H^{-1/2}(\Gamma)$, problem (12.3.12) has precisely one solution belonging to H^1 on $\Omega^\pm$. It is expressed as*

$$u = \mathcal{B}\varphi - \mathcal{A}\psi. \tag{12.3.13}$$

Proof. If $[u]_\Gamma$ and $[Tu]_\Gamma$ are zero, then it follows from (12.3.4) that the solution is zero. On the other hand, defining a solution by (12.3.13), we see from the jump relations (12.3.10) that $[u]_\Gamma = \varphi$ and $[Tu]_\Gamma = \psi$, so that (12.3.13) is the required solution.   □

This result can be extended to the case of the inhomogeneous equation by using the unique solvability of the system $Lu = f$ on the torus.

Remark. Note that the unique solvability of the Neumann problems have not been assumed in this subsection.

12.4 Operators on Γ and the Calderón Projections

Now, following [258], we set

$$B = \frac{1}{2}(\gamma^-\mathcal{B} + \gamma^+\mathcal{B}) \quad \text{and} \quad \widehat{B} = \frac{1}{2}(T^-\mathcal{A} + T^+\mathcal{A}). \tag{12.4.1}$$

We *call B* the *direct value of the double-layer potential*. This is a bounded operator in $H^{1/2}(\Gamma)$, and $\widehat{B}$ is a bounded operator in $H^{-1/2}(\Gamma)$.

These definitions and the jump relations imply the following relations for the boundary values of the potentials and their conormal derivatives:

$$T^{\pm}\mathcal{A}\psi = \pm\frac{1}{2}\psi + \widehat{B}\psi \quad \text{and} \quad \gamma^{\pm}\mathcal{B}\varphi = \mp\frac{1}{2}\varphi + B\varphi. \tag{12.4.2}$$

Next, (12.3.8) and (12.3.9) imply

$$u^{\pm} = \mp Bu^{\pm} + \frac{1}{2}u^{\pm} \pm AT^{\pm}u, \tag{12.4.3}$$

$$T^{\pm}u = \pm Hu^{\pm} \pm \widehat{B}T^{\pm}u + \frac{1}{2}T^{\pm}u. \tag{12.4.4}$$

Here we have managed without analytical calculation of the limits for the solution on the boundary and computation of its conormal derivatives. But again, in the case of additional smoothness, these relations agree with the known ones.

We set

$$P^{+} = \begin{pmatrix} \frac{1}{2}I - B & A \\ H & \frac{1}{2}I + \widehat{B} \end{pmatrix} \quad \text{and} \quad P^{-} = \begin{pmatrix} \frac{1}{2}I + B & -A \\ -H & \frac{1}{2}I - \widehat{B} \end{pmatrix}. \tag{12.4.5}$$

Let X denote the direct product of the spaces $H^{1/2}(\Gamma)$ and $H^{-1/2}(\Gamma)$. It contains two subspaces X^{+} and X^{-}: the former consists of the columns Q^{+} of the Cauchy data (Dirichlet and Neumann) for the system $Lu = 0$ in Ω^{+} and the latter, of the columns Q^{-} of the Cauchy data for this system in Ω^{-}. Under our assumption of the unique solvability of the Dirichlet problems, the elements $(\varphi, \psi)'$ of each of these subspaces are parametrized by their first components: $u^{\pm} = \varphi$ can be chosen arbitrarily, and $T^{\pm}u = \psi$ is determined uniquely. Uniqueness for the two Dirichlet problems implies uniqueness for the two Cauchy problems.

It is easy to derive from (12.2.5), (12.3.11), and (12.4.1) that the operators P^{-} and P^{+} are bounded on X and take each column $(\varphi, \psi)'$ in X to the columns of the Cauchy data of the solution (12.3.13) in Ω^{-} and the solution with the minus sign in Ω^{+}, respectively:

$$P^{-}(\varphi,\psi)' = (u^{-}, T^{-}u)', \quad P^{+}(\varphi,\psi)' = -(u^{+}, T^{+}u)'.$$

It is seen from formulas (12.4.3) and (12.4.4) that P^{-} leaves the columns $(u^{-}, T^{-}u)'$ fixed and annihilates the columns $(u^{+}, T^{+}u)'$, while P^{+} leaves the columns $(u^{+}, T^{+}u)'$ fixed and annihilates the columns $(u^{-}, T^{-}u)'$. Therefore,

$$(P^{\pm})^{2} = P^{\pm}, \quad P^{+}P^{-} = P^{-}P^{+} = 0. \tag{12.4.6}$$

Moreover, the sum $P^{+} + P^{-}$ equals the identity operator.

We see that X is *the direct sum of the subspaces* X^{+} *and* X^{-}, *and* $P^{\pm}$ *are complementary projections onto these subspaces*. These projections are called the *Calderón projections*: such operators were first introduced in the general theory of elliptic problems by Calderón [82].

It is easy to show that each of the relations in (12.4.6) is equivalent to the relations

$$BA = A\widehat{B}, \quad HB = \widehat{B}H, \quad \frac{1}{4}I - B^2 = AH, \quad \frac{1}{4}I - (\widehat{B})^2 = HA; \quad (12.4.7)$$

cf. [258].

12.5 Strong Coercivity of Forms and Invertibility of the Operators A and H

First, note that relations (12.4.3) and (12.4.4) for the Cauchy data of the solutions of the homogeneous system with the top signs can be rewritten as

$$\left(\frac{1}{2}I + B\right)u^+ = AT^+u \quad (12.5.1)$$

and

$$Hu^+ = \left(\frac{1}{2}I - \widehat{B}\right)T^+u, \quad (12.5.2)$$

and the same relations with the bottom signs can be rewritten as

$$\left(\frac{1}{2}I - B\right)u^- = -AT^-u \quad (12.5.3)$$

and

$$-Hu^- = \left(\frac{1}{2}I + \widehat{B}\right)T^-u. \quad (12.5.4)$$

Theorem 12.5.1. *The form of the operator A satisfies the strong coercivity inequality*

$$\|\psi\|^2_{H^{-1/2}(\Gamma)} \leq C_1 \operatorname{Re}(A\psi, \psi)_\Gamma, \quad \psi \in H^{-1/2}(\Gamma), \quad (12.5.5)$$

so that A is invertible as an operator from $H^{-1/2}(\Gamma)$ to $H^{1/2}(\Gamma)$.

Proof. Let $u = \mathcal{A}\psi$. This function has equal traces $u^\pm$ on Γ and belongs to $H^1(\mathbb{T})$, and we have the strong coercivity inequality on the torus for it. The Green identities for $Lu = 0$ in $\Omega^\pm$ and the second jump relation in (12.3.10) imply (cf. estimates in Section 11.3)

$$\|\psi\|_{H^{-1/2}(\Gamma)} \leq C_2\|u\|_{H^1(\mathbb{T})}.$$

Therefore, we have

$$\|\psi\|^2_{H^{-1/2}(\Gamma)} \leq C_2^2 \|u\|^2_{H^1(\mathbb{T})}$$
$$\leq C_3 \operatorname{Re} \Phi_{\mathbb{T}}(u,u) = C_3 \operatorname{Re} \Phi_{\Omega^+}(u,u) + C_3 \operatorname{Re} \Phi_{\Omega^-}(u,u)$$
$$= C_3 \operatorname{Re}[(T^+u, u^+)_\Gamma - (T^-u, u^-)_\Gamma] = C_3 \operatorname{Re}(\psi, A\psi)_\Gamma,$$

again by virtue of the Green identities in $\Omega^\pm$ and the second jump relation in (12.3.10). This gives (12.5.5). The invertibility of A follows now by the Lax–Milgram theorem. The corresponding form is $(A\psi, \psi)_\Gamma$.   □

Theorem 12.5.2. *The strong coercivity of the forms $\Phi_{\Omega^\pm}$ on $H^1(\Omega^\pm)$ implies the strong coercivity inequality*

$$\|\varphi\|^2_{H^{1/2}(\Gamma)} \leq C_2 \operatorname{Re}(H\varphi, \varphi)_\Gamma, \quad \varphi \in H^{1/2}(\Gamma), \tag{12.5.6}$$

for the form of the operator H and the invertibility of H as an operator from $H^{1/2}(\Gamma)$ to $H^{-1/2}(\Gamma)$.

Proof. The proof is similar to the proof of Theorem 12.5.1. We set $u = \mathcal{B}\varphi$. By virtue of the third jump relation, the norm $\|\varphi\|_{H^{1/2}(\Gamma)}$ is dominated by $\|u^\pm\|_{H^{1/2}(\Gamma)}$ and, hence, by $\|u\|_{H^1(\Omega^\pm)}$. The Green identities and the third and fourth jump relations imply

$$\Phi_{\Omega^+}(u,u) + \Phi_{\Omega^-}(u,u) = (H\varphi, \varphi)_\Gamma.   □$$

Corollary 12.5.3. *The solutions of the Dirichlet problems for the homogeneous system $Lu = 0$ in $\Omega^\pm$ with conditions $u^\pm = g$ can be constructed in the form of a single-layer potential by the formula*

$$u = \mathcal{A}\psi, \quad \text{where } A\psi = g, \quad \text{i.e., } \psi = A^{-1}g, \tag{12.5.7}$$

and the solutions of the Neumann problems in $\Omega^\pm$ with conditions $T^\pm u = h$ can be constructed in the form of a double-layer potential by the formula

$$u = \mathcal{B}\varphi, \quad \text{where } H\varphi = -h, \quad \text{i.e., } \varphi = -H^{-1}h, \tag{12.5.8}$$

under the assumption of the strong coercivity of the forms $\Phi_{\Omega^\pm}$ on $H^1(\Omega^\pm)$.

We also mention that *inequalities (12.5.5) and (12.5.6) can be rewritten with inverse operators as*

$$\|\varphi\|^2_{H^{1/2}(\Gamma)} \leq C_4 \operatorname{Re}(A^{-1}\varphi, \varphi)_\Gamma, \tag{12.5.9}$$

$$\|\psi\|^2_{H^{-1/2}(\Gamma)} \leq C_5 \operatorname{Re}(H^{-1}\psi, \psi)_\Gamma. \tag{12.5.10}$$

If the surface Γ and the coefficients of L are infinitely differentiable, then A and H are strongly elliptic pseudodifferential operators on Γ of orders -1 and 1, respectively; cf. [103] and [186].

Theorem 12.5.4. *If the forms $\Phi_{\Omega^\pm}$ are strongly coercive on $H^1(\Omega^\pm)$, then the operators $\frac{1}{2}I \pm B$ in $H^{1/2}(\Gamma)$ and $\frac{1}{2}I \pm \widehat{B}$ in $H^{-1/2}(\Gamma)$ are invertible.*

Proof. Let us verify the invertibility of $\frac{1}{2}I + B$. Suppose that the kernel of this operator contains a function $\varphi \in H^{1/2}(\Gamma)$. We set $u^+ = \varphi$. By virtue of (12.5.1), we have $AT^+u = 0$. But the operator A is invertible. Therefore, $T^+u = 0$. Uniqueness for the Neumann problem in Ω^+ implies $u = 0$ and $\varphi = 0$.

From the same formula (12.5.1) it is seen that the range of the operator $\frac{1}{2}I + B$ is the entire space $H^{1/2}(\Gamma)$.

We have shown that the operator $\frac{1}{2}I + B$ is invertible. The invertibility of $\frac{1}{2}I - B$ and $\frac{1}{2}I \pm \widehat{B}$ is verified in a similar way. $\qquad\square$

Corollary 12.5.5. *If the forms $\Phi_{\Omega^\pm}$ are strongly coercive on $H^1(\Omega^\pm)$, then the solutions of the Dirichlet problems can also be constructed by using relations (12.4.2) in the form of double-layer potentials by the formula*

$$u = \mathcal{B}\varphi, \quad where \ (\mp\frac{1}{2}I + B)\varphi = g, \quad i.e. \ \varphi = (\mp\frac{1}{2}I + B)^{-1}g, \qquad (12.5.11)$$

and the solutions of the Neumann problems can be constructed in the form of single-layer potentials by the formula

$$u = \mathcal{A}\psi, \quad where \ (\pm\frac{1}{2}I + \widehat{B})\psi = h, \quad i.e. \ \psi = (\pm\frac{1}{2}I + \widehat{B})^{-1}h. \qquad (12.5.12)$$

If the surface Γ and the coefficients of the operator L are infinitely differentiable, then $\frac{1}{2}I \pm B$ and $\frac{1}{2}I \pm \widehat{B}$ are elliptic pseudodifferential operators of order zero.

Remark 12.5.6. There is yet another approach to the proof of the invertibility of the operators $\frac{1}{2}I \pm B$ and $\frac{1}{2}I \pm \widehat{B}$. It consists in combining the relations (12.5.1)–(12.5.4) and the results for A and H just obtained with results for the operators D and N obtained in Section 11.3.

For example, consider the operator $\frac{1}{2}I - \widehat{B}$. Relation (12.5.2) implies

$$(Hu^+, u^+)_\Gamma = \left(\left(\frac{1}{2}I - \widehat{B}\right)T^+u, u^+\right)_\Gamma = \left(\left(\frac{1}{2}I - \widehat{B}\right)D^+u^+, u^+\right)_\Gamma.$$

From (12.5.6) we obtain the *strong coercivity inequality* for the form of the operator $\left(\frac{1}{2}I - \widehat{B}\right)D^+$:

$$Re\left(\left(\frac{1}{2}I - \widehat{B}\right)D^+\varphi, \varphi\right)_\Gamma \geq c\|\varphi\|^2_{H^{1/2}(\Gamma)}, \qquad (12.5.13)$$

$c > 0$. Therefore, this is an invertible operator from $H^{1/2}(\Gamma)$ to $H^{-1/2}(\Gamma)$. But D^+ is invertible as well. Hence $\frac{1}{2}I - \widehat{B}$ is invertible in $H^{-1/2}(\Gamma)$.

This can be compared with (12.6.1) below and with results of [103, p. 59].

3 The spaces H^s and strongly elliptic systems in Lipschitz domains

12.6 Relations between Operators on the Boundary

Proposition 12.6.1. *The following relations hold*:

$$A^{-1} = D^+ + D^-, \quad H^{-1} = N^+ + N^-, \quad N^{\pm} = \left(\frac{1}{2}I \pm B\right)^{-1} A,$$

$$D^{\pm} = \left(\frac{1}{2}I \mp \widehat{B}\right)^{-1} H. \tag{12.6.1}$$

More precisely, the first of them follows from the strong coercivity of the form $\Phi_{\mathbb{T}}$ on $H^1(\mathbb{T})$, and the others hold if the forms $\Phi_{\Omega^{\pm}}$ are strongly coercive on $H^1(\Omega^{\pm})$.

Proof. For coinciding $u^{\pm}$, (12.3.4) implies

$$u^{\pm} = -A[Tu]_{\Gamma} = A(D^+ + D^-)u^{\pm}. \tag{12.6.2}$$

Therefore, $D^+ + D^-$ is a right inverse of the operator A. But A is invertible; hence $D^+ + D^-$ is also its left inverse. Although, the relation

$$(D^+ + D^-)A\psi = \psi \tag{12.6.3}$$

can also be derived from the second jump relation in (12.3.10).
    Similarly, for coinciding $T^{\pm}u$, (12.3.4) implies

$$T^{\pm}u = -H[u]_{\Gamma} = H(N^+ + N^-)T^{\pm}u. \tag{12.6.4}$$

Therefore, $N^+ + N^-$ is a right inverse of H. It is also a left inverse of H, because the operator H is invertible. Again, the relation

$$(N^+ + N^-)H\varphi = \varphi \tag{12.6.5}$$

can also be derived from the definition of H and the third jump relation.
    The remaining relations in (12.6.1) follow from (12.5.1)–(12.5.4). $\qquad\square$

In particular, we see that the operators $N^+ + N^-$ and $D^+ + D^-$ are invertible. The latter is invertible by virtue of the strong coercivity of the form $\Phi_{\mathbb{T}}$ on $H^1(\mathbb{T})$. Curiously, the invertibility of the operator $D^+ + D^-$ does not require the existence of N^+ and N^-.

12.7 Duality Relations on Γ

In this subsection we derive three useful relations. The operators $\widetilde{A}$, $\widetilde{H}$, and $\widetilde{B}$ in these relations correspond to the formal adjoint $\widetilde{L}$ of L.
    Setting $u = \mathcal{A}\psi_1$ and $v = \widetilde{\mathcal{A}}\psi_2$, $\psi_j \in H^{-1/2}(\Gamma)$, we obtain

$$(A\psi_1, \psi_2)_\Gamma = (\psi_1, \widetilde{A}\psi_2)_\Gamma \qquad (12.7.1)$$

from the second Green identities in $\Omega^\pm$ and the first two jump relations in (12.3.10). Similarly, setting $u = \mathcal{B}\varphi_1$ and $v = \widetilde{\mathcal{B}}\varphi_2$, $\varphi_j \in H^{1/2}(\Gamma)$, we obtain

$$(H\varphi_1, \varphi_2)_\Gamma = (\varphi_1, \widetilde{H}\varphi_2)_\Gamma \qquad (12.7.2)$$

from the same Green identities and the last two jump relations.

Now we prove the relation

$$(\widehat{B}\psi, \varphi)_\Gamma = (\psi, \widetilde{B}\varphi)_\Gamma \qquad (\psi \in H^{-1/2}(\Gamma), \quad \varphi \in H^{1/2}(\Gamma)). \qquad (12.7.3)$$

For this purpose, we set $u = \mathcal{A}\psi$ and $v = \widetilde{\mathcal{B}}\varphi$. The second Green identity in Ω^+ implies

$$(T^+u, v^+)_\Gamma = (u^+, \widetilde{T}^+v)_\Gamma,$$

which gives (by virtue of relations which the reader can find in the preceding sections)

$$\left(\frac{1}{2}\psi + \widehat{B}\psi, -\frac{1}{2}\varphi + \widetilde{B}\varphi\right)_\Gamma = (A\psi, -\widetilde{H}\varphi)_\Gamma. \qquad (12.7.4)$$

Similarly, the second Green identity in Ω^- yields

$$-(T^-u, v^-)_\Gamma = -(u^-, \widetilde{T}^-v)_\Gamma,$$

which gives

$$-\left(-\frac{1}{2}\psi + \widehat{B}\psi, \frac{1}{2}\varphi + \widetilde{B}\varphi\right)_\Gamma = -(A\psi, -\widetilde{H}\varphi)_\Gamma. \qquad (12.7.5)$$

Summing (12.7.4) and (12.7.5), we obtain (12.7.3).

12.8 Problems with Boundary Conditions on a Nonclosed Surface

In this subsection we consider the system $Lu = 0$ on the torus $\mathbb{T}$ with boundary conditions on a nonclosed $(n-1)$-dimensional Lipschitz surface $\Gamma_1 \subset \mathbb{T}$ with $(n-2)$-dimensional Lipschitz boundary $\partial\Gamma_1$ $(n \geq 2)$, which is not included in Γ_1. To be more precise, we assume that Γ_1 is an open part of a closed Lipschitz surface Γ that divides the torus into two domains $\Omega^\pm$ and, furthermore, that the boundary $\partial\Gamma_1$ divides Γ into two domains Γ_1 and Γ_2. There is an obvious arbitrariness in the choice of the complementary part Γ_2 of the surface Γ. We denote the sides of Γ facing $\Omega^\pm$ by $\Gamma^\pm$. In a similar way we define $\Gamma_1^\pm$. Boundary and transmission conditions will be set on $\Gamma_1^\pm$. As previously, we assume the normal vector at points of Γ to be directed to Ω^-.

Let $\Omega_0 = \mathbb{T} \setminus \overline{\Gamma}_1$. Note that this domain is not Lipschitz.

First, we define the space $H^1(\Omega_0)$ in a standard way.

Definition 1. The *space* $H^1(\Omega_0)$ consists of functions u belonging to L_2 on the domain Ω_0 and such that all of their first derivatives $\partial_j u$ in the sense of distributions on this domain also belong to L_2. This space is endowed with the standard norm

$$\|u\|^2_{H^1(\Omega_0)} = \|u\|^2_{L_2(\Omega_0)} + \sum \|\partial_j u\|^2_{L_2(\Omega_0)}. \tag{12.8.1}$$

Now we give the second definition.

Definition 2. The *space* $H^1(\Omega_0)$ consists of functions belonging to $L_2(\Omega_0)$ such that the restrictions of these functions to $\Omega^\pm$ belong to $H^1(\Omega^\pm)$ and their traces on $\Gamma^\pm$ (which belong to $H^{1/2}(\Gamma)$) coincide on Γ_2. The norm on this space is defined by

$$\|u\|^2_{H^1(\Omega_0)} = \|u\|^2_{H^1(\Omega^+)} + \|u\|^2_{H^1(\Omega^-)}. \tag{12.8.2}$$

Proposition 12.8.1. *Definitions* 1 *and* 2 *are equivalent.*

Proof. Obviously, $H^1(\Omega_0)$ in the sense of the first definition is contained in $H^1(\Omega_0)$ in the sense of the second. The converse follows from the formula for integration by parts over a Lipschitz domain (see Proposition 9.3.3): if we write this formula in $\Omega^\pm$ for a function u belonging to $H^1(\Omega_0)$ in the sense of the second definition and for a test function in $C_0^\infty(\Omega_0)$ and sum these formulas, then the terms on Γ_2 will cancel each other. Therefore, the first derivatives on $\Omega^\pm$ together determine the first derivatives on Ω_0; cf. Section 3.5 on gluing together functions on half-spaces.

The norms (12.8.1) and (12.8.2) coincide, provided that the norms on $H^1(\Omega^\pm)$ are of the form (12.8.1). $\qquad\square$

It is seen from the first definition that the space $H^1(\Omega_0)$ does not depend on the choice of the surface Γ_2 and from the second definition that the jump $[u]_\Gamma$ of a function $u \in H^1(\Omega_0)$ vanishes on Γ_2, i.e., belongs to $\widetilde{H}^{1/2}(\Gamma_1)$.

Proposition 12.8.2. *Let u be a solution of the system $Lu = 0$ in Ω_0 belonging to $H^1(\Omega_0)$. Then the jump $[Tu]_\Gamma$ vanishes on Γ_2 and, therefore, belongs to $\widetilde{H}^{-1/2}(\Gamma_1)$.*

Proof. Let Ω^0 be a small ball centered at a point of Γ_2 and contained inside Ω_0, and let v be a function in $\widetilde{H}^1(\Omega^0)$ extended by zero outside Ω^0. Then $(Lu, v)_{\Omega^0} = 0$, and the boundary term in the Green identity for Ω^0 vanishes, so that $\Phi_{\Omega^0}(u, v) = 0$. Let us now write the Green identities for these functions in $\Omega^\pm \cap \Omega^0$. Summing them, we obtain $([Tu]_\Gamma, v^+)_{\Gamma_2} = 0$. By virtue of the arbitrariness in the choice of v on the intersection of the ball with Γ_2, we have $[Tu]_\Gamma = 0$ on this intersection. Therefore, this jump vanishes on Γ_2. $\qquad\square$

We consider the following four problems for the system $Lu = 0$ in Ω_0.

I The Dirichlet problem with conditions

$$u^\pm = g^\pm \text{ on } \Gamma_1^\pm, \tag{12.8.3}$$

where $g^\pm \in H^{1/2}(\Gamma_1)$ and $[g]_{\Gamma_1} = g^- - g^+ \in \widetilde{H}^{1/2}(\Gamma_1)$.

II The Neumann problem with conditions

$$T^{\pm}u = h^{\pm} \text{ on } \Gamma_1^{\pm}, \tag{12.8.4}$$

where $h^{\pm} \in H^{-1/2}(\Gamma_1)$ and $[h]_{\Gamma_1} \in \widetilde{H}^{-1/2}(\Gamma_1)$.

III The transmission problem with conditions

$$[u]_{\Gamma_1} = g, \quad [Tu]_{\Gamma_1} = h \text{ on } \Gamma_1, \tag{12.8.5}$$

where $g \in \widetilde{H}^{1/2}(\Gamma_1)$ and $h \in \widetilde{H}^{-1/2}(\Gamma_1)$.

IV The mixed problem with conditions

$$u^{+} = g \text{ on } \Gamma_1^{+}, \quad T^{+}u = h \text{ on } \Gamma_1^{-}, \tag{12.8.6}$$

where $g \in H^{1/2}(\Gamma_1)$ and $h \in H^{-1/2}(\Gamma_1)$.

In these settings the assumptions about the data on Γ_1 are essential.

Such problems arise in electrostatics, electrodynamics, and acoustics; in this case, L is the Laplace or the Helmholtz operator and Γ_1 is a nonclosed screen. They arise also in the theory of elasticity; in this case, $Lu = 0$ is one of the systems of elasticity theory, and Γ_1 models a crack.

We begin with problem III.

Theorem 12.8.3. *Problem III is uniquely solvable, and its solution is expressed by*

$$u = \mathcal{B}[u]_{\Gamma_1} - \mathcal{A}[Tu]_{\Gamma_1}. \tag{12.8.7}$$

Proof. This is essentially a mere corollary of Theorem 12.3.7. If a solution of problem III is given, then it can be expressed by (12.3.13) with jumps on Γ_1 (extended by zero to Γ_2), i.e., by (12.8.7).

Vice versa, if u is the function defined by (12.8.7), then this is a solution of the system $Lu = 0$ outside $\overline{\Gamma}_1$ belonging to $H^1(\Omega_0)$: it belongs to $H^1(\Omega^{\pm})$, and its jumps vanish on Γ_2 (see Definition 2 and Proposition 12.8.2). □

In the rest of this section we can use expression (12.8.7) instead of (12.3.13).

Our further considerations are similar to those of mixed problems in Section 11.4, where the operators D_1 and N_1 were used.

Consider the Dirichlet problem I.

We introduce the operators

$$A_1\psi = (A\psi)|_{\Gamma_1} \quad \text{and} \quad B_1\varphi = (B\varphi)|_{\Gamma_1}$$
$$(\psi \in \widetilde{H}^{-1/2}(\Gamma_1), \ \varphi \in \widetilde{H}^{1/2}(\Gamma_1)). \tag{12.8.8}$$

Obviously, A_1 is a bounded operator from $\widetilde{H}^{-1/2}(\Gamma_1)$ to $H^{1/2}(\Gamma_1)$ and B_1 is a bounded operator from $\widetilde{H}^{1/2}(\Gamma_1)$ to $H^{1/2}(\Gamma_1)$.

Passing to Γ_1 from its two sides in (12.8.7), summing the resulting relations, and dividing by 2, we obtain the following equation for $\psi = [Tu]_{\Gamma_1}$ (see (12.4.2)):

$$g = B_1\varphi - A_1\psi, \quad \text{where} \quad g = \frac{1}{2}(g^+ + g^-) \quad \text{and} \quad \varphi = g^- - g^+. \tag{12.8.9}$$

Theorem 12.8.4. *The form of the operator A_1 satisfies the strong coercivity inequality*

$$\|\psi\|^2_{\widetilde{H}^{-1/2}(\Gamma_1)} \le C_1 \operatorname{Re}(A_1\psi, \psi)_{\Gamma_1}, \quad \psi \in \widetilde{H}^{-1/2}(\Gamma_1), \tag{12.8.10}$$

and this operator is invertible.

Proof. The proof is similar to those of Theorems 11.4.2 and 11.4.3. The strong coercivity inequality for the form of the operator A (see (12.5.5)) is inherited by the form of A_1. Invertibility follows by the Lax–Milgram theorem.   □

Theorem 12.8.5. *The Dirichlet problem in setting* I *has precisely one solution belonging to $H^1(\Omega_0)$.*

Indeed, the jump $\psi = [Tu]_{\Gamma_1}$ is determined from the Dirichlet data by using equation (12.8.9), after which the solution is constructed by (12.8.7). For the zero Dirichlet data, this solution is zero.

Now we proceed to the Neumann problem II. Assuming that we have inequality (12.5.6), we introduce the operators

$$H_1\varphi = (H\varphi)|_{\Gamma_1} \quad \text{and} \quad \widehat{B}_1\psi = (\widehat{B}\psi)_{\Gamma_1}$$
$$(\varphi \in \widetilde{H}^{1/2}(\Gamma_1), \ \psi \in \widetilde{H}^{-1/2}(\Gamma_1)). \tag{12.8.11}$$

Here H_1 is a bounded operator from $\widetilde{H}^{1/2}(\Gamma_1)$ to $H^{-1/2}(\Gamma_1)$, and $\widehat{B}_1$ is a bounded operator from $\widetilde{H}^{-1/2}(\Gamma_1)$ to $H^{-1/2}(\Gamma_1)$.

Calculating the conormal derivatives of both sides in (12.8.7) on the two sides of Γ_1, summing them, and dividing by 2, we obtain the following equation for $\varphi = [u]_{\Gamma_1}$ (see (12.4.2)):

$$h = -H_1\varphi - \widehat{B}_1\psi, \quad \text{where} \quad h = \frac{1}{2}(h^+ + h^-) \quad \text{and} \quad \psi = h^- - h^+. \tag{12.8.12}$$

Theorem 12.8.6. *The form of the operator H_1 satisfies the strong coercivity inequality*

$$\|\varphi\|^2_{\widetilde{H}^{1/2}(\Gamma_1)} \le C_2 \operatorname{Re}(H_1\varphi, \varphi)_{\Gamma_1}, \quad \varphi \in \widetilde{H}^{1/2}(\Gamma_1), \tag{12.8.13}$$

and this operator is invertible.

Proof. The proof is similar to that of Theorem 12.8.4. It uses inequality (12.5.6) and the Lax–Milgram theorem.   □

The next result is obtained similarly to Theorem 12.8.5.

Theorem 12.8.7. *The Neumann problem in setting* II *has precisely one solution belonging to $H^1(\Omega_0)$.*

Finally, consider problem IV. To construct its solution u, we seek the jumps

$$\varphi = [u]_{\Gamma_1} \in \widetilde{H}^{1/2}(\Gamma_1) \quad \text{and} \quad \psi = [Tu]_{\Gamma_1} \in \widetilde{H}^{-1/2}(\Gamma_1). \tag{12.8.14}$$

Calculating the conormal derivative of the function (12.8.7) on Γ_1^- (see (12.3.11) and (12.4.2)) and the boundary value of this function on Γ_1^+ (see (12.4.2)), we obtain the equations

$$-H_1\varphi - \left(-\frac{1}{2}I + \widehat{B}_1\right)\psi = h,$$

$$\left(-\frac{1}{2}I + B_1\right)\varphi - A_1\psi = g. \tag{12.8.15}$$

It is convenient to write them in this order. We follow the paper [134] on the equations of anisotropic elasticity.

The operator

$$\mathcal{T} = \begin{pmatrix} H_1 & -\frac{1}{2}I + \widehat{B}_1 \\ \frac{1}{2}I - B_1 & A_1 \end{pmatrix} \tag{12.8.16}$$

acts boundedly from the space

$$\widetilde{\mathcal{H}} = \widetilde{H}^{1/2}(\Gamma_1) \times \widetilde{H}^{-1/2}(\Gamma_1) \tag{12.8.17}$$

to the space

$$\mathcal{H} = H^{-1/2}(\Gamma_1) \times H^{1/2}(\Gamma_1). \tag{12.8.18}$$

These two spaces are dual with respect to the extension of the inner product $(\varphi_1, \psi_1)_{\Gamma_1} + (\psi_2, \varphi_2)_{\Gamma_1}$ to their direct product. For a column $U = (\varphi, \psi)'$, we have

$$(\mathcal{T}U, U) = (H_1\varphi, \varphi)_{\Gamma_1}$$
$$+ \left(\left(-\frac{1}{2}I + \widehat{B}_1\right)\psi, \varphi\right)_{\Gamma_1} + \left(\left(\frac{1}{2}I - B_1\right)\varphi, \psi\right)_{\Gamma_1} + (A_1\psi, \psi)_{\Gamma_1}. \tag{12.8.19}$$

Suppose that L is a formally self-adjoint operator on the torus. Then $\widehat{B} = B$, and it is seen from (12.7.3) and (12.8.19) that

$$\text{Re}(\mathcal{T}U, U) = (H_1\varphi, \varphi)_{\Gamma_1} + (A_1\psi, \psi)_{\Gamma_1}; \tag{12.8.20}$$

the other terms cancel (the sign Re on the right-hand side is not needed now). Using Theorems 12.8.4 and 12.8.6, we obtain

$$\|U\|_{\widetilde{\mathcal{H}}}^2 \leq C\,\text{Re}(\mathcal{T}U, U).$$

We see that the equation $\mathcal{T}U = F$, where F is a column $(h, g)'$, is uniquely solvable by the Lax–Milgram theorem.

This proves the following theorem.

Theorem 12.8.8. *If the operator L is formally self-adjoint, then problem IV is uniquely solvable.*

We also mention that relations (12.7.1) and (12.7.2) imply

$$(A_1\psi_1, \psi_2)_{\Gamma_1} = (\psi_1, \widetilde{A}_1\psi_2)_{\Gamma_1}, \qquad (12.8.21)$$

where $\psi_j \in \widetilde{H}^{-1/2}(\Gamma_1)$, and

$$(H_1\varphi_1, \varphi_2)_{\Gamma_1} = (\varphi_1, \widetilde{H}_1\varphi_2)_{\Gamma_1}, \qquad (12.8.22)$$

where $\varphi_j \in \widetilde{H}^{1/2}(\Gamma_1)$.

12.9 Problems with a Spectral Parameter in Transmission Conditions

Here we briefly discuss the following spectral problems.

$7°\quad Lu = 0$ in $\Omega^\pm$, $\quad [u]_\Gamma = 0$, $\quad [Tu]_\Gamma = -\lambda u^\pm$ on Γ. $\qquad (12.9.1)$

It is seen from (12.3.12), (12.3.13), and the second jump relation in (12.3.10) that, for the *eigenfunctions*, this problem is equivalent to the equation $A^{-1}\psi = \lambda\psi$, where $\psi = [Tu]_\Gamma$ and $u = -\mathcal{A}\psi$.

$8°\quad Lu = 0$ in $\Omega^\pm$, $\quad [Tu]_\Gamma = 0$, $\quad T^\pm u = -\lambda[u]_\Gamma$ on Γ. $\qquad (12.9.2)$

For the *eigenfunctions*, this problem is equivalent to the equation $H\varphi = \lambda\varphi$, where $\varphi = [u]_\Gamma$ and $u = \mathcal{B}\varphi$. This follows from (12.3.12)–(12.3.13) and the third jump formula in (12.3.10).

Similar problems can be stated with transmission conditions on Γ_1.

$9°\quad Lu = 0$ in Ω_0, $\quad [u]_{\Gamma_1} = 0$, $\quad [Tu]_{\Gamma_1} = -\lambda u^\pm$ on Γ_1. $\qquad (12.9.3)$

For the *eigenfunctions*, this problem is equivalent to the equation $A_1^{-1}\psi = \lambda\psi$, where $\psi = [Tu]_{\Gamma_1}$.

$10°\quad Lu = 0$ in Ω_0, $\quad [Tu]_{\Gamma_1} = 0$, $\quad T^\pm u = -\lambda[u]_{\Gamma_1}$ on Γ_1. $\qquad (12.9.4)$

For the *eigenfunctions*, this problem is equivalent to the equation $H_1\varphi = \lambda\varphi$, where $\varphi = [u]_{\Gamma_1}$.

The spectral properties of the operators A, H, A_1, and H_1 are similar to those of the operators N, D, N_1, and D_1, respectively. But for the eigenvalues of A and A_1, asymptotic relations can be obtained without the assumption that the Lipschitz surface is almost smooth. This was done by Rozenblum and Tashchiyan [316].

In the case of the operator A, if L is nearly self-adjoint, we can estimate the order of $A - A_0$ as follows. We have (see (12.2.5))

$$A - A_0 = \gamma(L^{-1} - L_0^{-1})\gamma^*.$$

The operator γ^* acts boundedly from $H^{-1/2}(\Gamma)$ to $H^{-1}(\mathbb{T})$. The operator

$$L^{-1} - L_0^{-1} = L^{-1}(L_0 - L)L_0^{-1}$$

acts boundedly from $H^{-1}(\mathbb{T})$ to $H^2(\mathbb{T})$ (see Theorem 12.1.1). But everything is limited by the operator γ, which we know to act boundedly from $H^{3/2-\varepsilon}(\mathbb{T})$ to $H^{1-\varepsilon}(\Gamma)$ for arbitrarily small $\varepsilon > 0$. Therefore, $A - A_0$ is a bounded operator from $H^{-1/2}(\Gamma)$ to $H^{1-\varepsilon}(\Gamma)$.

12.10 More General Transmission Problems

The transmission problems considered in the preceding subsections of this section refer to a *single* strongly elliptic system on the torus. Now suppose given generally *different* strongly elliptic operators $L_\pm$ on the domains $\Omega^\pm$. This means that their coefficients $a_{j,k}$ and b_j with the same subscripts do not necessarily coincide on the boundary. We assume that the corresponding forms $\Phi_{\Omega^\pm}$ are strongly coercive on $H^1(\Omega^\pm)$. We are interested in the problem of finding a pair of functions u on $\Omega^\pm$ satisfying the homogeneous (for simplicity) systems

$$L_+ u = 0 \text{ in } \Omega^+, \quad L_- u = 0 \text{ in } \Omega^- \tag{12.10.1}$$

and transmission conditions in which the jumps $[u]_\Gamma$ and $[Tu]_\Gamma$ are given, but the conormal derivatives defining the latter are determined by *different* systems in $\Omega^\pm$. This problem reduces to two problems: in one of them only the jump $[Tu]_\Gamma$ is nonzero, and in the other only $[u]_\Gamma$ is nonzero. Thus, we consider the following two problems.

I $[u]_\Gamma = 0$, $[Tu]_\Gamma = h$.

II $[u]_\Gamma = g$, $[Tu]_\Gamma = 0$.

Here $g \in H^{1/2}(\Gamma)$ and $h \in H^{-1/2}(\Gamma)$.

Theorem 12.10.1. *Under the above assumptions, these problems are uniquely solvable.*

Proof. Using the Poincaré–Steklov operators $N^\pm$ and $D^\pm$ associated with $L_\pm$, we obtain

$$N^\pm(T^\pm u) = \pm u^\pm \quad \text{and} \quad D^\pm(u^\pm) = \pm T^\pm u$$

(see (12.1.1)). In problem I, for $u^\pm = \varphi$, we have the equation

$$(D^+ + D^-)\varphi = -h. \tag{12.10.2}$$

In problem II, for $T^\pm u = \psi$, we have the equation

$$(N^+ + N^-)\psi = -g. \tag{12.10.3}$$

The operators on the left-hand sides have strongly coercive forms:

$$\|\varphi\|^2_{H^{1/2}(\Gamma)} \leq C_1 \operatorname{Re}((D^+ + D^-)\varphi, \varphi)_\Gamma, \tag{12.10.4}$$

$$\|\psi\|^2_{H^{-1/2}(\Gamma)} \leq C_2 \operatorname{Re}((N^+ + N^-)\psi, \psi)_\Gamma \tag{12.10.5}$$

(this follows from the consideration of the operators D and N in Section 11.3). Thus, we can apply the Lax–Milgram theorem. We obtain the invertibility of the first operator as an operator from $H^{1/2}(\Gamma)$ to $H^{-1/2}(\Gamma)$ and of the second one as an operator from $H^{-1/2}(\Gamma)$ to $H^{1/2}(\Gamma)$. The solutions of the initial problems are now constructed as solutions of the Dirichlet or Neumann problems in $\Omega^\pm$. $\square$

We can also consider the spectral problems

$7^{\circ\circ}$ $L_\pm u = 0$ in $\Omega^\pm$, $[u]_\Gamma = 0$, $[Tu]_\Gamma = -\lambda u^\pm$ on Γ $\qquad$ (12.10.6)

and

$8^{\circ\circ}$ $L_\pm u = 0$ in $\Omega^\pm$, $[Tu]_\Gamma = 0$, $T^\pm u = -\lambda[u]_\Gamma$ on Γ. $\qquad$ (12.10.7)

Problem. Show that, for the eigenfunctions, problem $7^{\circ\circ}$ is equivalent to the equation

$$(D^+ + D^-)\varphi = \lambda\varphi, \quad \text{where } \varphi = u^\pm, \tag{12.10.8}$$

and problem $8^{\circ\circ}$ is equivalent to the equation

$$\psi = \lambda(N^+ + N^-)\psi, \quad \text{where } \psi = T^\pm u. \tag{12.10.9}$$

These results are consistent with results of Section 12.9 on the problems considered there by virtue of relations of Section 12.6. But now we cannot derive the invertibility of the operator $D^+ + D^-$ from the strong coercivity of the forms $\Phi_{\Omega^\pm}$ only on $\widetilde{H}^1(\Omega^\pm)$.

We can also prove a usual set of spectral properties for the operators $D^+ + D^-$ and $N^+ + N^-$, including spectral asymptotics (provided that the surface Γ is almost smooth and the systems in $\Omega^\pm$ are formally self-adjoint or have formally self-adjoint leading parts), the location of eigenvalues, and the basis property or the completeness of generalized eigenfunctions. We do not dwell on this.

Chapter 4
More General Spaces and Their Applications

13 Elements of Interpolation Theory

13.1 Contents of the Section. The Spaces L_p

1. Theorems of interpolation theory are applied, in particular, in situations of the following type (we describe them very roughly at the moment). Suppose given two pairs of Banach spaces (X_0, X_1) and (Y_0, Y_1) with properties specified later on. From these pairs scales of spaces $\{X_\theta\}$ and $\{Y_\theta\}$, $0 < \theta < 1$, are constructed in a special way. Next, suppose that there is a linear operator T acting continuously from X_0 to Y_0 and from X_1 in Y_1. It turns out that T acts continuously from X_θ to Y_θ for $0 < \theta < 1$. Moreover, sometimes, under certain assumptions, an additional information about properties (e.g., invertibility) of T as an operator from X_0 to Y_0 and from X_1 to Y_1 implies that T has similar properties as an operator from X_θ to Y_θ. Such results often simplify analytic considerations. In some cases, this is the only known method for obtaining needed results, and in some other cases, this method is the simplest one.

This theory emerged in the past century. The first was a theorem of M. Riesz (1926) and Thorin (1938) about operators in L_p spaces. We shall say a few words about these theorems in the next subsection. The foundations of the abstract theory were laid in the late 1950s–early 1960s by Calderón, S. Krein, Lions, Peetre, Aronszajn, Gagliardo, and other mathematicians. We mention only some of the numerous works on this topic. The first monographs [60], [223], and [376] appeared in 1976–1979. The theory continued to develop, and now this is a separate direction of functional analysis with important applications, in particular, to the theory of general Banach and function spaces, the theory of differential and integral equations, and approximation theory. In the extensive survey [76], bibliography includes 786 entries. We also mention the monographs [57], [77], [80], and [246] and the survey [130].

In this section, we prove only a few theorems (particularly important in our opinion), but we shall try to comment on all the material presented. Many proofs which

© Springer International Publishing Switzerland 2015
M.S. Agranovich, *Sobolev Spaces, Their Generalizations and Elliptic Problems
in Smooth and Lipschitz Domains*, Springer Monographs in Mathematics,
DOI 10.1007/978-3-319-14648-5_4

we omit can be found in [60], [376], and [223], but some theorems were proved later, and in these cases, we refer to journal articles.

We warn the reader that we do not aim at maximum generality. We selected a material which, according to the author's experience, is surely useful for applications to elliptic partial differential equations. Strengthenings, versions, and other additional material can be found in the cited literature. Our brief survey only touches on the foundations of interpolation theory; cf. [258, Appendix B].

2. The spaces L_p were touched on in the preceding sections. Although the reader most likely knows elementary facts on the L_p spaces very well, we briefly summarize these facts below (for references, the book [137, vol. I] can be used).

If U is a domain in $\mathbb{R}^n$, the *space* $L_p(U)$ with $1 \leq p < \infty$ is the separable Banach space of measurable complex-valued (for definiteness) functions $u(x)$ on U with finite norm

$$\|u\|_{L_p(U)} = \left(\int\limits_U |u(x)|^p \, dx \right)^{1/p}. \tag{13.1.1}$$

The triangle inequality for this norm is called the *Minkowski inequality*. Now let $1 < p < \infty$ and $p + p' = pp'$. Then for any functions $u \in L_p(U)$ and $v \in L_{p'}(U)$, *Hölder's inequality*

$$\left| \int\limits_U u(x)v(x) \, dx \right| \leq \|u\|_{L_p(U)} \|v\|_{L_{p'}(U)} \tag{13.1.2}$$

holds. For $1 < p < \infty$, the spaces $L_p(U)$ and $L_{p'}(U)$ are dual with respect to the extension of the inner product in $L_2(U)$ to the direct product of these spaces.

The *space* $L_\infty(U)$ consists of bounded measurable functions on U with finite norm

$$\|u\|_{L_\infty(U)} = \inf_X \sup_{x \in X} |u(x)|, \tag{13.1.3}$$

where the suprema are over subsets $X \subset U$ of full measure. For $p = 1$, p' is set to ∞. The space $L_\infty(U)$ is Banach and dual to $L_1(U)$ as well, but not vice versa; $L_\infty(U)$ is not separable, and these two spaces are not reflexive. The linear manifold $C_0^\infty(U)$ is dense in $L_p(U)$ for $1 \leq p < \infty$.

The operator of multiplication by a function in $L_\infty(U)$ is a multiplier on any $L_p(U)$.

Instead of a domain U, we can consider a space U with positive measure (we do not dwell on details). This approach covers, in particular, the space l_p of numerical sequences $\{u_k\}_1^\infty$ with norm

$$\|u\|_{l_p} = \left(\sum_{k=1}^\infty |u_k|^p \right)^{1/p} \tag{13.1.4}$$

for $1 \leq p < \infty$. Here the measure is concentrated at the points $k \in \mathbb{N}$. For these p, the finite sequences (in which the number of nonzero members u_k is finite) are dense in l_p. The space l_∞ consists of bounded sequences with norm $\|u\|_{l_\infty} = \sup |u_k|$. Note also that, instead of the spaces $L_p(U)$ of numerical functions, the spaces $L_p(U, X)$ of functions on U taking values in a Banach space X can be considered. These two possible generalizations should be borne in mind below in this section, where we touch on L_p.

The spaces L_p can also be considered for $0 < p < 1$, but in this case, they are only *quasi-normed*, or, to be more precise, *quasi-Banach* (they are complete). A *quasi-norm* $\|u\|$ differs from a norm by an additional factor on the right-hand side of the triangle inequality:

$$\|u + v\| \leq C(\|u\| + \|v\|). \tag{13.1.5}$$

3. We tell about fundamental facts of interpolation theory on the basis of examples in the framework of the spaces H^s, whose theory has already been constructed in the preceding sections, and L_p. In fact, interpolation theory is the more informative the larger is the class of spaces under consideration, which is often "fitted" to the analytical problem being studied. Recall that, in Section 14, we shall briefly discuss the spaces H_p^s and B_p^s. In that section, we shall give interpolation relations for these spaces.

In all instances of the coincidence or equality of spaces in this section, we mean the equivalence of norms on them, although it is sometimes possible to prove that these norms are equal. All embeddings of spaces are assumed to be continuous. All Banach spaces are considered over the field of complex numbers.

13.2 Basic Definitions and the Complex Interpolation Method

13.2.1

Let $X = (X_0, X_1)$ be a pair of Banach spaces. It is said to be an *interpolation pair* if both spaces are continuously and linearly embedded in a Hausdorff[1] topological vector space $\mathcal{X}$. (Other names for such a pair are a *Banach pair* and *compatible Banach spaces*.) This allows us to consider linear combinations of elements of X_0 and X_1. The vector spaces $\Sigma(X) = X_0 + X_1$ and $\Delta(X) = X_0 \cap X_1$ are endowed with the norms

$$\|x\|_{\Sigma(X)} = \inf\{\|x_0\|_{X_0} + \|x_1\|_{X_1} : x = x_0 + x_1, \; x_0 \in X_0, \; x_1 \in X_1\} \tag{13.2.1}$$

and

$$\|x\|_{\Delta(X)} = \max\{\|x\|_{X_0}, \|x\|_{X_1}\}. \tag{13.2.2}$$

It is easy to verify that both spaces thus obtained are Banach. Obviously, $\Delta(X) \subset X_j \subset \Sigma(X)$. It is frequently assumed that $\Delta(X)$ is dense in X_0 and X_1. *We make this assumption unless otherwise specified.* If X_1 is (continuously) embedded in X_0, then $X_1 = \Delta(X)$ and $X_0 = \Sigma(X)$, and for the ambient space we can take X_0. In the general case, $\Sigma(X)$ is the minimal ambient space.

---

[1] Any two different points have disjoint neighborhoods.

First, we must construct intermediate, or interpolation, spaces. The main methods for this are the complex and the real interpolation method; we describe them below. Thus, let $X = (X_0, X_1)$ be an interpolation pair.

13.2.2

The *complex interpolation method* is as follows. It was proposed by Calderón [84] and Lions [234]; a close form of this method was also suggested by S. Krein (in terms of "analytic scales"; see [223] and references therein). We note beforehand that, for functions $f(z)$ of a scalar variable with values in a Banach space, the notions of boundedness, continuity, and analyticity are defined: it suffices to replace absolute values by norms. In particular, for analytic, or holomorphic, functions, the Cauchy integral formula and the maximum modulus principle can be proved; see [137, Vol. I, Chap. III, Sec. 4] or [179, Chap. III].

Let S and S_0 be the closed and open vertical strips of width 1 in the complex plane:

$$S = \{z : 0 \le \operatorname{Re} z \le 1\}, \quad S_0 = \{z : 0 < \operatorname{Re} z < 1\}. \tag{13.2.3}$$

By $F(X)$ we denote the space of functions f on S with values in $\Sigma(X)$ that

(1) are holomorphic in S_0 (i.e., have local Taylor series expansions converging in the norm of $\Sigma(X)$);

(2) are continuous and bounded on S;

(3) take values in X_0 on the left boundary of the strip and in X_1 on the right boundary, are continuous there with respect to $\operatorname{Im} z$, and tend to zero as $\operatorname{Im} z \to \pm\infty$.

Obviously, $F(X)$ is a vector space. A norm on it is defined by

$$\|f\|_{F(X)} = \max\{\sup_t \|f(it)\|_{X_0}, \ \sup_t \|f(1+it)\|_{X_1}\}. \tag{13.2.4}$$

(Here max can be written instead of sup.) It can be verified that the space thus obtained is Banach.

The *complex interpolation space* $X_\theta = [X_0, X_1]_\theta$, $0 \le \theta \le 1$, is defined as the space of all $x \in \Sigma(X)$ such that $x = f(\theta)$ for some function $f \in F(X)$. The norm on X_θ is defined by

$$\|x\|_{X_\theta} = \inf\{\|f\|_{F(X)} : f \in F(X), \ f(\theta) = x\}. \tag{13.2.5}$$

It can be verified that this space is Banach as well.

In [60], also another version of the complex method is described.

All spaces $[X_0, X_1]_\theta$, $0 \le \theta \le 1$, contain $\Delta(X)$ and are contained in $\Sigma(X)$. Under our assumption that $\Delta(X)$ is dense in X_0 and in X_1, the spaces $[X_0, X_1]_0$ and $[X_0, X_1]_1$ coincide with the initial spaces X_0 and X_1, respectively.

We also mention the following properties of the spaces X_θ.

1. The equality $[X_0, X_1]_\theta = [X_1, X_0]_{1-\theta}$ holds.

2. The space $\Delta(X)$ is dense in all X_θ.

3. If $X_1 \subset X_0$ and $0 < \theta_0 < \theta_1 < 1$, then $X_1 \subset X_{\theta_1} \subset X_{\theta_0} \subset X_0$; moreover, each space is dense in the next one.

4. If X_0 and X_1 coincide, then all X_θ coincide.

Below we present the fundamental theorem on the complex interpolation method. Let $X = (X_0, X_1)$ and $Y = (Y_0, Y_1)$ be two interpolation pairs, and let T be a bounded linear operator from $\Sigma(X)$ to $\Sigma(Y)$. We denote its restriction to X_θ by T_θ. This notation is also used in what follows.

Theorem 13.2.1. *Let T_0 be a bounded operator from X_0 to Y_0 with norm M_0, and let T_1 be a bounded operator from X_1 to Y_1 with norm M_1. Then T_θ is a bounded operator from X_θ to Y_θ $(0 < \theta < 1)$. Moreover, its norm M_θ satisfies the inequality*

$$M_\theta \le M_0^{1-\theta} M_1^\theta. \tag{13.2.6}$$

Proof. Let x be an element of X_θ. Suppose that $y = Tx$, ε is an arbitrarily small positive number, and $f(z)$ is a function from $F(X)$ such that $f(\theta) = x$ and $\|f\|_{F(X)} \le \|x\|_{X_\theta} + \varepsilon$. Consider the function

$$g(z) = M_0^{z-\theta} M_1^{\theta-z} T f(z), \quad z \in S. \tag{13.2.7}$$

(In general, proofs of fundamental statements about the complex interpolation method usually require selecting suitable functions of a complex variable.) It is easy to verify that it belongs to $F(Y)$. Moreover,

$$g(\theta) = T f(\theta) = Tx = y.$$

We have

$$\|g(it)\|_{Y_0} \le M_0^{-\theta} M_1^\theta M_0 \|f(it)\|_{X_0} \le M_0^{1-\theta} M_1^\theta (\|x\|_{X_\theta} + \varepsilon),$$
$$\|g(1+it)\|_{Y_1} \le M_0^{1-\theta} M_1^{\theta-1} M_1 \|f(1+it)\|_{X_1} \le M_0^{1-\theta} M_1^\theta (\|x\|_{X_\theta} + \varepsilon).$$

Thus, $y \in Y_\theta$ and estimate (13.2.6) holds. $\qquad\square$

The passage from an interpolation pair to intermediate spaces is called an *interpolation functor*. (A precise definition can be found in [60] or [376].) In the case under consideration, this is the passage from the pair X to the spaces X_θ. If

$$\|T_\theta\| \le C \max(\|T_0\|, \|T_1\|), \tag{13.2.8}$$

where the constant C does not depend on T, then the interpolation functor is said to be *uniform*; in the case under consideration, we have a uniform functor with $C = 1$. The validity of (13.2.6) can be expressed in words as "the functor is *exact of type* θ." This estimate means the logarithmic convexity of the norm $\|T_\theta\|$ (which means that the graph of the logarithm of this norm is convex downward).

Remark. Situations where T is initially given and bounded only as an operator from X_0 to Y_0 and from X_1 to Y_1 are often encountered. Suppose that these are operators T_0 and T_1. To define their extension to a bounded operator from $\Sigma(X)$ to $\Sigma(Y)$, we must assume that $T_0 = T_1$ on $\Delta(X)$. This assumption is called the *compatibility condition* on the operators T_0 and T_1. It is sufficient to assume the coincidence of these operators on a dense subset of $\Delta(X)$. If $X_1 \subset X_0$ and $Y_1 \subset Y_0$, then the compatibility condition reduces to $T_1 = T_0|_{X_1}$.

Indeed, under the compatibility condition, the operator T is well defined on $\Sigma(X)$ by

$$Tx = T_0 x_0 + T_1 x_1 \quad (x = x_0 + x_1, \ x_0 \in X_0, \ x_1 \in X_1).$$

This operator acts from $\Sigma(X)$ to $\Sigma(Y)$, and its norm does not exceed $\max(\|T_0\|, \|T_1\|)$; this can be shown by selecting summands x_0 and x_1 for x so that

$$\|x_0\|_{X_0} + \|x_1\|_{X_1} \leq \|x\|_{\Sigma(X)} + \varepsilon$$

for arbitrarily small $\varepsilon > 0$ given in advance.

13.2.3

In this subsection, we give two important examples. Here the role of U can be played, in particular, by $\mathbb{R}^n$, the half-space, a bounded domain (e.g., with Lipschitz boundary), or a compact manifold with or without boundary; it is these cases which are particularly interesting for us.

Theorem 13.2.2. 1°. *For any real $s_0 \neq s_1$, the spaces $H^{s_0}(U)$ and $H^{s_1}(U)$ form an interpolation pair, and*

$$[H^{s_0}(U), H^{s_1}(U)]_\theta = H^s(U), \tag{13.2.9}$$

where

$$s = (1 - \theta)s_0 + \theta s_1. \tag{13.2.10}$$

In particular,

$$[H^0(U), H^1(U)]_\theta = H^\theta(U). \tag{13.2.11}$$

2°. *The spaces $L_{p_0}(U)$ and $L_{p_1}(U)$ with any $p_0 \neq p_1$ in $[1, \infty]$ form an interpolation pair, and*

$$[L_{p_0}(U), L_{p_1}(U)]_\theta = L_p(U), \tag{13.2.12}$$

where

$$\frac{1}{p} = \frac{1 - \theta}{p_0} + \frac{\theta}{p_1}. \tag{13.2.13}$$

Note that in the case 2°, the density assumption on $\Delta(X)$ does not hold if one of the subscripts is infinity.

Proof of (13.2.11) *for $U = \mathbb{R}^n$.* As we know, the spaces $H^s = H^s(\mathbb{R}^n)$ are isomorphic to $\widehat{H}^s = \widehat{H}^s(\mathbb{R}^n)$ (see Section 1.1). Therefore, it suffices to prove a similar relation for the spaces $\widehat{H}^s$. (This reduces the proof to considering the simpler weighted

L_2 spaces.) For simplicity, we assume that $s_0 = 0$ and $s_1 = 1$. Consider the pair $X = (X_0, X_1)$, where $X_0 = \widehat{H}^0$ and $X_1 = \widehat{H}^1$. It is sufficient to verify the inequalities

$$\|v(\xi)\|_{X_\theta} \le \|v(\xi)\|_{\widehat{H}^\theta} \quad \text{and} \quad \|v(\xi)\|_{\widehat{H}^\theta} \le \|v(\xi)\|_{X_\theta} \qquad (13.2.14)$$

for functions $v(\xi)$ in $\widehat{H}^\theta$ and X_θ, respectively.

Suppose that $v(\xi)$ belongs to $\widehat{H}^\theta$, i.e., $(1+|\xi|^2)^{\theta/2}v(\xi) \in \widehat{H}^0 = L_2$. Taking arbitrarily small $\varepsilon > 0$, we set

$$f(z) = f(z, \xi) = \exp(\varepsilon z^2 - \varepsilon \theta^2)(1+|\xi|^2)^{-z/2}(1+|\xi|^2)^{\theta/2}v(\xi). \qquad (13.2.15)$$

This function is defined so that $f(\theta) = v(\xi)$, and it belongs to $F(X)$. Furthermore,

$$\|f(it)\|_{L_2} \le \exp(-\varepsilon t^2)\|(1+|\xi|^2)^{\theta/2}v(\xi)\|_{L_2} \le \|v\|_{\widehat{H}^\theta},$$
$$\|f(1+it)\|_{\widehat{H}^1} \le \exp(\varepsilon(1-t^2))\|(1+|\xi|^2)^{(\theta-1+1)/2}v(\xi)\|_{L_2} \le e^\varepsilon \|v\|_{\widehat{H}^\theta}.$$

This leads to the inequality

$$\|v(\xi)\|_{X_\theta} \le e^\varepsilon \|v(\xi)\|_{\widehat{H}^\theta}$$

and, thereby, to the first inequality in (13.2.14).

The proof of the second inequality requires certain preparations. We need the following theorem.

Theorem 13.2.3 (on three lines). *Suppose that a numerical function $h(z)$ is holomorphic in the strip S_0 and continuous and bounded on the strip S. Let m_θ be the least upper bound of its absolute value on the vertical line with abscissa θ. Then*

$$m_\theta \le m_0^{1-\theta} m_1^\theta \quad (0 \le \theta \le 1). \qquad (13.2.16)$$

This theorem is similar to Hadamard's well-known three-circles theorem. Its proof can be found, e.g., in [60, Sec. 1.1] and [407, Chap. XII, Sec. 1].

Note also that, *for any nonnegative numbers m_0 and m_1,*

$$m_0^{1-\theta} m_1^\theta \le \max(m_0, m_1), \qquad (13.2.17)$$

which is easy to check separately for $m_0 > m_1$ and $m_0 < m_1$.

We proceed to prove the second inequality in (13.2.14). Take a function $v(\xi) \in X_\theta$. For arbitrarily small $\varepsilon > 0$, there exists a function $f(z) = f(z, \xi) \in F(X)$ such that $f(\theta, \xi) = v(\xi)$ and $\|f\|_{F(X)} \le \|v\|_{X_\theta} + \varepsilon$. Consider the function

$$g(z) = g(z, \xi) = (1+|\xi|^2)^{z/2}f(z). \qquad (13.2.18)$$

We set

$$h(z) = \int g(z, \xi)\varphi(\xi)\, d\xi, \qquad (13.2.19)$$

where $\varphi(\xi)$ is any compactly supported smooth function on $\mathbb{R}^n$ with L_2-norm 1. The idea is that, for any z, we have

$$\|g(z,\xi)\|_{L_2} = \sup|h(z)|, \tag{13.2.20}$$

where the least upper bound is over all such φ.

To the numerical function $h(z)$ we apply the three-lines theorem and the remark after it. We obtain

$$|h(\theta)| \le (\sup|h(it)|)^{1-\theta}(\sup|h(1+it)|)^\theta$$
$$\le \max(\sup|h(it)|, \sup|h(1+it)|). \tag{13.2.21}$$

By virtue of the Schwarz inequality, we have

$$|h(it)| \le \|g(it)\|_{L_2} \quad \text{and} \quad |h(1+it)| \le \|g(1+it)\|_{L_2}. \tag{13.2.22}$$

It follows from (13.2.20)–(13.2.22) that

$$\|g(\theta)\|_{L_2} \le \max(\sup\|g(it)\|_{L_2}, \sup\|g(1+it)\|_{L_2}),$$

i.e.,

$$\|f(\theta)\|_{\widehat{H}^\theta} \le \max(\sup\|f(it)\|_{L_2}, \sup\|f(1+it)\|_{\widehat{H}^1}),$$

so that

$$\|v\|_{\widehat{H}^\theta} \le \|v\|_{X_\theta} + \varepsilon.$$

We obtain the second inequality in (13.2.14). $\square$

Later on, we shall explain how to transfer this result to other U. Its generalizations to the spaces H_p^s and B_p^s will be given in Section 14.

Assertion 2° is verified for all U specified above simultaneously.

Proof of 2° for $1 < p_j < \infty$.[2] We use the notation

$$X_0 = L_{p_0}(U), \quad X_1 = L_{p_1}(U), \quad X_\theta = [X_0, X_1]_\theta, \quad L_p = L_p(U). \tag{13.2.23}$$

We must verify that the norms of a function in the spaces L_p and X_θ dominate each other. Here by x we denote a point of U.

Suppose that a function $u(x)$ belongs to L_p. Let us normalize it: $\|u\|_{L_p} = 1$. Consider the function

$$f(z) = \exp(\varepsilon z^2 - \varepsilon\theta^2)|u|^{p/p(z)}u/|u|, \tag{13.2.24}$$

where ε is a small positive number and $1/p(z) = (1-z)/p_0 + z/p_1$. We remove the set of points x at which $u(x) = 0$ from U. The function $f(z)$ belongs to $F(X)$ and equals u at $z = \theta$. Simple estimates show that

$$\|f(it)\|_{L_{p_0}} \le 1 \quad \text{and} \quad \|f(1+it)\|_{L_{p_1}} \le e^\varepsilon, \tag{13.2.25}$$

---

[2] The first part of the proof follows the book [60]. There the proof is given for $1 \le p_j \le \infty$.

so that $\|f\|_{F(X)} \le e^{\varepsilon}$. This leads to the inequality $\|u\|_{X_\theta} \le \|u\|_{L_p}$.

Let us prove the reverse inequality. First we note that (13.2.13) implies

$$\frac{1}{p'} = \frac{1-\theta}{p'_0} + \frac{\theta}{p'_1}. \tag{13.2.26}$$

Indeed, this relation is obtained by subtracting both sides of (13.2.13) from 1. Let $v(x)$ be a function in $C_0^\infty(U)$ with $L_{p'}$-norm 1. We set

$$g(z) = \exp(\varepsilon z^2 - \varepsilon\theta^2)|v|^{p'/p'(z)}v/|v|, \tag{13.2.27}$$

where ε is again a small positive number and $1/p'(z) = (1-z)/p'_0 + z/p'_1$; the points x at which $v(x) = 0$ are removed from U. This function belongs to $F(X')$, where $X' = (L_{p'_0}, L_{p'_1})$, and equals v at $z = \theta$. Similarly to (13.2.25), we have

$$\|g(it)\|_{L_{p'_0}} \le 1 \quad \text{and} \quad \|g(1+it)\|_{L_{p'_1}} \le e^{\varepsilon}. \tag{13.2.28}$$

Now we take a function $u(x) \in X_\theta$ and normalize it for simplicity: $\|u\|_{X_\theta} = 1$. Let us find $f(z) \in F(X)$ such that $f(\theta) = u$ and $\|f\|_{F(X)} \le 1 + \varepsilon$, i.e.,

$$\|f(it)\|_{L_{p_0}} \le 1 + \varepsilon \quad \text{and} \quad \|f(1+it)\|_{L_{p_1}} \le 1 + \varepsilon. \tag{13.2.29}$$

We set

$$h(z) = \int_U f(z,x)g(z,x)\,dx. \tag{13.2.30}$$

For $z = \theta$, we have

$$h(\theta) = \int_U u(x)v(x)\,dx; \tag{13.2.31}$$

therefore,

$$\|u\|_{L_p} = \sup|h(\theta)|, \tag{13.2.32}$$

where the least upper bound is over all v specified above. By the three-lines theorem and the remark after it, we have

$$|h(\theta)| \le \max(\sup|h(it)|, \sup|h(1+it)|). \tag{13.2.33}$$

Applying Hölder's inequality to the integral (13.2.30) and using (13.2.29) and (13.2.28), we obtain

$$|h(it)| \le 1 + \varepsilon \quad \text{and} \quad |h(1+it)| \le (1+\varepsilon)e^{\varepsilon}. \tag{13.2.34}$$

These inequalities, (13.2.32), and (13.2.33) imply

$$\|u\|_{L_p} \le (1+\varepsilon)e^{\varepsilon},$$

which yields the required inequality $\|u\|_{L_p} \leq \|u\|_{X_\theta}$.   □

Assertion $2°$ can also be proved for spaces U with measure and for functions with values in Banach spaces. Moreover, the following assertion is valid. If $X = (X_0, X_1)$ is a Banach pair and $0 < \theta < 1$, then

$$[L_{p_0}(U, X_0), L_{p_1}(U, X_1)]_\theta = L_p(U, X_\theta), \tag{13.2.35}$$

where p is the same as in (13.2.13).

13.2.4

As a corollary of Theorems 13.2.1 and 13.2.2 for the spaces L_p, the *Riesz–Thorin theorem* can be obtained. Its direct proof is given, e.g., in [60, Sec. 1.1] and [407, Chap. 12, Sec. 1].

Theorem 13.2.4. *If a linear operator T acts boundedly from $L_{p_j}(U)$ to $L_{q_j}(V)$ for $j = 0, 1$, where $p_0 \neq p_1$, $q_0 \neq q_1$, and p_j, $q_j \in [1, \infty]$, then it acts boundedly from $L_p(U)$ to $L_q(V)$, where*

$$\frac{1}{p} = \frac{1-\theta}{p_0} + \frac{\theta}{p_1} \quad and \quad \frac{1}{q} = \frac{1-\theta}{q_0} + \frac{\theta}{q_1}, \quad 0 < \theta < 1. \tag{13.2.36}$$

Moreover, an estimate of the form (13.2.6) holds.

Thorin used functions holomorphic in a strip and thus suggested the general construction of the complex interpolation method.

From the Riesz–Thorin theorem we derive two classical inequalities for the convolution

$$h(x) = (f * g)(x) = \int_U f(x-y)g(y)\,dy \tag{13.2.37}$$

(cf. [60]). Suppose that a function $f \in L_1(U)$ is fixed and consider (13.2.37) as an operator taking g to h. If $g \in L_1(U)$, then the inequality

$$\|h\|_{L_1(U)} \leq \|f\|_{L_1(U)}\|g\|_{L_1(U)} \tag{13.2.38}$$

follows from Fubini's theorem. If $g \in L_\infty(U)$, then, obviously,

$$\|h\|_{L_\infty(U)} \leq \|f\|_{L_1(U)}\|g\|_{L_\infty(U)}. \tag{13.2.39}$$

We conclude that, *for $f \in L_1(U)$, the convolution (13.2.37) is a bounded operator on any space $L_p(U)$ and*

$$\|h\|_{L_p(U)} \leq \|f\|_{L_1(U)}\|g\|_{L_p(U)} \quad (1 \leq p \leq \infty). \tag{13.2.40}$$

Now we suppose that f belongs to some $L_r(U)$ and try to treat (13.2.37) as an operator from $L_p(U)$ to $L_q(U)$. For $p = 1$, inequality (13.2.40) with interchanged f

and g yields

$$\|h\|_{L_r(U)} \le \|g\|_{L_1(U)}\|f\|_{L_r(U)} \quad (1 \le r \le \infty). \tag{13.2.41}$$

For $p = r'$, where

$$\frac{1}{r} + \frac{1}{r'} = 1, \tag{13.2.42}$$

Hölder's inequality implies

$$\|h\|_{L_\infty(U)} \le \|f\|_{L_r(U)}\|g\|_{L_{r'}(U)}. \tag{13.2.43}$$

We conclude that the operator under consideration acts boundedly from $L_p(U)$ to $L_q(U)$ for

$$\frac{1}{p} = \frac{1-\theta}{1} + \frac{\theta}{r'} \quad \text{and} \quad \frac{1}{q} = \frac{1-\theta}{r} + \frac{\theta}{\infty}, \tag{13.2.44}$$

$0 < \theta < 1$, and obtain the inequality

$$\|h\|_{L_q(U)} \le \|f\|_{L_r(U)}\|g\|_{L_p(U)}, \tag{13.2.45}$$

which is known as *Young's inequality*. Eliminating θ from (13.2.44), we see that *Young's inequality is valid for*

$$\frac{1}{q} = \frac{1}{p} - \frac{1}{r'}, \quad 1 < p < r'. \tag{13.2.46}$$

The Riesz–Thorin theorem readily implies also the following classical *Hausdorff–Young theorem*.

Theorem 13.2.5. *The Fourier transform F is a continuous operator from $L_p(\mathbb{R}^n)$ to $L_{p'}(\mathbb{R}^n)$ for $1 \le p \le 2$, $p + p' = pp'$. More precisely, for these p,*

$$\|Fu\|_{L_{p'}} \le (2\pi)^{n/p'}\|f\|_{L_p}. \tag{13.2.47}$$

Proof. As is known, F is a continuous operator from $L_2(\mathbb{R}^n)$ to $L_2(\mathbb{R}^n)$ (and even an isomorphism up to a multiplier: see Parseval's identity (1.1.5)) and from $L_1(\mathbb{R}^n)$ to $L_\infty(\mathbb{R}^n)$ (which is easy to verify). This implies the assertion of the theorem and estimate (13.2.47). $\qquad\square$

There is a similar theorem about Fourier series, also due to Hausdorff and Young; see [407, Chap. XII, Sec. 2]. In [407], a more general theorem of F. Riesz for the case of a uniformly bounded orthonormal basis instead of a trigonometric system is also proved.

Theorem 13.2.6. *Suppose that $1 < p \le 2$ and $p + p' = pp'$.*

$1°$. *Let $u(t)$ be a function from $L_p(0, 2\pi)$ with Fourier coefficients*

$$c_k = \frac{1}{2\pi} \int_0^{2\pi} u(t)e^{-ikt}dt \quad (k \in \mathbb{Z}). \tag{13.2.48}$$

Then the sequence $c = \{c_k\}_{-\infty}^{\infty}$ *belongs to* $l_{p'}$, *and*

$$\|c\|_{l_{p'}} \leq (2\pi)^{-1/p}\|u\|_{L_p}. \tag{13.2.49}$$

2°. *If a sequence* $c = \{c_k\}_{-\infty}^{\infty}$ *belongs to* l_p, *then there exists a function* $u \in L_{p'}(0, 2\pi)$ *with Fourier coefficients* (13.2.48), *and the inequality*

$$(2\pi)^{-1/p'}\|u\|_{L_{p'}} \leq \|c\|_{l_p} \tag{13.2.50}$$

is valid.

The function $u(t)$ is constructed from its Fourier coefficients as the limit of the partial sums

$$\sum_{|k|\leq N} c_k e^{ikt}.$$

13.2.5

We supplement our exposition of the complex method with the following theorem, which provides a better understanding of properties of the space $F(X)$. Its proof is given in [223, Chap. IV, Sec. 1].

Theorem 13.2.7. 1°. *At* $z = s + it \in S_0$, *the functions in* $F(X)$ *can be represented as*

$$f(z) = \int_{-\infty}^{\infty} f(i\tau)\mu_0(z,\tau)\,d\tau + \int_{-\infty}^{\infty} f(1+i\tau)\mu_1(z,\tau)\,d\tau, \tag{13.2.51}$$

where

$$\mu_j(z,\tau) = \frac{1}{2} \frac{\sin \pi s}{\cosh \pi(\tau - t) - (-1)^j \cos \pi s} \quad (j = 0, 1). \tag{13.2.52}$$

2°. *For* $z \in S$,

$$\|f(z)\|_{\Sigma(X)} \leq \|f\|_{F(X)}. \tag{13.2.53}$$

3°. *In* $F(X)$ *the set of functions of the form*

$$g(z) = e^{\delta z^2} \sum_{1}^{N} x_k e^{\lambda_k z} \tag{13.2.54}$$

is dense, where $x_k \in \Delta(X)$, *the* λ_k *are real, and* $\delta > 0$. *Therefore, the set of functions with values in* $\Delta(X)$ *is dense in* $F(X)$.

We shall use the last assertion in Section 13.7.

13.2.6

In this subsection, we give the simplest statement of the important *Marcinkiewicz theorem*. This theorem, as well as the Riesz–Thorin theorem, was one of the first theorems of interpolation theory.

Suppose for simplicity that U and V are domains in $\mathbb{R}^n$ and f is a scalar real- or complex-valued function on U. Its distribution function $m(\sigma, f)$, $\sigma > 0$, is defined as the Lebesgue measure of the set of points at which $|f(x)| > \sigma$. The space $L_p(U)$ ($1 \le p < \infty$) is contained in the *weak L_p space* $L_p^*(U)$, which consists of those f for which

$$\|f\|_{L_p^*(U)} = \sup_\sigma \sigma m(\sigma, f)^{1/p} < \infty.$$

This is a quasi-norm. The space $L_\infty^*(U)$ is defined as $L_\infty(U)$.

A linear operator T from $L_p(U)$ to $L_q^*(V)$ is said to be bounded if $\|Tf\|_{L_q^*(V)} \le C\|f\|_{L_p(U)}$ for some C; the greatest lower bound for such C is called its norm.

Theorem 13.2.8. *Let $p_0 \ne p_1$, and let T be a bounded operator from $L_{p_0}(U)$ to $L_{q_0}^*(V)$ with norm M_0^* and from $L_{p_1}(U)$ to $L_{q_1}^*(V)$ with norm M_1^*. Suppose that θ and p are the same as in (13.2.36) and $p \le q$. Then T is a bounded operator from $L_p(U)$ to $L_q(V)$ with norm not exceeding $C_\theta (M_0^*)^{1-\theta} (M_1^*)^\theta$.*

This is, to a certain extent, a generalization of the Riesz–Thorin theorem, because, at the extreme index values, the operator is not required to be bounded as acting from $L_p(U)$ to $L_q(V)$; furthermore, real-valued functions are allowed. On the other hand, the assumption $p \le q$ is added (and is essential).

More general statements of the Marcinkiewicz theorem and their applications are discussed, e.g., in [60].

13.3 The Real Method

The real method in various forms was proposed by Lions, Peetre (see, e.g., [233], [238], and [297]), and other mathematicians; detailed references can be found in [60], [376], and [77]. We give one of the possible definitions, namely, describe the so-called *K-method* [297].

Let again (X_0, X_1) be an interpolation pair. Following Peetre, we define the numerical function $K(t, x) = K(t, x, X)$ for $t \ge 0$ and $x \in \Sigma(X)$ as

$$K(t, x) = \inf\{\|x_0\|_{X_0} + t\|x_1\|_{X_1} : x_0 \in X_0, \ x_1 \in X_1, \ x = x_0 + x_1\}. \qquad (13.3.1)$$

It is called the *K-functional*. For each $t > 0$, this is a norm on $\Sigma(X)$ equivalent to the initial one. For fixed x, this is a continuous increasing function convex upward [376]. The functions

$$K(t, x, X) \quad \text{and} \quad t^{-1}K(t, x, X) \qquad (13.3.2)$$

are bounded above. Moreover, as shown, e.g., in [77, p. 296], the former function has a finite limit as $t \to \infty$ and the latter, as $t \to 0$. Note also that

$$t^{-1}K(t,x,(X_0,X_1)) = K(t^{-1},x,(X_1,X_0)).\qquad(13.3.3)$$

Now we define the *real interpolation space* $X_{\theta,q} = (X_0,X_1)_{\theta,q}$ with $0 < \theta < 1$ and $1 \leq q < \infty$ as the space of all $x \in \Sigma(X)$ for which

$$\|x\|_{X_{\theta,q}} = \left(\int_0^\infty (t^{-\theta}K(t,x))^q \frac{dt}{t}\right)^{1/q} < \infty.\qquad(13.3.4)$$

Expression (13.3.4) is a norm, and the space $X_{\theta,q}$ with this norm is Banach. For $q = \infty$, the norm is defined by

$$\|x\|_{X_{\theta,\infty}} = \sup_{t>0} t^{-\theta}K(t,x).\qquad(13.3.5)$$

All these spaces contain $\Delta(X)$ and are contained in $\Sigma(X)$. The following assertions are valid.

1. The equality $(X_0,X_1)_{\theta,q} = (X_1,X_0)_{1-\theta,q}$ holds.
2. The space $\Delta(X)$ is dense in $X_{\theta,q}$ for $q < \infty$.
3. If $X_1 \subset X_0$, then $X_{\theta_1,q} \subset X_{\theta_0,q}$ for $\theta_0 < \theta_1$.
4. If $q_1 < q_2$, then $X_{\theta,q_1} \subset X_{\theta,q_2}$.
5. If X_0 and X_1 coincide, then all $X_{\theta,q}$ coincide with them.

Below we give the main theorem about the K-method, which is an analogue of Theorem 13.2.1. Suppose again that $X = (X_0,X_1)$ and $Y = (Y_0,Y_1)$ are two interpolation pairs and T is a bounded linear operator from $\Sigma(X)$ to $\Sigma(Y)$. We denote its restrictions to X_j and $X_{\theta,q}$ by T_j (as before) and $T_{\theta,q}$, respectively. This notation will also be used in what follows.

Theorem 13.3.1. *Let T_j be bounded operators from X_j to Y_j with norms $\|T_j\| = M_j$ $(j = 0, 1)$. Then $T_{\theta,q}$ is a bounded operator from $X_{\theta,q}$ to $Y_{\theta,q}$ $(0 < \theta < 1, 1 \leq q \leq \infty)$ with norm $M_{\theta,q}$ satisfying the inequality*

$$M_{\theta,q} \leq M_0^{1-\theta}M_1^{\theta}.\qquad(13.3.6)$$

Proof. Let $x \in X_{\theta,q}$. We have

$$K(t,Tx,Y) \leq \|T_0\|K\left(\frac{\|T_1\|}{\|T_0\|}t,x,X\right).$$

Therefore,

$$\|Tx\|_{Y_{\theta,q}} \leq \|T_0\|\left(\int_0^\infty \left(t^{-\theta}K\left(\frac{\|T_1\|}{\|T_0\|}t,x,X\right)\right)^q \frac{dt}{t}\right)^{1/q}.$$

Making the change $\tau = (\|T_1\|/\|T_0\|)t$ in the integral, we obtain

$$\|Tx\|_{Y_{\theta,q}} \le \|T_0\| \left(\frac{\|T_1\|}{\|T_0\|} \right)^\theta \|x\|_{X_{\theta,q}}.$$

Thus, $T_{\theta,q}x$ belongs to $Y_{\theta,q}$, and the required estimate holds. □

In this case, we can say again that the interpolation functor $X \mapsto X_{\theta,q}$ is exact of order θ.

There also exist other real methods, such as the J-method (Peetre), the method of means (Lions and Peetre), and the trace method (Lions; this method was suggested by trace theorems for Sobolev-type spaces). However, it can be proved that they are equivalent; this is usually done in monographs (see, e.g., [60] and [376]). These versions of the real method can be useful in situations where it is required to investigate the interpolation of new spaces. We do not dwell on these and other methods, as well as on the general notion of the interpolation functor. We strongly recommend that the reader read, at least, about the trace method in Lions and Magenes [237] or in Triebel [376].

The following theorem exemplifies real interpolation (this is an analogue of the second part of Theorem 13.2.2, but we give it without proof).

Theorem 13.3.2. *For $0 < \theta < 1$ and $1 \le p_0 < p_1 \le \infty$,*

$$(L_{p_0}(U), L_{p_1}(U))_{\theta,p} = L_p(U). \tag{13.3.7}$$

Here U, θ, and p are the same as in Theorem 13.2.2.

Note that relation (13.3.7) can be generalized as

$$(L_{p_0}(U), L_{p_1}(U))_{\theta,q} = L_{p,q}(U), \tag{13.3.8}$$

where $q > p_0$. Here the space on the right-hand side is the *Lorentz space*, and relation (13.3.8) can be taken for one of its definitions. Other definitions are given in [60] and [376]. The spaces L_p^* in Section 13.2.6 are a special case of Lorentz spaces.

Below we state a theorem on the coincidence of the real and the complex method.

Theorem 13.3.3. *If H_1 and H_2 are Hilbert spaces and one of them is continuously and densely embedded in the other, then, for $0 < \theta < 1$,*

$$(H_1, H_2)_{\theta,2} = [H_1, H_2]_\theta. \tag{13.3.9}$$

This fact can be found in [376, Sec. 1.18.10, Remark 3]. This coincidence was essentially known to Lions [236].

The assumptions made in [49] are somewhat relaxed: it is only required there that the spaces in question form an interpolation pair and their intersection be dense in each of them.

In particular, for $0 < \theta < 1$ and $s_0 \ne s_1$, we have

$$(H^{s_0}(U), H^{s_1}(U))_{\theta,2} = H^s(U), \tag{13.3.10}$$

where s is defined by (13.2.10).

Interpolation relations for the real method and the spaces H_p^s and B_p^s will be listed in Section 14.

13.4 Retractions and Coretractions

The theorems stated in the next three sections are well known; their proofs can be found in [60] and [376]. We want, in particular, to explain how the circle of spaces admitting interpolation can be enlarged in practice.

We need the following definition. Let X and Y be two Banach spaces. A continuous operator R from X to Y is called a *retraction* if it has a continuous right inverse S: $RS = I$, where I is the identity operator on Y. The space Y is then called a *retract* of X, and the operator S is called a *coretraction*.

Obviously, $P = SR$ is a *projection* in X, i.e., $P^2 = P$. Indeed, $(SR)^2 = S(RS)R = SR$.

A subspace X' in X is said to be *complemented* if there is another subspace X'' such that X is the direct sum $X' \dotplus X''$. The existence of a continuous projection P onto X' implies that this subspace is complemented:

$$X = PX + (I - P)X. \tag{13.4.1}$$

Here I is the identity operator on X.

The converse is also true: if a space X is the direct sum of its subspaces X' and X'', then there exist continuous projections from X onto X' and X''. Namely, if $X \ni x = x' + x''$, where $x' \in X'$ and $x'' \in X''$, then we can set $Px = x'$ and $(I - P)x = x''$.

The space X'' can be chosen in different ways, and the projection P depends on this choice. It is possible to say that P projects X onto X' *parallel to X''*.

Finite-dimensional spaces are always complemented; see Proposition 18.1.1. Subspaces of Hilbert spaces are always complemented as well: they have orthogonal complements.

Thus, $X_1 = SY$ is a complemented subspace of X. The retract Y of X turns out to be isomorphic to it: R and S are mutually inverse transformations $X_1 \to Y$ and $Y \to X_1$.

Now we change notation. Suppose given two interpolation pairs $X = (X_0, X_1)$ and $Y = (Y_0, Y_1)$ and continuous maps R: $\Sigma(X) \to \Sigma(Y)$ and S: $\Sigma(Y) \to \Sigma(X)$. Suppose also that Y_0 is a retract of X_0, and the corresponding retraction R_0 and coretraction S_0 are the restrictions of the given maps R and S to X_0 and Y_0, respectively. Similarly, suppose that Y_1 is a retract of X_1, and the corresponding retraction R_1 and coretraction S_1 are the restrictions of the given maps R and S to X_1 and Y_1, respectively. The operators R_i and S_i are assumed to be continuous.

Theorem 13.4.1. *Under the above assumptions, each* Y_θ, $0 < \theta < 1$, *is a retract of* X_θ. *The corresponding retraction* R_θ *and coretraction* S_θ *are the restrictions of the maps* R *and* S *to* X_θ *and* Y_θ, *respectively.*

Similarly, each $Y_{\theta,q}$ *is a retract of* $X_{\theta,q}$, $0 < \theta < 1$, $1 \le q \le \infty$. *The corresponding retraction* $R_{\theta,q}$ *and coretraction* $S_{\theta,q}$ *are the restrictions of the given maps* R *and* S *to* $X_{\theta,q}$ *and* $Y_{\theta,q}$, *respectively.*

This theorem is used in situations where X_θ or $X_{\theta,q}$ is known. For this reason, the following assertion is convenient.

Corollary 13.4.2. *Under the assumptions of Theorem* 13.4.1, $R_\theta X_\theta$ *coincides with* Y_θ *and* $R_{\theta,q} X_{\theta,q}$ *coincides with* $Y_{\theta,q}$.

An important example of the application of this assertion is as follows. The restriction of the functions in $H^s(\mathbb{R}^n)$ to a bounded domain Ω with smooth or Lipschitz boundary Γ (or to a half-space) is a retraction of this space onto $H^s(\Omega)$: the corresponding coretraction is the extension of functions in $H^s(\Omega)$ to functions in $H^s(\mathbb{R}^n)$ (see Section 10). Since we already know that (13.2.9) holds for the spaces $H^s(\mathbb{R}^n)$, *a similar result for the spaces* $H^s(\Omega)$ *follows as a corollary*. Note that, without the universal extension operator constructed in Section 10, which applies also to negative s, we could not obtain such a result in full generality. The situation with spaces in $\mathbb{R}^n_+$ is similar. Thus, we see, in particular, that *the spaces* $H^s(\Omega)$ *with fractional indices can be defined by interpolating spaces with integer indices.*

In a similar way, these relations can be partially obtained for a closed manifold being the smooth or Lipschitz boundary Γ of a bounded domain Ω. In this case, we have a retraction $H^{s+1/2}(\Omega) \to H^s(\Gamma)$, namely, the passage to the trace of a function in $H^{s+1/2}(\Omega)$ on $H^s(\Gamma)$, where $s > 0$ (for a smooth boundary) or $0 < s < 1$ (for a Lipschitz boundary). The corresponding coretraction is the reconstruction of a function (to be more precise, of one of functions) from its trace. Since relation (13.2.9) holds on Ω, we can derive its validity on Γ for s in the range specified above.

But these limitations can be removed: see Triebel's book [378], where local charts are used, and his paper [379]. The result can also be extended to function spaces on a part of a manifold (accordingly, with smooth or Lipschitz boundary).

The following theorem about the interpolation of (complemented) subspaces is derived from Theorem 13.4.1.

Theorem 13.4.3. *Let* (X_0, X_1) *be an interpolation pair, and let* Z *be a complemented subspace in* $\Sigma(X)$ *such that the restrictions of the corresponding projection to* X_0 *and* X_1 *are continuous. Then* $(X_0 \cap Z, X_1 \cap Z)$ *is an interpolation pair as well, and*

$$[X_0 \cap Z, X_1 \cap Z]_\theta = [X_0, X_1]_\theta \cap Z, \tag{13.4.2}$$

$$(X_0 \cap Z, X_1 \cap Z)_{\theta,q} = (X_0, X_1)_{\theta,q} \cap Z. \tag{13.4.3}$$

The interpolation of quotient spaces is discussed in [376, Sec. 1.17.2].

13.5 Duality

We begin with the following general fact (see [60, Sec. 2.7]).

Theorem 13.5.1. *Let $X = (X_0, X_1)$ be an interpolation pair, and let X_j^*, $j = 0, 1$, be the spaces dual to X_j. Then $X^* = (X_0^*, X_1^*)$ is an interpolation pair as well, and*

$$[\Delta(X)]^* = \Sigma(X^*), \quad [\Sigma(X)]^* = \Delta(X^*). \tag{13.5.1}$$

Now we state a duality theorem for the real and the complex method (see [60, Secs. 3.7 and 4.5]).

Theorem 13.5.2. *Let $X = (X_0, X_1)$ be an interpolation pair. Then, for $0 < \theta < 1$ and $1 \le q < \infty$,*

$$(X_0, X_1)_{\theta, q}^* = (X_0^*, X_1^*)_{\theta, q'}, \quad \frac{1}{q} + \frac{1}{q'} = 1. \tag{13.5.2}$$

If at least one of the spaces X_0 and X_1 is reflexive, then

$$[X_0, X_1]_\theta^* = [X_0^*, X_1^*]_\theta, \quad 0 < \theta < 1, \tag{13.5.3}$$

and the norms of these spaces are equal.

Let us illustrate this theorem by examples.

1. Obviously,

$$[H^{s_0}(\mathbb{R}^n), H^{s_1}(\mathbb{R}^n)]_\theta^* = [H^{-s_0}(\mathbb{R}^n), H^{-s_1}(\mathbb{R}^n)]_\theta, \tag{13.5.4}$$

$$[L_{p_0}(U), L_{p_1}(U)]_\theta^* = [L_{p_0'}(U), L_{p_1'}(U)]_\theta. \tag{13.5.5}$$

2. Using the duality of the spaces $H^s(\Omega)$ and $\widetilde{H}^{-s}(\Omega)$, we obtain the following corollary for a bounded Lipschitz domain Ω from relation (13.2.9).

Corollary 13.5.3. *A relation of the form* (13.2.9) *is valid for the spaces* $\widetilde{H}^s(\Omega)$.

Remark 13.5.4. Usually, in cases of interest to us, the forms determining duality were extensions of the same form to the direct product of the corresponding spaces. In the situations which we considered, these were extensions of the inner product in L_2 spaces on various sets ($\mathbb{R}^n$, compact manifolds, $\mathbb{R}_+^n$, bounded domains, closed surfaces, or their parts, sometimes after certain continuations) to direct products. These forms are also natural extensions of forms that define the action of a distribution on a test function (which is a more general possible point of view).

It should be taken into account that, in these cases, the form satisfies the generalized Schwarz inequality

$$|(u, v)| \le C\|u\|\|v\|^*, \tag{13.5.6}$$

where the constant C may differ from 1. If $C = 1$, then we can say that (13.5.6) is the *exact Schwarz inequality*. In Theorems 13.5.1 and 13.5.2, the exact Schwarz inequality is meant. To pass from a generalized Schwarz inequality to the exact one, it suffices to perform renormalization: e.g., replace the norm $\|v\|^*$ by $\|v\|' = \|v\|^*/C$.

13.6 Iterated Interpolation

In interpolation theory, *reiteration theorems* are often useful. See [60] and [376]. Their meaning is as follows. If we have two spaces obtained by interpolation from spaces X_0 and X_1 and again perform interpolation, departing from the obtained spaces, then we obtain spaces which might be obtained by interpolation from X_0 and X_1. Below we give a statement for complex interpolation; the relationship between the indices does not need comments.

Theorem 13.6.1. *Let $\{X_\theta\}$ be a scale of spaces obtained by complex interpolation from spaces X_0 and X_1, and let*

$$\theta = (1-\eta)\theta_0 + \eta\theta_1, \tag{13.6.1}$$

where η, θ_0, and θ_1 are strictly between 0 and 1 and $\theta_0 \neq \theta_1$. Then

$$[X_{\theta_0}, X_{\theta_1}]_\eta = X_\theta, \tag{13.6.2}$$

and the norms of these spaces are equal.

The statement of a similar result for real interpolation is somewhat more complicated [76]:

Theorem 13.6.2. *Let $X = (X_0, X_1)$ be an interpolation pair, and let*

$$0 < \theta_j < 1, \quad 0 < \eta < 1 \quad (j = 0,1), \quad \theta = (1-\eta)\theta_0 + \eta\theta_1, \quad 1 \leq q \leq \infty. \tag{13.6.3}$$

Then, for $\theta_0 \neq \theta_1$ and any $q_j \in [1,\infty]$,

$$(X_{\theta_0,q_0}, X_{\theta_1,q_1})_{\eta,q} = X_{\theta,q}. \tag{13.6.4}$$

If $\theta_0 = \theta_1 = \theta$, then relation (13.6.4) is valid for

$$\frac{1}{q} = \frac{1-\eta}{q_0} + \frac{\eta}{q_1}. \tag{13.6.5}$$

For real and complex or complex and real interpolations performed successively, the following relations hold (see [60, Sec. 4.7]).

Theorem 13.6.3. *Let $X = (X_0, X_1)$ be an interpolation pair,*

$$0 < \theta_0 < \theta_1 < 1, \quad 0 < \eta < 1, \quad 1 \leq q_j \leq \infty,$$
$$\theta = (1-\eta)\theta_0 + \eta\theta_1, \quad and \quad \frac{1}{q} = \frac{1-\eta}{q_0} + \frac{\eta}{q_1}. \tag{13.6.6}$$

Then

$$[X_{\theta_0,q_0}, X_{\theta_1,q_1}]_\eta = X_{\theta,q}. \tag{13.6.7}$$

Furthermore, for $1 \leq q \leq \infty$,

$$(X_{\theta_0}, X_{\theta_1})_{\eta,q} = X_{\theta,q}. \tag{13.6.8}$$

13.7 Interpolation and Extrapolation of Invertibility

The results of this section are especially useful in studying partial differential equations.

13.7.1

Theorem 13.7.1. *Let* $X = (X_0, X_1)$ *and* $Y = (Y_0, Y_1)$ *be two interpolation pairs, and let* T *be a continuous operator from* $\Sigma(X)$ *to* $\Sigma(Y)$. *Suppose that the operators* T_θ *are continuous and invertible for* $\theta = 0$ *and* 1. *Suppose also that the following compatibility condition holds: the restriction of the operators* T_0 *and* T_1 *to* $\Delta(X)$ *coincide, and the inverse operators coincide on* $\Delta(Y)$. *Then all* $T_\theta \colon X_\theta \to Y_\theta$ *are invertible operators.*

A similar assertion is also valid for real interpolation.

The simplest case is where $X_1 \subset X_0$ and $T_1 = T_0|_{X_1}$. A similar theorem was stated and used in [155] in the case $X = Y$.

Proof of Theorem 13.7.1. We have the operators $S_0 = T_0^{-1}$ on Y_0 and $S_1 = T_1^{-1}$ on Y_1. Their extension S to $\Sigma(Y)$ can be well defined, and this is a bounded operator from $\Sigma(Y)$ to $\Sigma(X)$ (see the remark in Section 13.2.2). It is easy to verify that this is a two-sided inverse of T. For example, if

$$y = y_0 + y_1, \quad y_j \in Y_j,$$

then

$$TSy = T(S_0y_0 + S_1y_1) = T_0S_0y_0 + T_1S_1y_1 = y_0 + y_1 = y,$$

so that TS is the identity operator; ST is the identity operator for similar reasons.

It remains to apply Theorems 13.2.1 and 13.3.1.   □

There is also a strong theorem about the *extrapolation of invertibility* for complex interpolation, which implies results that are difficult to obtain without it in some particular situations. Its applications are considered in Sections 16 and 17.

Theorem 13.7.2. *Let* $X = (X_0, X_1)$ *and* $Y = (Y_0, Y_1)$ *be two interpolation pairs, and let* T *be a continuous operator from* $\Sigma(X)$ *to* $\Sigma(Y)$. *Suppose that its restriction* $T_\theta \colon X_\theta \to Y_\theta$ *is invertible for some* $\theta = \theta_0 \in (0, 1)$. *Then* T_θ *is invertible for all* θ *in some neighborhood of* θ_0.

This fact is also known as the *stability of the invertibility* of an operator.

This theorem was obtained by Shneiberg [347]; his results have drawn a wide response in the literature. This theorem resembles the classical fact of functional analysis that an operator close to an invertible operator in norm is invertible, but it is deeper, it refers to spaces with different norms, and its proof is nontrivial. Similar theorems for real interpolation were proved in [404] (about extrapolation of invertibility with respect to θ; a certain analytic structure in the real interpolation method is specified) and in [90] (about extrapolation of invertibility with respect to (θ, q)); see also references in [76, Section 12].

Shneiberg's estimates are interesting and useful. We begin with the result which has the simplest formulation.

If T_t is an invertible operator, then we denote the norm of its inverse by $\beta(t)$. Let $M = \max \|T_j\|$ $(j = 0, 1)$.

Theorem 13.7.3. *If $T_{1/2}$ is invertible, then T_t is invertible for*

$$q_0(t) < \left[M\beta\left(\tfrac{1}{2}\right)\right]^{-1}, \quad where \ \ q_0(t) = \left| \tan\left[\tfrac{\pi}{2}\left(t - \tfrac{1}{2}\right)\right]\right|; \qquad (13.7.1)$$

moreover, for these t we have

$$\beta(t) \le \frac{M\beta\left(\tfrac{1}{2}\right) - q_0(t)}{M\left[1 - q_0(t)M\beta\left(\tfrac{1}{2}\right)\right]}. \qquad (13.7.2)$$

Let us comment this. We have

$$\beta\left(\tfrac{1}{2}\right) \ge \frac{1}{\|T_{1/2}\|} \ge \frac{1}{M}$$

by virtue of (13.2.6); thus, $M\beta\left(\tfrac{1}{2}\right) \ge 1$, and $q_0(t) < 1$, so that the numerator on the right-hand side in (13.7.2) is positive.

In the same papers theorems about the extrapolation (stability) of the Fredholm property with the preservation of the index of operators were proved (see definitions in Section 18.1). It was also shown there that if the kernel or the cokernel of the initial operator is trivial, then this property is stable as well. See also [198].

13.7.2

In the rest of this section, we give more complete results of Shneiberg and outline their proofs, which are not very simple. We elaborate on some details but under the additional assumption that at least one of the spaces X_j is reflexive. This allows us to use relation (13.5.3).

The reader may skip this material, if desired.

We introduce the notation

$$\gamma(\theta) = \gamma(\theta, T) = \inf \frac{\|Tx\|_{Y_\theta}}{\|x\|_{X_\theta}} \qquad (13.7.3)$$

and denote the set of points θ at which $\gamma(\theta) > 0$ by K_T. Let $s \in K_T$. Shneiberg's main result is a lower bound for $\gamma(t)$ at other points t, which makes it possible to estimate the size of the neighborhood of s contained in K_T. For $s, t \in (0, 1)$, we set

$$q(s,t) = \left| \frac{\tan\left[\frac{\pi}{2}\left(s - \frac{1}{2}\right)\right] - \tan\left[\frac{\pi}{2}\left(t - \frac{1}{2}\right)\right]}{1 - \tan\left[\frac{\pi}{2}\left(s - \frac{1}{2}\right)\right]\tan\left[\frac{\pi}{2}\left(t - \frac{1}{2}\right)\right]} \right|. \tag{13.7.4}$$

In particular, $q(1/2, t) = q_0(t)$. Note that $q(s, t) < 1$, because

$$\tan\alpha\,(1 + \tan\beta) < 1 + \tan\beta$$

for $-\pi/4 < \beta \le \alpha < \pi/4$.

Theorem 13.7.4. *The following inequality holds*:

$$\gamma(t) \ge M \frac{\gamma(s) - q(s,t)M}{M - q(s,t)\gamma(s)}. \tag{13.7.5}$$

Proof. We need the following generalization of Schwarz' well-known lemma for a function $\varphi(z)$ holomorphic and satisfying inequality $|\varphi(z)| \le M_1$ inside the unit disk (see [170]): for any points z and λ in this disk,

$$|\varphi(z)| \ge M_1 \frac{|\varphi(\lambda)| - \left|\frac{\lambda - z}{1 - \bar{z}\lambda}\right| M_1}{M_1 - \left|\frac{\lambda - z}{1 - \bar{z}\lambda}\right| |\varphi(\lambda)|}. \tag{13.7.6}$$

From this inequality, applying a conformal transformation, we derive an inequality for functions $\varphi(z)$ holomorphic and satisfying the condition $|\varphi(z)| \le M_1$ in the vertical strip S_0. We write it for real $s, t \in (0, 1)$:

$$|\varphi(t)| \ge M_1 \frac{|\varphi(s)| - q(s,t)M_1}{M_1 - q(s,t)|\varphi(s)|} = \frac{|\varphi(s)| - q(s,t)M_1}{1 - q(s,t)\frac{|\varphi(s)|}{M_1}}. \tag{13.7.7}$$

Note that the value of this expression decreases with increasing M_1. (If a worse upper bound is given, then a worse lower bound is obtained.)

This implies the following estimate for $f \in F(X)$.

Lemma 13.7.5. *The following inequality holds*:

$$\|f(t)\|_{X_t} \ge \|f(z)\|_{F(X)} \frac{\|f(s)\|_{X_s} - q(s,t)\|f(z)\|_{F(X)}}{\|f(z)\|_{F(X)} - q(s,t)\|f(s)\|_{X_s}}. \tag{13.7.8}$$

Proof of the lemma. It suffices to obtain this inequality for functions $f \in F(X)$ with values in $\Delta(X)$ (instead of $\Sigma(X)$), because the set of such functions is dense in $F(X)$. (This was mentioned in assertion 3° of Theorem 13.2.7.) Thus, suppose that the values $f(z)$ belong to $\Delta(X)$. We use the first relation in (13.5.1). It means that there exists a bilinear (not sesquilinear) form $\langle f, g \rangle$ on $\Delta(X) \times \Sigma(X^*)$ which expresses the general form of a continuous linear functional on $\Delta(X)$. This form is continuous in the norms of the spaces $\Delta(X)$ and $\Sigma(X^*)$ and satisfies Schwarz' exact inequality

$$|\langle f, g \rangle| \le \|f\|_{\Delta(X)} \|g\|_{\Sigma(X^*)}. \tag{13.7.9}$$

If $g(z) \in F(X^*)$, then the function

$$\varphi(z) = \langle f(z), g(z) \rangle \tag{13.7.10}$$

is holomorphic in S_0 and bounded on S, so that we can use inequality (13.7.7). Now let us choose $g(z)$.

Since $f(s)$ belongs to X_s and the dual of this space is X_s^* (see Theorem 13.5.2), there exists a functional $g \in X_s^*$ which takes the value $\|f(s)\|_{X_s}$ at $f(s)$ and has norm $\|g\|_{X_s^*} = 1$ (see, e.g., [214, Chap. IV, Sec. 1, Subsec. 3]). It remains a continuous linear functional on $\Delta(X)$ and can be written as $\langle f, g \rangle$ with the same g. Now we take a small $\varepsilon > 0$ and find a function $g(z) \in F(X^*)$ such that $g(s) = g$ and $\|g(z)\|_{F(X^*)} < 1 + \varepsilon$. This defines the function (13.7.10). On the boundaries of the strip S, we have

$$|\varphi(z)| \le \|f(z)\|_{F(X)}(1 + \varepsilon). \tag{13.7.11}$$

According to the three-lines theorem and the remark after it, this inequality holds in the entire strip S. At the points s and t, we have

$$|\varphi(s)| = \|f(s)\|_{X_s} \quad \text{and} \quad |\varphi(t)| \le \|f(t)\|_{X_t}(1 + \varepsilon). \tag{13.7.12}$$

Combining (13.7.7), (13.7.11), and (13.7.12) and reducing by $1 + \varepsilon$, we obtain the inequality

$$\|f(t)\|_{X_t} \ge \|f(z)\|_{F(X)} \frac{\|f(s)\|_{X_s} - q(s,t)\|f(z)\|_{F(X)}(1 + \varepsilon)}{\|f(z)\|_{F(X)}(1 + \varepsilon) - q(s,t)\|f(s)\|_{X_s}}. \tag{13.7.13}$$

The passage to the limit as $\varepsilon \to 0$ yields (13.7.8). $\qquad\square$

We return to the proof of Theorem 13.7.4. Let $x \in X_t$, $\|x\|_{X_t} = 1$. Then, for arbitrarily small $\varepsilon > 0$, there exists a function $f(z) \in F(X)$ such that $\|f\|_{F(X)} \le 1 + \varepsilon$ and $f(t) = x$. According to the lemma (with interchanged s and t),

$$\|f(s)\|_{X_s} \ge (1 + \varepsilon) \frac{1 - q(s,t)(1 + \varepsilon)}{1 + \varepsilon - q(s,t)} = r(\varepsilon). \tag{13.7.14}$$

Consider the function $h(z) = Tf(z)$. It belongs to $F(Y)$. We have

$$\|h(z)\|_{F(Y)} \le M(1 + \varepsilon) \quad \text{and} \quad \|h(s)\|_{Y_s} \ge \gamma(s)r(\varepsilon). \tag{13.7.15}$$

Again applying the lemma, we obtain

$$\|h(t)\|_{Y_t} \ge M(1 + \varepsilon) \frac{\gamma(s)r(\varepsilon) - q(s,t)M(1 + \varepsilon)}{M(1 + \varepsilon) - q(s,t)\gamma(s)r(\varepsilon)}. \tag{13.7.16}$$

As $\varepsilon \to 0$, the quantity $r(\varepsilon)$ tends to 1, so that (13.7.16) implies

$$\|h(t)\|_{Y_t} \geq M \frac{\gamma(s) - q(s,t)M}{M - q(s,t)\gamma(s)}.$$

This leads to inequality (13.7.5).   □

Corollary 13.7.6. *If $s \in K_T$ and*

$$\gamma(s) > Mq(s,t), \qquad\qquad (13.7.17)$$

then $t \in K_T$.

For $s = 1/2$, inequality (13.7.17) takes the form $\gamma(s) > Mq_0(t)$.

Now, instead of K_T, consider the set L_T of points s in the interval $(0,1)$ at which the operator T_s is invertible. For these points, we have

$$\gamma(s,T) = \|T_s^{-1}\|^{-1} = [\beta(s)]^{-1}. \qquad\qquad (13.7.18)$$

Lemma 13.7.7. *If L_T is nonempty, then $L_T = K_T$.*

Proof. For $t \in K_T$, we have the a priori estimate

$$\|x\|_{X_t} \leq \gamma^{-1}(t)\|Tx\|_{Y_t},$$

which ensures that the operator $T_t \colon X_t \to Y_t$ has trivial kernel and closed range (see the remark to Proposition 18.1.7 in Section 18.1). If there is a point $s \in L_T$, then the whole set $\Delta(Y)$ in contained in the image of $T_s \colon X_s \to Y_s$ and, hence, of $T \colon \Sigma(X) \to \Sigma(Y)$; therefore, $\Delta(Y)$ is contained in the image of the operator $T_t \colon X_t \to Y_t$ for any $t \in K_T$. But $\Delta(Y)$ is dense in all Y_t. Thus, $Y_t = TX_t$ for all $t \in K_T$.   □

This lemma is absent in [347].

Lemma 13.7.7, relation (13.7.5), and Corollary 13.7.6 imply the following assertion.

Corollary 13.7.8. *If $s \in L_T$ and $q(s,t) < [M\beta(s)]^{-1}$, then $t \in L_T$. Moreover,*

$$\beta(t) \leq \frac{M\beta(s) - q(s,t)}{M(1 - q(s,t)M\beta(s))}. \qquad\qquad (13.7.19)$$

Here, as well as in (13.7.2), the positivity of the denominator implies that of the numerator.

In particular, we obtain the assertion of Theorem 13.7.3.

13.8 Further Results

13.8.1

In the context of our book, it is useful to dwell on spaces constructed by using fractional powers of operators.

Let A be an unbounded self-adjoint positive operator in a Hilbert space H. Using its real powers A^γ, we can construct a scale of Hilbert spaces H_t so that A^γ is an isomorphism of H_t onto $H_{t-\gamma}$ for any t; see Section 18.3.3.

Theorem 13.8.1. *Any two spaces H_{t_1} and H_{t_2} form an interpolation pair, and for $0 < \theta < 1$,*

$$[H_{t_1}, H_{t_2}]_\theta = H_{(1-\theta)t_1 + \theta t_2}. \tag{13.8.1}$$

A proof of this theorem can be found, e.g., in [223].

Consider examples. On $L_2(\mathbb{R}^n)$, for A we can take the operator Λ (see Section 1); its powers generate the scale of spaces $H^t(\mathbb{R}^n)$. On a closed smooth Riemannian manifold M, for A we can be take the similar operator $(-\Delta + I)^{1/2}$ (where Δ is the Beltrami–Laplace operator); see Section 6.4. This operator has discrete spectrum, and its powers generate the scale of spaces $H^t(M)$. On a half-space, a bounded domain, and a manifold with boundary, we can consider powers of operators associated with certain elliptic problems.

In particular, what is said above is very useful in the case of the torus $\mathbb{T} = \mathbb{T}^n$. In this case, the operator $A = (-\Delta + I)^{1/2}$ has the orthonormal basis in $L_2(\mathbb{T})$ consisting of the eigenfunctions $(2\pi)^{-n/2} e^{i\alpha \cdot x}$; the corresponding eigenvalues are $(|\alpha|^2 + 1)^{1/2}$. Therefore, the powers A^s ($s \in \mathbb{R}$) act on smooth (for simplicity) functions

$$u(x) = \sum_{\alpha \in \mathbb{Z}^n} c_\alpha(u) e^{i\alpha \cdot x}$$

as

$$A^s u(x) = \sum_{\alpha \in \mathbb{Z}^n} (|\alpha|^2 + 1)^{s/2} c_\alpha(u) e^{i\alpha \cdot x}. \tag{13.8.2}$$

Since A^s is an isomorphism of $H^s(\mathbb{T})$ onto $L_2(\mathbb{T})$, it turns out that the norm on $H^s(\mathbb{T})$ is equivalent to the norm (2.2.9):

$$\left(\sum (1 + |\alpha|^2)^s |c_\alpha(u)|^2 \right)^{1/2}. \tag{13.8.3}$$

We also mention that both the theory of partial differential equations and abstract operator theory often have "their own," intrinsic, means for determining whether given operators acting on scales of Banach spaces are, say, bounded, invertible, or Fredholm without appealing to interpolation theory. We have seen this already in Chapters 2 and 3 and shall see in further sections. But in these means, mechanisms of interpolation theory can often be revealed. This theory provides a useful point of view on these means.

13.8.2

Now, we state some assertions about the extrapolation of the compactness of operators (see [105] and the references cited therein and in [76, Sec. 7, Subsec. 1]).

Theorem 13.8.2. *Let* $X = (X_0, X_1)$ *and* $Y = (Y_0, Y_1)$ *be two Banach pairs, and let* $T: \Sigma(X) \to \Sigma(Y)$ *be a continuous operator. Then the following assertions hold.*

1. *If the operator* $T_\theta: X_\theta \to Y_\theta$ *is compact for some* $\theta \in (0, 1)$*, then this is true for all* $\theta \in (0, 1)$.

2. *If the operator* $T_1: X_1 \to Y_1$ *is compact, then so are the operators* $T_{\theta,q}: X_{\theta,q} \to Y_{\theta,q}$ *for all* $\theta \in (0, 1)$ *and* $q \in [1, \infty]$.

Note that, in theorems on elliptic operators on bounded domains or on closed manifolds, compactness usually follows directly from that the space containing the image of the operator is compactly embedded in its domain. But in some works on partial differential equations (for example, in [252]), results of [105] were essentially used.

13.8.3

There is a series of papers with investigation of the dependence of the spectrum of an operator on the indices of a space constructed from an interpolation pair. There are examples in which the spectrum depends on these indices. One of the simplest is the spectrum of the *Cesàro operator*.

$$Tu(t) = \frac{1}{t} \int_0^t u(\tau) d\tau$$

on $L_p(0, \infty)$, $1 < p < \infty$, which depends on p (see [70]). Below we cite a positive result [308, p. 460].

Theorem 13.8.3. *Let* (X_0, X_1) *be an interpolation pair, and let* T *be a continuous operator in* $\Sigma(X)$. *Suppose that the spectrum* $\sigma(T_{\theta_0})$ *is countable for some* $\theta_0 \in [0, 1]$. *Then the spectra* $\sigma(T_\theta)$ *coincide with it for all* $\theta \in (0, 1)$.

This result does not extend to $[0, 1]$.

Note, however, that in classical elliptic problems in bounded domains and on closed manifolds, the countability of the spectrum and its independence of the indices of the space readily follows as a rule from regularity theorems. We saw this in Chapter 2. See also Section 17.2 below; there embeddings will also be used.

13.8.4

Below we give yet another useful theorem. For definiteness, we state it for the complex interpolation method. Suppose given a set of Banach spaces $\{X_t\}$, where the subscript t ranges over some interval I. We say that they form an *interpolation scale* for the complex method if any two spaces X_{t_0} and X_{t_1} with $t_0 < t_1$ form an interpolation pair and all spaces X_t with $t \in (t_0, t_1)$ are obtained from them by complex interpolation:

$$X_t = [X_{t_0}, X_{t_1}]_\theta, \quad \text{where } t = t_0(1 - \theta) + t_1\theta. \tag{13.8.4}$$

Theorem 13.8.4. *Suppose that Banach spaces X_i ($i = 1, 2, 3, 4$) are continuously embedded in a Hausdorff space. If*

$$[X_1, X_3]_\theta = X_2 \quad and \quad [X_2, X_4]_\vartheta = X_3, \quad where \ \theta, \vartheta \in (0, 1), \tag{13.8.5}$$

then

$$[X_1, X_4]_\xi = X_2 \quad and \quad [X_1, X_4]_\eta = X_3, \tag{13.8.6}$$

where

$$\xi = \frac{\theta\vartheta}{1 - \theta + \theta\vartheta} \quad and \quad \eta = \frac{\vartheta}{1 - \theta + \theta\vartheta}. \tag{13.8.7}$$

This theorem is due to Wolf [399]; see also [189]. It has a version for the real interpolation method. This theorem essentially means that if two interpolation scales coincide on the intersection of the corresponding intervals of variation of the index, then they can be united into one scale. Also, these scales can "branch."

In the context of this book, Theorem 13.8.4 is useful for understanding variational elliptic problems. Namely, we know that the spaces $H^s(\Omega)$ and $\widetilde{H}^s(\Omega)$ with $|s| < 1/2$ coincide (up to equivalence of norms). This allows us to compose interpolation scales from the spaces $\widetilde{H}^s(\Omega)$ on the left (or on the right) of some point $s_0 \in (-1/2, 1/2)$ and the spaces $H^s(\Omega)$ on the right (respectively, on the left) of this point. The solution space and the space of right-hand sides turn out to be in the same interpolation scale. We mentioned this in Section 8.1. But the two norms on these spaces with $0 \neq s \in (-1/2, 1/2)$ are only equivalent (do not coincide). The situation with operators on a part of the boundary is similar (it was considered in Section 12; see also Section 17).

13.8.5

Below we state a theorem which we shall use in Section 16.4. We borrowed it from [325], where this theorem is given with reference to [60, Sec. 3.5b].

Theorem 13.8.5. *Let X, Y, and Z be three Banach spaces such that X is continuously embedded in Y, and let T be a bounded linear operator from X to Z. Suppose that*

$$\|Tu\|_Z \leq C\|u\|_X^{1-\sigma}\|u\|_Y^\sigma \quad (u \in X) \tag{13.8.8}$$

for constant $C > 0$ and $\sigma \in (0, 1)$. Then T can be continuously extended to a bounded operator from $(X, Y)_{\sigma,1}$ to Z.

13.8.6

We call the reader's attention to results on the interpolation of analytic families of operators obtained by Stein [360] and included in Stein and Weiss' book [361]. We do not cite them here, but they have drawn a noticeable response in the literature.

13.9 Positive Operators and Their Fractional Powers

Let X be a Banach space with norm $\|\cdot\|$, and let A be a closed linear operator in X with dense domain $D(A) \subset X$. We say that A is *positive* if the closed negative half-axis $\overline{\mathbb{R}}_-$ is contained in its resolvent set and, on this half-axis, the norm of the resolvent $R(\lambda) = (A - \lambda I)^{-1}$ satisfies the inequality

$$\|R(\lambda)\| \le \frac{C}{1 + |\lambda|}. \tag{13.9.1}$$

This implies that the resolvent set contains a sector with bisector $\mathbb{R}_-$ and a neighborhood of the origin (see, e.g., [222]); in a sector with a slightly smaller opening angle and the same bisector and in a slightly smaller neighborhood, an inequality of the form (13.9.1) remains valid.

For simplicity, we consider only positive operators with discrete spectrum.

Below by Λ we denote a fixed closed sector with bisector $\mathbb{R}_+$ outside which estimate (13.9.1) holds. It contains the spectrum of the operator.

Examples of positive operators are the operator A_1 on H_{-1} and the operator A_2 on H_0 considered in Section 11.9. In these cases, there is (a common) sector Λ with opening angle less than π. Examples of positive operators given in Sections 6 and 7 are, of course, systems elliptic with parameter along $\mathbb{R}_-$ and problems with homogeneous boundary conditions uniquely solvable at all points of the closure of this ray.

Given a positive operator A, its powers A^z with any complex z can be defined. The powers $A^{-\alpha}$ with $\operatorname{Re}\alpha > 0$ are defined by

$$A^{-\alpha} = -\frac{1}{2\pi i} \int_\gamma \lambda^{-\alpha} R(\lambda)\, d\lambda. \tag{13.9.2}$$

Here the infinite contour γ encloses the spectrum of A. It largely goes along the rays bounding Λ, but, near the origin, passes around the origin on the right and has no common points with $\overline{\mathbb{R}}_-$. The traverse direction is downward from infinity (i.e., negative). This is a bounded operator. The unbounded operator A^α is defined as the inverse of $A^{-\alpha}$. The precise definitions of the powers A^α for general complex numbers α can be found, e.g., in the books [376] and [222] and also in [37] and [176].

For the multiplication of two powers, the usual rule of addition of exponents applies. If $0 \le \alpha < \beta$, then the domain $D(A^\beta)$ is densely embedded in $D(A^\alpha)$. Furthermore, if A acts on a Hilbert space, then $(A^\alpha)^* = (A^*)^\alpha$ [222].

In the literature, much attention is paid to determining the domains of positive fractional powers.

We mention Kato's theorem in [200] (which strengthens a 1959 result of Krasnosel'sky and Sobolev; see [222]). In particular, this theorem contains the following assertion.

Theorem 13.9.1. *If A and B are positive operators in Hilbert spaces H and H', respectively, and T is a bounded operator from H to H' such that $TD(A) \subset D(B)$ and $\|BTu\|_{H'} \le C\|Au\|_H$ on $D(A)$, then $TD(A^\theta) \subset D(B^\theta)$ and $\|B^\theta Tu\|_{H'} \le C_\theta \|A^\theta u\|_H$ on $D(A^\theta)$, $0 < \theta < 1$.*

However, of special interest is the following theorem, which was proved in Triebel's book [376, Sec. 1.15].

Theorem 13.9.2. *Let A be a positive operator in a Hilbert space $X = H$ such that its purely imaginary powers A^{it} are bounded operators for sufficiently small $|t|$, whose norms are bounded uniformly in t. Suppose that $0 \le \operatorname{Re}\alpha < \operatorname{Re}\beta < \infty$ and $0 < \theta < 1$. Then*

$$[D(A^\alpha), D(A^\beta)]_\theta = D(A^{\alpha(1-\theta)+\beta\theta}). \tag{13.9.3}$$

Arendt mentioned in [37, Sec. 4.4.10] that the converse is also true. The simplest case in which all purely imaginary powers are bounded is the case of a self-adjoint positive operator A with a discrete spectrum in a Hilbert space H. Partly repeating what is said in Section 18.3.3, we remind the reader that, in this case, there exists an orthonormal basis consisting of the eigenvectors e_j of A ($j = 1, 2, \ldots$), and if the λ_j are the corresponding eigenvalues, then any vector in H expands in the series

$$u = \sum_1^\infty (u, e_j)e_j, \quad \|u\|^2 = \sum_1^\infty |(u, e_j)|^2, \tag{13.9.4}$$

and the operator A^α acts as

$$A^\alpha u = \sum_1^\infty \lambda_j^\alpha (u, e_j)e_j. \tag{13.9.5}$$

This operator is unbounded for $\operatorname{Re}\alpha > 0$: the series

$$\|A^\alpha u\|^2 = \sum_1^\infty |\lambda_j^\alpha|^2 |(u, e_j)|^2$$

cannot converge for all u. But, obviously, for purely imaginary $\alpha = it$, this is a uniformly bounded operator.

Thus, the local boundedness of the purely imaginary powers of a positive operator is a necessary and sufficient condition for relation (13.9.3) to hold, as in the case of a self-adjoint operator. This condition is not always satisfied; see, e.g., [215, p. 342].

According to Triebel, little is known about the boundedness of the purely imaginary powers of positive operators. For the operators corresponding to "smooth" problems elliptic with parameter outside a sector with opening angle less than π, a positive result was obtained by Seeley in [338].

The deep study of the Kato problem, which we touch on in Section 16.6, brought positive results for all operators A_1 corresponding, in particular, to the problems in Lipschitz domains considered in this book. We shall be able to explain this in Section 16.6 but with references to certain implications in works on the Kato problem.

By way of preparation, we state a result contained in Yagi's short paper [401].

Theorem 13.9.3. *Suppose that a positive operator A in a Hilbert space H satisfies the relation*

$$D(A^{\alpha}) = D(A^{*\alpha}) \tag{13.9.6}$$

for $0 < \alpha < \varepsilon$ with sufficiently small ε. Then

$$D(A^{\alpha}) = [H, D(A)]_{\alpha} \quad and \quad D(A^{*\alpha}) = [H, D(A^*)]_{\alpha} \quad (0 \le \alpha \le 1). \tag{13.9.7}$$

Moreover, the purely imaginary powers of A are locally bounded.

14 The Spaces W_p^s, H_p^s, and B_p^s

In Sections 14.1–14.5, we briefly describe the facts of the theory of the spaces W_p^s with $s \ge 0$ and of the spaces H_p^s and B_p^s with $-\infty < s < \infty$ and $1 < p < \infty$.

As we already mentioned, for $p = 2$, the last two spaces coincide with H^s. In the general case, like in the case $p = 2$, the superscript s is related to the smoothness of elements of H_p^s or B_p^s. The subscript p indicates their connection with the space L_p.

In Section 14.6, we, in addition, give the definitions of these spaces for $p = 1$ and $p = \infty$. In Section 14.7, we define the more general Triebel–Lizorkin spaces $F_{p,q}^s$ and the general Besov spaces $B_{p,q}^s$. A detailed consideration of the properties of the spaces defined in these two sections is beyond the scope of this book. But we mention some properties of the spaces $B_{p,q}^s$ to be used in Section 16.4.

14.1 Spaces of Functions on $\mathbb{R}^n$

In this subsection, we use largely Triebel's book [376].

As in Section 1, we need the operator

$$\Lambda^t = F^{-1}(1 + |\xi|^2)^{t/2} F, \quad t \in \mathbb{R}, \tag{14.1.1}$$

where F is the Fourier transform in the sense of distributions.

We assume that the numbers p and p' are larger than 1 and related by

$$\frac{1}{p} + \frac{1}{p'} = 1. \tag{14.1.2}$$

The *Sobolev space* $W_p^s(\mathbb{R}^n)$ with nonnegative integer $s = m$ and $1 < p < \infty$ is defined as the space of functions $u(x)$ which belong to $L_p(\mathbb{R}^n)$ together with their derivatives (in the sense of distributions) of order up to m and have finite norm

$$\|u\|_{W_p^m(\mathbb{R}^n)} = \left(\sum_{|\alpha| \leq m} \int |D^\alpha u(x)|^p \, dx \right)^{1/p}, \tag{14.1.3}$$

where we can keep only $\alpha = (0,\ldots,0)$ and $(m,0,\ldots,0),\ldots,(0,\ldots,0,m)$.

The *Slobodetskii space* $W_p^s(\mathbb{R}^n)$ with positive noninteger $s = m + \theta$, $m = 0,1,\ldots$, $0 < \theta < 1$, $1 < p < \infty$, is defined as the space of functions $u(x)$ which belong to $L_p(\mathbb{R}^n)$ together with their derivatives (in the sense of distributions) up to order $m = [s]$ and have finite norm

$$\|u\|_{W_p^s(\mathbb{R}^n)} = \|u\|_{W_p^m(\mathbb{R}^n)} + \sum_{|\alpha|=m} \left(\iint \frac{|D^\alpha u(x) - D^\alpha u(y)|^p}{|x-y|^{n+p\theta}} \, dx \, dy \right)^{1/p}. \tag{14.1.4}$$

In his main work [350], Slobodetskii dealt with the case $p = 2$, but he also considered these spaces with $p \neq 2$ [351].

The space $H_p^s(\mathbb{R}^n)$, usually called the *Bessel potential space*, is defined for any real s and $1 < p < \infty$ as the space of all distributions in S' with finite norm

$$\|u\|_{H_p^s(\mathbb{R}^n)} = \|\Lambda^s u\|_{L_p(\mathbb{R}^n)}. \tag{14.1.5}$$

In the literature, these spaces are also referred to as the *Lebesgue spaces*, *Liouville spaces*, and *generalized Sobolev spaces*. For a nonnegative integer s, $H_p^s(\mathbb{R}^n)$ coincides with the Sobolev space $W_p^s(\mathbb{R}^n)$. For $p = 2$ and any s, it coincides with $H^s(\mathbb{R}^n)$. As previously, by coincidence we always mean coincidence up to equivalence of norms.

The *Besov space* $B_p^s(\mathbb{R}^n)$ can be defined for all s and $1 < p < \infty$ as the space of all distributions in S' with finite norm

$$\|u\|_{B_p^s(\mathbb{R}^n)} = \|\Lambda^{s-\sigma} u\|_{W_p^\sigma(\mathbb{R}^n)} \tag{14.1.6}$$

for some fixed positive noninteger σ. This space does not depend on the choice of σ. Other definitions are also possible; see, e.g., [376]. For noninteger $s > 0$, $B_p^s(\mathbb{R}^n)$ coincides with the Slobodetskii space $W_p^s(\mathbb{R}^n)$, and for $p = 2$ and any s, it again coincides with $H^s(\mathbb{R}^n)$.

Note that, for integer $s = m + 1 > 0$, the norm on $B_p^s(\mathbb{R}^n)$ can be defined by expression (14.1.4) with $\theta = 1$:

$$\|u\|_{B_p^{m+1}(\mathbb{R}^n)} = \|u\|_{W_p^m(\mathbb{R}^n)} + \sum_{|\alpha|=m} \left(\iint \frac{|D^\alpha u(x) - D^\alpha u(y)|^p}{|x-y|^{n+p}} \, dx \, dy \right)^{1/p}; \tag{14.1.7}$$

see [197, p. 7].

The spaces $H_p^s(\mathbb{R}^n)$ and $B_p^s(\mathbb{R}^n)$ coincide only for $p = 2$.

Theorem 14.1.1. *All of the spaces listed above are Banach; in particular, they are complete. The linear manifolds $C_0^\infty(\mathbb{R}^n)$ and $S(\mathbb{R}^n)$ are dense in each of them. The operator Λ^t maps isomorphically $H_p^s(\mathbb{R}^n)$ onto $H_p^{s-t}(\mathbb{R}^n)$ and $B_p^s(\mathbb{R}^n)$ onto $B_p^{s-t}(\mathbb{R}^n)$ for any s, t, and $1 < p < \infty$.*

In the initial definition of the spaces $H^s(\mathbb{R}^n)$, we described these spaces in terms of isomorphisms with weighted L_2-spaces $\widehat{H}^s(\mathbb{R}^n)$. For $p \neq 2$, there are no similar descriptions. In particular, instead of the theorem that the Fourier transform maps isomorphically $L_2(\mathbb{R}^n)$ onto itself, we have only the Hausdorff–Young theorem (Theorem 13.2.5). But the operator (14.1.1) retains its importance. There are also discrete norms (similar to those described in Section 1.14) useful in some technical aspects. We describe them in Section 14.7.

Below we indicate some inclusions. The spaces $W_p^s(\mathbb{R}^n)$, $B_p^s(\mathbb{R}^n)$, and $H_p^s(\mathbb{R}^n)$ expand with decreasing the superscript, and the embedding operators are continuous. Moreover, for these spaces, we have (see, e.g., [376, Sec. 2.3.3])

$$H_p^{s+\varepsilon}(\mathbb{R}^n) \subset B_p^s(\mathbb{R}^n) \subset H_p^s(\mathbb{R}^n) \quad \text{for } 1 < p \leq 2,$$
$$B_p^{s+\varepsilon}(\mathbb{R}^n) \subset H_p^s(\mathbb{R}^n) \subset B_p^s(\mathbb{R}^n) \quad \text{for } 2 \leq p. \tag{14.1.8}$$

Here ε is an arbitrarily small positive number. All embeddings are continuous and dense.

Now we formulate two deep *embedding theorems*. In [376, Sec. 2.8.1], they were proved by using discrete representations of functions. We shall touch on these representations in Section 14.7; see also [60, Sec. 6.5].

Theorem 14.1.2. *If*

$$1 < p \leq q < \infty \quad \text{and} \quad s - \frac{n}{p} \geq t - \frac{n}{q}, \tag{14.1.9}$$

then the following continuous dense embeddings hold:

$$W_p^s(\mathbb{R}^n) \subset W_q^t(\mathbb{R}^n) \quad (s, t \geq 0),$$
$$H_p^s(\mathbb{R}^n) \subset H_q^t(\mathbb{R}^n), \quad B_p^s(\mathbb{R}^n) \subset B_q^t(\mathbb{R}^n). \tag{14.1.10}$$

In particular, $W_p^s(\mathbb{R}^n)$ is continuously embedded in $L_q(\mathbb{R}^n)$ for $0 < n/p - s \leq n/q$.

Theorem 14.1.3. *For*

$$s > \frac{n}{p} + t, \tag{14.1.11}$$

where $t > 0$, the space $W_p^s(\mathbb{R}^n)$ is continuously embedded in $C_b^t(\mathbb{R}^n)$. This embedding also holds for $s = n/p + t$ provided that t is noninteger.

In particular, $W_p^s(\mathbb{R}^n)$ is continuously embedded in $C_b(\mathbb{R}^n)$ for $s > n/p$.

For $p = 2$ and $s > n/2 + t$, this embedding was proved in Section 1. For other indices, means of interpolation theory are used in the proof in [376].

Some additional remarks to the embedding theorems will be given in Section 14.4.

The following theorem is about duality.

Theorem 14.1.4. *The spaces $H_p^s(\mathbb{R}^n)$ and $H_{p'}^{-s}(\mathbb{R}^n)$ with $1 < p < \infty$ are dual to each other with respect to the extension of the inner product $(u,v)_{0,\mathbb{R}^n}$ to their direct product. Moreover, the generalized Hölder–Schwarz inequality*

$$|(u,v)_{0,\mathbb{R}^n}| \leq C\|u\|_{H_p^s(\mathbb{R}^n)}\|v\|_{H_{p'}^{-s}(\mathbb{R}^n)} \tag{14.1.12}$$

is valid. The spaces $B_p^s(\mathbb{R}^n)$ and $B_{p'}^{-s}(\mathbb{R}^n)$ are dual as well.

The proof of the first assertion is reduced, by using the operators $\Lambda^{\pm s}$, to the case $s = 0$, in which the duality between L_p and $L_{p'}$ and the usual Hölder inequality are employed.

In particular, by virtue of the above statements, the dual of $W_p^s(\mathbb{R}^n)$ with integer $s \geq 0$ is the space $H_{p'}^{-s}(\mathbb{R}^n)$ and the dual of $W_p^s(\mathbb{R}^n)$ with noninteger $s > 0$ is $B_{p'}^{-s}(\mathbb{R}^n)$.

Theorem 14.1.4 implies the reflexivity of the spaces $W_p^s(\mathbb{R}^n)$, $H_p^s(\mathbb{R}^n)$, and $B_p^s(\mathbb{R}^n)$.

Below we give some interpolation relations; see, e.g., [60, Sec. 6.4].

Theorem 14.1.5. *Suppose that $\theta \in (0,1)$, $p > 1$, $q > 1$, $s \neq t$, and*

$$\frac{1}{r} = \frac{1-\theta}{p} + \frac{\theta}{q}. \tag{14.1.13}$$

Then the spaces of functions on $\mathbb{R}^n$ satisfy the relations

$$[H_p^s, H_q^t]_\theta = H_r^{(1-\theta)s+\theta t}, \quad (H_p^s, H_q^s)_{\theta,r} = H_r^s,$$
$$(H_p^s, H_p^t)_{\theta,q} = B_q^{(1-\theta)s+\theta t}, \tag{14.1.14}$$

and

$$[B_p^s, B_q^t]_\theta = B_r^{(1-\theta)s+\theta t}, \quad (B_p^s, B_q^t)_{\theta,r} = B_r^{(1-\theta)s+\theta t}. \tag{14.1.15}$$

Not going into details, we only mention that these results can be derived from other definitions of these spaces, with discrete norms (see Section 14.7), by using interpolation theorems for L_p (and l_p) spaces with values in certain Banach spaces, for which our spaces turn out to be retracts; see [60] or [376] for details.

Here the following circumstances attract attention. The Bessel potential spaces and the Besov spaces form generally different interpolation scales. The Sobolev spaces belong to the former scale, and the Slobodetskii spaces belong to the latter. Under the real interpolation of Bessel potential spaces, Besov spaces can arise.

As to multipliers, we restrict ourselves by the simplest assertions, which are similar to those proved in Section 1.9 and as easy to verify.

Theorem 14.1.6. *In $L_p(\mathbb{R}^n)$, the operator of multiplication by a bounded measurable function is bounded. In the Sobolev spaces $W_p^m(\mathbb{R}^n)$ with positive integer m, the operator of multiplication by a function $a(x)$ is bounded if this function belongs to $C_b^{m-1,1}(\mathbb{R}^n)$. In the Slobodetskii spaces $W_p^s(\mathbb{R}^n)$ with noninteger $s = m + \theta > 0$, $0 < \theta < 1$, the operator of multiplication by a function $a(x)$ is bounded if this function belongs to $C_b^{m,\vartheta}(\mathbb{R}^n)$ with $\vartheta \in (\theta, 1)$. Similar assertions are valid for the dual spaces $H_{p'}^{-m}(\mathbb{R}^n)$ and $B_{p'}^{-s}(\mathbb{R}^n)$.*

In particular, the estimates of Section 1.9 can be generalized to these spaces. Deeper facts can be found in [369], [254], and [318].

We should also mention Fourier multipliers. A function $\psi(\xi)$ is called a *Fourier multiplier* for a given space of functions of x on $\mathbb{R}^n$ if the operator

$$u(x) \mapsto ((F^{-1}\psi) * u)(x), \tag{14.1.16}$$

where F is the Fourier transform in the sense of distributions, is bounded in this space. We denote the space of Fourier multipliers for $L_p(\mathbb{R}^n)$ by M_p. For example, $M_2 = L_\infty$. A classical result concerning the M_p spaces is due to Mikhlin [263]; it has many generalizations, which can be found, e.g., in [182], [239], [60], and [376,377]. Below we cite a theorem from [60, Sec. 6.1].

Theorem 14.1.7. *Let $\rho(\xi)$ be a function continuously differentiable up to order l, where $l > n/2$, and let*

$$|\xi^\alpha||\partial^\alpha \rho(\xi)| \le A \quad (|\alpha| \le l). \tag{14.1.17}$$

Then $\rho(\xi) \in M_p$ for all $p \in (1, \infty)$, and the norm of the operator (14.1.16) does not exceed $C_p A$.

A theorem about the trace operator, which is, as previously, defined for smooth functions by

$$\mathrm{Tr}\colon f(x) \mapsto f(x', 0), \tag{14.1.18}$$

is stated as follows; see, e.g., [60, Sec. 6.6], [376, Sec. 2.9], and the references therein.

Theorem 14.1.8. *For $s > 1/p$, the operator* Tr *can be extended to a bounded operator*
(1) from $H_p^s(\mathbb{R}^n)$ to $B_p^{s-1/p}(\mathbb{R}^{n-1})$,
(2) from $B_p^s(\mathbb{R}^n)$ to $B_p^{s-1/p}(\mathbb{R}^{n-1})$,
(3) and, hence, from $W_p^s(\mathbb{R}^n)$ to $B_p^{s-1/p}(\mathbb{R}^{n-1})$.
These operators have bounded right inverses.

Recall that $B_p^{s-1/p}(\mathbb{R}^{n-1})$ coincides with $W_p^{s-1/p}(\mathbb{R}^{n-1})$ for noninteger $s - 1/p > 0$. Note that *the right inverse is the same operator*, i.e., it acts from $B_p^{s-1/p}(\mathbb{R}^{n-1})$ to the intersection of the spaces $H_p^s(\mathbb{R}^n)$ and $B_p^s(\mathbb{R}^n)$. This is ascertained in the book [197].

In particular, the traces of functions belonging to H spaces belong to B spaces.

A theorem about the Cauchy data operator on a hyperplane and its right inverse is also valid: see [376].

14.2 Spaces on Smooth Manifolds

If M is a closed compact C^∞ manifold, then the norms on all spaces

$$W_p^s(M), \quad s \ge 0, \quad H_p^s(M), \quad \text{and} \quad B_p^s(M), \quad s \in \mathbb{R}, \quad 1 < p < \infty,$$

can be naturally defined by using a sufficiently fine partition of unity on M and the corresponding norms on $\mathbb{R}^n$, as in Section 2. It is easy to formulate properties of these spaces implied by similar properties of function spaces on $\mathbb{R}^n$; we shall not dwell on them. If M is of finite smoothness, then the range of admissible s narrows. These spaces expand with decreasing the superscript and with decreasing the subscript; the embedding operators are continuous. The embeddings of $H_p^\sigma(M)$ in $H_p^s(M)$ and of $B_p^\sigma(M)$ in $B_p^s(M)$ for $\sigma > s$ are compact. The same is true for function spaces on a bounded domain, which are considered in the next subsection.

14.3 Function Spaces on $\mathbb{R}_+^n$ and on Smooth Domains

In this subsection, Ω denotes the half-space $\mathbb{R}_+^n$ or a smooth bounded domain in $\mathbb{R}^n$.

For $-\infty < s < \infty$ and $1 < p < \infty$, the *spaces $H_p^s(\Omega)$ and $B_p^s(\Omega)$* are defined as the spaces of all restrictions u to Ω of functions (more precisely, distributions) w in the corresponding spaces on $\mathbb{R}^n$. The norm of such a function u is defined in a standard way as the greatest lower bound of the norms of its extensions w. This refers, in particular, to the Sobolev–Slobodetskii spaces $W_p^s(\Omega)$ with $s \ge 0$.

The norms on the Sobolev–Slobodetskii spaces are equivalent to norms of the form (14.1.3)–(14.1.4) with integration over Ω.

For positive integer m and $0 \le \theta < 1$, the space $H_p^{-m+\theta}(\Omega)$ consists of the derivatives of order up to m of functions in $H_p^\theta(\Omega)$ (in $L_p(\Omega)$ for $\theta = 0$) in the sense of distributions on Ω. For $0 < \theta < 1$, a similar assertion is true for the spaces $B_p^{-m+\theta}(\Omega)$ and $B_p^\theta(\Omega) = W_p^\theta(\Omega)$; see [278].

The Hestenes operator $\mathcal{E}_N$ (see Sections 3.1 and 5.1) can be used to extend functions $u \in W_p^s(\Omega)$, $s \ge 0$, to functions $w \in W_p^s(\mathbb{R}^n)$ for sufficiently large N. The Rychkov universal operator $\mathcal{E}$ [320] can be used to extend functions in $H_p^s(\Omega)$ to functions in $H_p^s(\mathbb{R}^n)$ and functions in $B_p^s(\Omega)$ to functions in $B_p^s(\mathbb{R}^n)$ for any s. It is bounded in the corresponding norms. In Section 10 we considered the Rychkov operator for $p = 2$.

Using definitions and the universal extension operator, we obtain the following theorem.

Theorem 14.3.1. *The $H_p^s(\Omega)$ and $B_p^s(\Omega)$ are Banach spaces. The linear manifold $C^\infty(\overline{\Omega})$ is dense in them. All inclusions and embedding theorems given in Section 14.1 remain valid for Ω instead of $\mathbb{R}^n$. In the case of a bounded domain and $\sigma > s$, the embeddings of $H_p^\sigma(\Omega)$ in $H_p^s(\Omega)$ and of $B_p^\sigma(\Omega)$ in $B_p^s(\Omega)$ are compact. The interpolation relations (14.1.14) and (14.1.15) are valid for Ω instead of $\mathbb{R}^n$.*

In particular, if m is a nonnegative integer and $s = m + \theta, 0 < \theta < 1$, then, in view of the first formula in (14.1.14),

$$H_p^s(\Omega) = [W_p^m(\Omega), W_p^{m+1}(\Omega)]_\theta. \tag{14.3.1}$$

The last assertion of the theorem follows from the fact that the restriction of functions to Ω turns out to be a retraction; see Theorem 13.4.1. All our statements concerning multipliers can be extended to function spaces on Ω as well.

Now we define spaces $\widetilde{H}_p^s(\Omega)$ and $\widetilde{B}_p^s(\Omega)$ for arbitrary s and $1 < p < \infty$. The former consists of all functions (or distributions) belonging to $H_p^s(\mathbb{R}^n)$ and supported in $\overline{\Omega}$. The norm is taken from $H_p^s(\mathbb{R}^n)$. The latter consists of all functions (or distributions) belonging to $B_p^s(\mathbb{R}^n)$ and supported in $\overline{\Omega}$. The norm is taken from $B_p^s(\mathbb{R}^n)$. In these new spaces, the functions from $C_0^\infty(\Omega)$ extended by zero outside Ω are dense. The interpolation relations extend to these spaces, too.

Theorem 14.3.2. *The spaces $H_p^s(\Omega)$ and $\widetilde{H}_{p'}^{-s}(\Omega)$ are dual to each other. The same is true for the spaces $B_p^s(\Omega)$ and $\widetilde{B}_{p'}^{-s}(\Omega)$. In particular, all these spaces are reflexive.*

A proof of this theorem with the definition of the corresponding duality form is given in [278]. Cf. [376] and [379].

The operator $\mathcal{E}_0$ of extension by zero (see Section 3.4) acts boundedly from $H_p^s(\Omega)$ to $H_p^s(\mathbb{R}^n)$ and from $B_p^s(\Omega)$ to $B_p^s(\mathbb{R}^n)$ for $0 \leq s < 1/p$. Taking into account Theorem 14.3.2, we see that this remains true for $-1/p' < s < 0$ and the following result is valid.

Theorem 14.3.3. *The spaces $H_p^s(\Omega)$ and $\widetilde{H}_p^s(\Omega)$ with $-1/p' < s < 1/p$ can be identified, their norms are equivalent. The same is true for the spaces $B_p^s(\Omega)$ and $\widetilde{B}_p^s(\Omega)$.*

By $\mathring{W}_p^s(\Omega)$ we denote the completion of $C_0^\infty(\Omega)$ in $W_p^s(\Omega)$. Similarly we define $\mathring{H}_p^s(\Omega)$ and $\mathring{B}_p^s(\Omega)$.

For a function u on Ω, by $\widetilde{u}$ we denote its extension to $\mathbb{R}^n$ by zero outside Ω. They are identified in the following theorem (see [173]).

Theorem 14.3.4. *The spaces $\widetilde{W}_p^s(\Omega)$ and $\mathring{W}_p^s(\Omega)$ with $s > 0$, $s \neq k + 1/p$ for a nonnegative integer k, can be identified. For other $s = k + 1/p > 0$, $k = [s]$, $\widetilde{W}_p^s(\Omega)$ can be considered as a linear subspace in $\mathring{W}_p^s(\Omega)$ with the stronger topology determined by the norm*

$$\|u\|_{W_p^s(\Omega)} + \sum_{|\alpha|=k} \left\|\rho^{1/2} D^\alpha u\right\|_{L_p(\Omega)} \tag{14.3.2}$$

(*cf.* (5.1.15)). *It is continuously embedded in* $\mathring{W}_p^s(\Omega)$ *but nonclosed.*

See also Triebel [376], Section 4.3.2.

Below we give a trace theorem. By Γ we denote the boundary of a domain Ω.

Theorem 14.3.5. *If Ω is a half-space or a smooth domain, then the operator of taking trace on Γ acts boundedly from $H_p^{s+1/p}(\Omega)$ to $B_p^s(\Gamma)$ and from $B_p^{s+1/p}(\Omega)$ to $B_p^s(\Gamma)$ for $s > 0$. There is a common bounded right inverse operator, which acts from $B_p^s(\Gamma)$ to both spaces $H_p^{s+1/p}(\Omega)$ and $B_p^{s+1/p}(\Omega)$, i.e., to their intersection.*

The spaces $\widetilde{H}_{p'}^{-s}(\Omega)$ and $\widetilde{B}_{p'}^{-s}(\Omega)$ with $s > 1/p$ contain continuous functionals on $H_p^s(\Omega)$ and $B_p^s(\Omega)$, respectively, supported on the boundary. They can be realized as functionals on the set of boundary values of functions in $H_p^s(\Omega)$ and $B_p^s(\Omega)$ and their derivatives.

14.4 Remarks to Embedding Theorems

Sobolev's original proofs of embedding theorems (see [354]) are very instructive. In essence, Sobolev used integral representations of functions obtained by using potential-type operators. His method was subsequently developed and perfected in the literature, including educational books. We mention the textbook [358] and borrow the following example from it.

Let Ω be a bounded domain, and let $u(x)$ be a function in $C_0^\infty(\Omega)$. Using a fundamental solution $E(x)$ of the Laplace equation, we can write $u(x)$ as the volume potential

$$u(x) = \int_\Omega E(x-y)\Delta u(x)\,dx. \tag{14.4.1}$$

Integrating by parts, we reduce this formula to the form

$$u(x) = c_n \sum_{k=1}^n \int_\Omega \frac{x_k - y_k}{|x-y|^n} \partial_k u(y)\,dy. \tag{14.4.2}$$

Here the first derivatives are subject to the action of integral operators with a weak singularity. This representation extends to functions in $\mathring{W}_p^1(\Omega)$. This makes it possible to show that $\mathring{W}_p^1(\Omega)$ is continuously embedded in $L_q(\Omega)$ if $1 \le p \le n$, $q < \infty$, and

$$1 \ge \frac{n}{p} - \frac{n}{q}. \tag{14.4.3}$$

If the inequality in (14.4.3) is strict, this embedding is compact. If $p > n$, a compact embedding in $C(\overline{\Omega})$ is obtained. Moreover, in the former case, conditions for the continuity and compactness of the embedding in $L_q(\Omega_m)$ are obtained; here Ω_m is the intersection of Ω with an m-dimensional hyperplane.

These results are generalized to $W_p^1(\Omega)$ by extending functions in this space to functions in $\mathring{W}_p^1(\Omega_1)$ for a larger bounded domain Ω_1. Thus, it suffices to assume the boundary of Ω to be Lipschitz. The results remain the same.

The results are generalized to the case of the space $\mathbb{R}^n$ by covering it by equal domains, e.g., balls. Only the results on the compactness of embeddings are lost.

The results can be generalized to $W_p^l(\Omega)$ with positive integer l (cf. Theorems 14.1.2 and 14.1.3), e.g., by induction on l.

There is a huge literature on embedding theorems. This topic includes embedding theorems in function spaces on submanifolds of lower dimension (the simplest example is trace theorems) and "converse" theorems on constructing extension operators. In addition to the approaches based on integral representations (not only in terms of derivatives but also in terms of differences), an approach using approximation of functions by trigonometric polynomials and by entire functions of exponential type was developed. For further reading we recommend, in the first place, the monographs [289] of Nikol'skii and [63] of Besov, Il'in, and Nikol'skii; see also references therein. As we have already mentioned, means of interpolation theory are also used.

14.5 The Spaces H_p^s and B_p^s on Lipschitz Domains and Lipschitz Surfaces

The references for this subsection are [278, 376, 379].

Here by Ω we denote a bounded domain with Lipschitz boundary Γ.

The definitions of the spaces $H_p^s(\Omega)$ and $B_p^s(\Omega)$ given in Section 14.3 remain the same in the case of a bounded Lipschitz domain for any s. In these spaces, the linear manifold $C^\infty(\overline{\Omega})$ remains dense, which is understood in the same sense as in Section 9.2. As previously, the fundamental case is that of $p = 2$, in which $H_2^s = B_2^s$ and the subscript 2 is omitted. The definitions of the spaces $\widetilde{H}_p^s(\Omega)$, $\widetilde{B}_p^s(\Omega)$, $\mathring{H}_p^s(\Omega)$, and $\mathring{B}_p^s(\Omega)$ are also retained.

It was the case of Lipschitz domains for which Rychkov constructed his universal operator of extension of functions from a domain to the entire space. This operator serves all spaces $H_p^s(\Omega)$ and $B_p^s(\Omega)$ and more general function spaces on Lipschitz domains, which will be described in Section 14.7.

Thanks to the existence of the extension operator, the W, H, and B spaces on Lipschitz domains retain many of the properties mentioned in the preceding sections. This relates to what was said before Theorem 14.3.5. In particular, norms of nonnegative integer order s on $H_p^s(\Omega)$ and norms of positive noninteger order s on $B_p^s(\Omega)$ can be defined by the Sobolev–Slobodetskii formulas, and these norms are equivalent to the norms in the initial definitions. Theorems 14.3.1–14.3.4 remain valid. Multipliers on $H_p^1(\Omega)$ and $B_p^1(\Omega)$ are the operators of multiplication by functions in $C^{0,1}(\overline{\Omega})$.

However, on the boundary Γ of a Lipschitz domain, the spaces $H_p^s(\Gamma)$ and $B_p^s(\Gamma)$ are now *invariantly* defined only for $|s| \leq 1$. These bounds cannot be violated, since transformations of local coordinates are generally only Lipschitz. The spaces $B_p^s(\Gamma)$ are more commonly used because of their role in the trace theorem.

Given such a surface Γ, norms on $B_p^s(\Gamma)$ with $s \in (0,1)$ can be defined locally by the Slobodetskii formulas and globally by using a partition of unity.

The embedding of $B_p^\sigma(\Gamma)$ in $B_p^s(\Gamma)$ remains continuous, compact, and dense for $-1 \leq s < \sigma \leq 1$.

For all spaces $B_p^s(\Gamma)$, the operator of multiplication by a Lipschitz function is a multiplier.

The theorem about the trace is stated as follows: as previously, this operator acts boundedly from $H_p^{s+1/p}(\Omega)$ and $B_p^{s+1/p}(\Omega)$ to $B_p^s(\Gamma)$, but *only for* $0 < s < 1$; see [197]. A corresponding not very simple example can be found in [195]. These operators have a common right inverse.

Let us say a few words about the very instructive book [197]. Its authors consider questions related to the traces of functions on (closed) subsets of $\mathbb{R}^n$ and to the possibility of extending functions on subsets of $\mathbb{R}^n$ to functions on $\mathbb{R}^n$ with given properties. The latter problem was studied by Whitney for smooth functions by using Taylor expansions [396, 397], and the authors of [197] follow his ideas, but their constructions are linked to the Besov spaces on subsets and can be applied in the case where these subsets are Lipschitz boundaries.

However, the definition of the *trace of a function* on a subset used in [197] somewhat differs from the definition used above (see also [359, Chap. VI, Sec. 4]). Given a function $f(x)$ locally belonging to, say, $L_1(\mathbb{R}^n)$, we can consider the limits

$$\widetilde{f}(x) = \lim_{r \to 0} \frac{1}{|B(x,r)|} \int\limits_{B(x,r)} f(y)\, dy.$$

Here $B(x,r)$ is a ball of radius r centered at x and $|B(x,r)|$ is its volume. According to Lebesgue's theorem, this limit exists and coincides with $f(x)$ almost everywhere. Redefining f on a set of measure zero, we can assume that this coincidence takes place at all points where the limit exists. After this, the function is referred to as strictly defined. Its restriction $f|_F$ to a subset $F \subset \mathbb{R}^n$ is defined pointwise. The membership of a trace in certain spaces is studied (in trace theorems) or assumed (in extension theorems).

The spaces $\widetilde{H}_{p'}^{-s}(\Omega)$ and $\widetilde{B}_{p'}^{-s}(\Omega)$ with $s > 1/p$ contain continuous functionals on $H_p^s(\Omega)$ and $B_p^s(\Omega)$, respectively, supported on the boundary of Γ. They can be realized as functionals of the boundary values of functions in these spaces.

The spaces $B_p^s(\Gamma)$ and $B_{p'}^{-s}(\Gamma)$ are dual.

The B spaces on nonclosed Lipschitz surfaces can also be considered. Suppose that a closed Lipschitz $(n-1)$-dimensional surface Γ is divided by a closed Lipschitz $(n-2)$-dimensional surface Γ_0 into two open parts, Γ_1 and Γ_2; for simplicity, we assume the parts to be simply connected. In this case, we can consider the spaces $B_p^s(\Gamma_1)$ and $B_p^s(\Gamma_2)$, $|s| \leq 1$. The space $B_p^s(\Gamma_1)$ ($|s| \leq 1$) can be defined as the space of the restrictions of elements of $B_p^s(\Gamma)$ to Γ_1 with the inf-norm. There is a bounded

operator of extension of functions in $B_p^s(\Gamma_1)$ to functions in $B_p^s(\Gamma)$: this is a version
of the Rychkov operator. For $-1/p' < s < 1/p$, the operator of extension by zero is
bounded. The space $\widetilde{B}_p^s(\Gamma_1)$ is defined as the subspace of $B_p^s(\Gamma)$ consisting of the
elements supported in $\overline{\Gamma}_1$. For $-1/p' < s < 1/p$, the spaces $B_p^s(\Gamma_1)$ and $\widetilde{B}_p^s(\Gamma_1)$ can
be identified. The spaces $B_p^s(\Gamma_1)$ and $\widetilde{B}_{p'}^{-s}(\Gamma_1)$ are dual.

All interpolation relations which we know remain valid in Lipschitz domains and
on Lipschitz surfaces within natural limits. In particular, formula (14.3.1) remains
valid.

14.6 Spaces with $p = 1$ and $p = \infty$

By analogy with the spaces considered above, the W_p^s, H_p^s, and B_p^s spaces can be
defined for $p = 1$ and $p = \infty$. However authors of monographs emphasize that one
should not expect that all facts known for $1 < p < \infty$ can be generalized without
losses.

The space $H_p^s(\mathbb{R}^n)$ with $p = 1$ or $p = \infty$ can be defined as the space of functions
f in $S'(\mathbb{R}^n)$ for which $\Lambda^s f \in L_p(\mathbb{R}^n)$. The space $W_1^k(\mathbb{R}^n)$ with positive integer k
is the space of distributions which belong to $L_1(\mathbb{R}^n)$ together with their generalized
derivatives of order up to k. Norms on these spaces are introduced in an obvious
way; see, e.g., [376, Sec. 2.3.3, Remark 5] and [60, Sec. 6.2]. But the spaces $H_p^k(\mathbb{R}^n)$
and $W_p^k(\mathbb{R}^n)$ with such p do not generally coincide. In this connection, Triebel refers
to [359]; see [359, Chap. V, Sec. 6.6]. The relation

$$[H_1^s(\mathbb{R}^n)]^* = H_\infty^{-s}(\mathbb{R}^n)$$

is valid [60, Sec. 6.2], but here the subscripts cannot be interchanged, these spaces
are nonreflexive.

The function space B_∞^s on $\mathbb{R}^n$ or Ω for noninteger $s > 0$ is the Hölder space C_b^s
introduced in Section 1. The space B_∞^{m+1} with integer $m \geq 0$ is $C_b^{m,1}$ introduced in
Section 1; see [197, 318, 376].

14.7 The Triebel–Lizorkin Spaces and the General Besov Spaces

Below we give definitions of even more general Sobolev-type spaces. For this purpose, we use the system of functions $\varphi_k(x)$, say, from Section 1.14. First, suppose
that $s \in \mathbb{R}$ and $1 \leq p < \infty$.

The *Triebel–Lizorkin space* $F_{p,q}^s(\mathbb{R}^n)$ is defined as the space of all distributions
$u \in S'$ with finite norm

$$\|u\|_{F_{p,q}^s(\mathbb{R}^n)} = \left\|\left(\sum_{k=0}^{\infty} |2^{ks}\varphi_k * u|^q\right)^{1/q}\right\|_{L_p(\mathbb{R}^n)}. \qquad (14.7.1)$$

In this expression, we first take the l_q-norm of the sequence $\{2^{ks}(\varphi_k * u)(x)\}$ and then the L_p-norm of this l_q-norm.

The *Besov space* $B_{p,q}^s(\mathbb{R}^n)$ is defined as the space of all distributions $u \in S'$ with finite norm

$$\|u\|_{B_{p,q}^s(\mathbb{R}^n)} = \left(\sum_{k=0}^{\infty} \|2^{ks}\varphi_k * u\|_{L_p(\mathbb{R}^n)}^q\right)^{1/q}. \qquad (14.7.2)$$

Here the L_p-norms of the same functions $2^{ks}(\varphi_k * u)(x)$ are first calculated, and then the l_q-norm of the resulting numerical sequence is taken.

For $p = \infty$ and/or $q = \infty$, the above expressions still make sense when naturally modified: the L_p-norms are replaced by the corresponding norms in L_∞, and/or the l_q-norm is replaced by the norm in l_∞. A modification of the definition is indeed needed for the $F_{\infty,q}^s$ spaces (see [377] and [318, p. 9]); we do not dwell on this here.

Note that

$$F_{p,2}^s(\mathbb{R}^n) = H_p^s(\mathbb{R}^n). \qquad (14.7.3)$$

In particular, $F_{2,2}^s(\mathbb{R}^n) = H^s(\mathbb{R}^n)$; see Section 1.14. Next,

$$B_{p,p}^s(\mathbb{R}^n) = B_p^s(\mathbb{R}^n). \qquad (14.7.4)$$

Moreover, $B_{p,p}^s(\mathbb{R}^n) = F_{p,p}^s(\mathbb{R}^n)$ and, in particular, $B_{2,2}^s(\mathbb{R}^n) = F_{2,2}^s(\mathbb{R}^n) = H^s(\mathbb{R}^n)$. It is easy to see that, in these cases, the right-hand sides of (14.7.1) and (14.7.2) coincide.

These new spaces are Banach as well. Their definitions can be extended to $p \in (0,1)$ and $q \in (0,1)$, but for such p and/or q, the spaces become quasi-Banach.

All of them were studied in detail, e.g., in [376, 377]; see also the summary of the properties of these spaces in [318].

In regard to the general Besov spaces $B_{p,q}^s(\mathbb{R}^n)$, we add that if $s > 0$, $s = m + \theta$, where $m \in \mathbb{Z}_+$ and $0 < \theta \le 1$, $1 \le p < \infty$, and $1 \le q < \infty$, then the norm on these spaces can be defined as

$$\|u\|_{W_p^m(\mathbb{R}^n)} + \sum_{|\alpha|=m} \left(\int \frac{\|D^\alpha u(x+y) - D^\alpha u(x)\|_{L_p(\mathbb{R}^n)}^q}{|y|^{n+\theta q}} dy\right)^{1/q}. \qquad (14.7.5)$$

For $q = \infty$, this expression is replaced by

$$\|u\|_{W_p^m(\mathbb{R}^n)} + \sum_{|\alpha|=m} \sup_{h\neq 0} \frac{\|D^\alpha u(x+h) - D^\alpha u(x)\|_{L_p(\mathbb{R}^n)}}{|h|^\theta}; \qquad (14.7.6)$$

see [197, p. 7] and [325].

Besov spaces with three indices arise, in particular, in the real interpolation of spaces with two indices: for $s \neq t$ and $0 < \theta < 1$, we have

$$(H_p^s, H_p^t)_{\theta,q} = B_{p,q}^{(1-\theta)s+\theta t} \quad \text{and} \quad (B_p^s, B_p^t)_{\theta,q} = B_{p,q}^{(1-\theta)s+\theta t}; \qquad (14.7.7)$$

see, e.g., [60, Sec. 6.4].

The spaces $F_{p,q}^s(\Omega)$ and $B_{p,q}^s(\Omega)$ on a domain Ω are, naturally, defined as consisting of the restrictions of the elements of the corresponding function spaces on $\mathbb{R}^n$ with the inf-norm. The Rychkov operator serves these spaces, too. His results for function spaces on special Lipschitz domains give expressions for norms on these spaces which do not involve extensions (these norms are new but equivalent to those specified above); these are (14.7.1) and (14.7.2) with $L_p(\Omega)$ instead of $L_p(\mathbb{R}^n)$. (Cf. [319].) The system of functions φ_j is the same as in Section 10.3. In particular, for $q = 2$ and $q = p$, the norms on $H_p^s(\Omega)$ and in $B_p^s(\Omega)$, respectively, are

$$\|u\|_{H_p^s(\Omega)} = \left\|\left(\sum_{j=0}^\infty 2^{2js}|\varphi_j * u|^2\right)^{1/2}\right\|_{L_p(\Omega)} \qquad (14.7.8)$$

and

$$\|u\|_{B_p^s(\Omega)} = \left(\sum_{j=0}^\infty 2^{jsp}\|\varphi_j * u\|_{L^p(\Omega)}^p\right)^{1/p}. \qquad (14.7.9)$$

Expression (14.7.8) is particularly interesting for noninteger $s > 0$ and for $s < 0$, and expression (14.7.9), for integer $s \geq 0$ (including $s = 0$) and for $s < 0$.

In Section 16.4, we shall use the spaces $B_{p,q}^s(\Omega)$ to prove an important theorem about the smoothness of solutions of the Dirichlet and Neumann problems in Lipschitz domains. That section contains an additional information on these spaces. The reader will have an opportunity to see how this information is used. Actually, only the cases $q = \infty$ and 1 are needed there.

15 Applications to the General Theory of Elliptic Equations and Boundary Value Problems

15.1 General Elliptic Problems in the Sobolev–Slobodetskii Spaces

We shall not dwell on equations on closed manifolds and consider at once the general scalar elliptic problems from Section 7 in a smooth bounded domain:

$$a(x, D)u(x) = f(x) \text{ in } \Omega, \qquad (15.1.1)$$

$$b_j(x, D)u(x) = g_j(x) \text{ on } \Gamma \quad (j = 1, \dots, l). \qquad (15.1.2)$$

These problems can be considered in spaces much more general than those used in Section 7, in particular, with

$$u \in W_p^s(\Omega), \quad f \in W_p^{s-2l}(\Omega), \text{ and } g_j \in W_p^{s-r_j-1/p}(\Gamma), \quad 1 < p < \infty, \quad (15.1.3)$$

and sufficiently large s. We shall consider only this setting. For simplicity, we assume that s is integer and the following analogue of condition (7.1.6) holds:

$$s \geq 2l, \quad s > \max r_j + 1/p. \tag{15.1.4}$$

The operator $\mathcal{A}: u \mapsto (f, g_1, \ldots, g_l)$ corresponding to this problem acts boundedly from the solution space to the space of right-hand sides. The main statement remains true: the ellipticity of the problem is equivalent to the Fredholm property of the operator. Let us write the a priori estimate:

$$\|u\|_{W_p^s(\Omega)} \leq C_1 \left[\|f\|_{W_p^{s-2l}(\Omega)} + \sum_1^l \|g_j\|_{W_p^{s-r_j-1/p}(\Gamma)} + \|u\|_{L_p(\Gamma)} \right]. \tag{15.1.5}$$

This estimate was proved in [9] by the method of freezing coefficients modified as follows: the solution of the problem in the half-space for operators with constant coefficients without lower-order terms is written in the x-representation, without applying the Fourier transform; for this purpose, the kernels of the corresponding integral operators are explicitly calculated and the required boundedness properties of these operators are verified. In particular, important theorems on the boundedness of singular integral operators [88] are used.

Note that in [9], the a priori estimates are first derived in the Hölder norms ($u \in C^s(\overline{\Omega})$ with *noninteger* sufficiently large s). Such estimates are named *Schauder estimates*: they first appeared in Schauder's works [326, 327].

In fact, a two-sided estimate holds (by virtue of the boundedness of $\mathcal{A}$ mentioned above). If the solution is unique, then the last term on the right-hand side of (15.1.5) can be omitted.

A theorem on the regularity of solutions is also valid; we shall not state it here. The problems considered in Section 7 and in the present subsection are essentially the same, and the meaning of the theorem is that the smoothness of the solution is adequately determined by that of the right-hand sides.

Finally, the result on the unique solvability of problem (7.1.19), (7.1.4) elliptic with parameter along a ray at large values of the parameter remains valid. The a priori estimate for $s = 2l > r_j$ and homogeneous boundary conditions has the form

$$\|u\|_{W_p^s(\Omega)} + |\lambda| \|u\|_{L_p(\Omega)} \leq C_2 \|f\|_{L_p(\Omega)}. \tag{15.1.6}$$

If the boundary and the coefficients are infinitely smooth, then so are the eigenfunctions and the generalized eigenfunctions of the spectral problems, and therefore they belong to all of the corresponding spaces. When problems with a spectral parameter in the equation are considered, these are the spaces $W_p^s(\Omega)$. The eigenvalues do not depend on s and p. There are results on the completeness of the generalized eigenfunctions in these spaces; see [7, 29, 36].

15.2 Generalizations

In [9, Part II], a priori estimates were extended to Douglis–Nirenberg elliptic systems. Volevich [393] proved these estimates, too; he also constructed a right parametrix and, thereby, completed the proof of the theorem on the equivalence of ellipticity and the Fredholm property for systems in the W_p^s spaces. In parallel, this was done by Solonnikov [355–357].

Generalizations of the theory on the Triebel–Lizorkin spaces and general Besov spaces can be found in [376] and [161], and a summary of results is given in [318, Chap. II] (where scalar problems with normal boundary conditions were considered); see also references therein.

A motivation for considering problems in such general spaces is most frequently a preparation for the passage to nonlinear elliptic problems; see [318]. In this case, it is often important to minimize smoothness assumptions on the coefficients.

16 Applications to Boundary Value Problems in Lipschitz Domains. 1

In this section we obtain new results (in comparison with those of Sections 11 and 12) in the framework of the spaces H^s, and in the next section, of more general spaces. In Section 17 we also touch on nonstationary (parabolic) problems.

16.1 Main Boundary Value Problems in More General H^s Spaces

Within the framework of the spaces H^s, we can substantially generalize first the Green identity (11.1.4) and then the settings of the main boundary value problems. We first assume that the coefficients $a_{j,k}(x)$ and $b_j(x)$ belong to $C^{1/2}(\overline{\Omega})$ and the coefficient $c(x)$ is bounded and measurable. But when a formally adjoint operator is needed, we assume b_j to be Lipschitz.

Proposition 16.1.1. *The form Φ_Ω is bounded on the direct product of the spaces* $H^{1+s}(\Omega)$ *and* $H^{1-s}(\Omega)$, $|s| < 1/2$:

$$|\Phi_\Omega(u,v)| \le C\|u\|_{H^{1+s}(\Omega)}\|v\|_{H^{1-s}(\Omega)}. \tag{16.1.1}$$

Proof. First, suppose that Φ_Ω contains only higher-order terms. If $u \in H^{1+s}(\Omega)$ and $v \in H^{1-s}(\Omega)$, then $\partial_k u \in H^s(\Omega)$, $a_{j,k}\partial_k u \in H^s(\Omega)$, and $\partial_j v \in H^{-s}(\Omega)$. For $|s| < 1/2$, these spaces coincide with the corresponding spaces $\widetilde{H}^{\pm s}(\Omega)$ and are dual with respect to the extension of the inner product in $L_2(\Omega)$ to their direct product. The zero-order term is treated as a product of functions in $L_2(\Omega)$. The intermediate terms are treated in the same way if $s > 0$. If $s < 0$, then we use the membership of $b_j \partial_j u$ in $H^s(\Omega)$ and of v in $\widetilde{H}^{-s}(\Omega)$. Thus, (16.1.1) is obtained.   □

Remark. Note that we can relax the smoothness assumptions on the coefficients by fixing s which we are interested in. In this case, it is sufficient that the $a_{j,k}$ and the b_j be multipliers in $H^s(\Omega)$. Moreover, as seen from the proof, for $s > 0$, it is sufficient that the coefficients b_j be bounded and measurable, if the adjoint problem is not needed.

This remark should be kept in mind in what follows.

Now let us write the Green identity:

$$(Lu, v)_\Omega = \Phi_\Omega(u, v) - (T^+u, v^+)_\Gamma. \tag{16.1.2}$$

Here $u \in H^{1+s}(\Omega)$ and $v \in H^{1-s}(\Omega)$, $|s| < 1/2$. In the general case, we assume that $Lu \in \widetilde{H}^{-1+s}(\Omega)$; the left-hand side then makes sense. Next, $v^+ \in H^{1/2-s}(\Gamma)$, and the condition on s follows from the conditions in the trace theorem. As previously, we define the conormal derivative by the Green identity.

Proposition 16.1.2. *The following inequality holds for* $|s| < 1/2$:

$$\|T^+u\|_{H^{-1/2+s}(\Gamma)} \le C[\|u\|_{H^{1+s}(\Omega)} + \|Lu\|_{\widetilde{H}^{-1+s}(\Omega)}]. \tag{16.1.3}$$

Proof. We have

$$|(T^+u, v^+)_\Gamma| \le |(Lu, v)_\Omega| + |\Phi_\Omega(u, v)|.$$

The right-hand side does not exceed

$$C_2[\|Lu\|_{\widetilde{H}^{-1+s}(\Omega)} + \|u\|_{H^{1+s}(\Omega)}]\|v\|_{H^{1-s}(\Omega)}.$$

This implies (16.1.3), because

$$\|T^+u\|_{H^{-1/2+s}(\Gamma)} \le C_3 \sup_{v^+ \ne 0} \frac{|(T^+u, v^+)_\Gamma|}{\|v^+\|_{H^{1/2-s}(\Gamma)}},$$

and the norm in the denominator is bounded below by $C_4\|v\|_{H^{1-s}(\Omega)}$ for appropriate v determined by v^+.   □

The Dirichlet and Neumann problems with homogeneous boundary conditions are now generalized as follows. As previously, they are determined by the Green identity without the boundary term. But now, in the Dirichlet problem,

$$u \in \widetilde{H}^{1+s}(\Omega), \quad Lu = f \in H^{-1+s}(\Omega), \quad v \in \widetilde{H}^{1-s}(\Omega), \tag{16.1.4}$$

and in the Neumann problem,

$$u \in H^{1+s}(\Omega), \quad Lu = f \in \widetilde{H}^{-1+s}(\Omega), \quad v \in H^{1-s}(\Omega). \tag{16.1.5}$$

Here $|s| < 1/2$; as mentioned above, this is dictated by the trace theorem. In both problems, f and v belong to dual spaces.

The question of for what s these problems are uniquely solvable cannot already be answered by using the Lax–Milgram theorem, because the solution and the right-hand side do not belong to mutually dual spaces for $s \neq 0$. We discuss this question in this and following subsections by using, in particular, some results of interpolation theory. The first of them is Shneiberg's theorem (see Section 13.7). It gives the following results.

Theorem 16.1.3. 1. *If the form Φ_Ω is strongly coercive on $\widetilde{H}^1(\Omega)$, then there exists an $\varepsilon = \varepsilon(L) \in (0, 1/2]$ such that the operator*

$$\mathcal{L}_D : \widetilde{H}^{1+s}(\Omega) \to H^{-1+s}(\Omega) \qquad\qquad (16.1.6)$$

corresponding to the Dirichlet problem with homogeneous boundary conditions is invertible for $|s| < \varepsilon$. If only strong ellipticity holds, then the operator $\mathcal{L}_D$ is Fredholm for $|s| < \varepsilon$, and the dimensions of the kernel and cokernel of this operator coincide and do not depend on s.

2. *If the form Φ_Ω is strongly coercive on $H^1(\Omega)$, then there exists an $\varepsilon = \varepsilon(L) \in (0, 1/2]$ such that the operator*

$$\mathcal{L}_N : H^{1+s}(\Omega) \to \widetilde{H}^{-1+s}(\Omega) \qquad\qquad (16.1.7)$$

corresponding to the Neumann problem with homogeneous boundary conditions is invertible for $|s| < \varepsilon$. If Φ_Ω is only coercive on $H^1(\Omega)$, then the operator $\mathcal{L}_N$ is Fredholm for $|s| < \varepsilon$, and the dimensions of the kernel and cokernel of this operator coincide and do not depend on s.

It must be obvious to the reader how to obtain these results. The operator corresponding to the problem acts from one interpolation scale of spaces to another, and it is invertible at $s = 0$; hence it remains invertible for sufficiently small $|s|$. The assertions concerning the Fredholm property follow from those concerning invertibility for the operator $L + \tau I$ with sufficiently large τ instead of L. The assertions about the independence of the dimensions of the kernel and the cokernel on s follow from the independence of their elements on s.

For simplicity, we usually assume that if numbers $\varepsilon(L)$ exist for both the Dirichlet and Neumann problems, then they coincide (are equal to the smaller one). We also assume that $\varepsilon(L) = \varepsilon(\widetilde{L})$, because the operators L and $\widetilde{L}$ with homogeneous Dirichlet or Neumann conditions remain related by the second Green identity.

All this can be generalized to the Dirichlet and Neumann problems with inhomogeneous boundary conditions, as in Section 11.1. Some details related to a more general situation are contained in Section 17.1.

Statements on unique solvability or the Fredholm property hold simultaneously with those on the regularity of solutions, or, which is the same, on increasing their smoothness. In particular, *for problems with homogeneous boundary conditions, if $-\varepsilon < s' < s'' < \varepsilon$, the solution belongs to the space with index $1 + s'$, and the right-hand side belongs to the space with index $-1 + s''$, then the solution belongs to the space with index $1 + s''$.* (We discussed a similar connection between statements in Sections 6 and 7.)

Statements concerning the smoothness of the eigenfunctions and generalized eigenfunctions of our spectral problems can also be obtained. Assertions on their completeness and the basis property can be generalized. We shall discuss this in a more general situation in Section 17.2.

For those s for which the unique solvability of the Dirichlet problem is obtained, the Weyl decomposition can be generalized to the space $H^{1+s}(\Omega)$. In the theorem stated below, it is taken into account that the test functions belong to $H^{1-s}(\Omega)$. We have to complicate slightly the notation of spaces in comparison with that used in Section 11.2. Note that $1 + s$ lies strictly between $1/2$ and $3/2$.

Theorem 16.1.4. *Suppose that the Dirichlet problem for a system with operator L is uniquely solvable for $|s| < \varepsilon$. Then the space $H^{1+s}(\Omega)$ decomposes into the direct sum $H_1(s) \dotplus H_2(s)$ of the subspace $H_1(s) = \widetilde{H}^{1+s}(\Omega) = \overset{\circ}{H}{}^{1+s}(\Omega)$ of functions with zero boundary values and the subspace $H_2(s) = H_2(s, L)$ of solutions of the homogeneous system $Lu = 0$ in Ω. The latter is parametrized by the elements of $H^{1/2+s}(\Gamma)$.*

. For the corresponding decomposition $u = u_1 + u_2$, we keep the name "the *Weyl decomposition* of u in $H^{1+s}(\Omega)$ corresponding to the operator L."

The Dirichlet problem for the formally adjoint system is also uniquely solvable for $|s| < \varepsilon$. We assume that $Lu_2 = 0$ and $\widetilde{L}v_2 = 0$ for the Weyl decomposition $v = v_1 + v_2$ in $H^{1-s}(\Omega)$ corresponding to the operator $\widetilde{L}$. Then, for these s, an expression of the form (11.2.3) remains valid, namely,

$$\Phi_\Omega(u, v) = (Lu_1, v_1)_\Omega + (T^+u_2, v_2^+)_\Gamma. \qquad (16.1.8)$$

This allows us to extend decomposition (11.2.4) to those s which we consider now. We write the generalization as

$$\widetilde{H}^{-1+s}(\Omega) = \widetilde{H}_1^{-1+s}(\Omega) \dotplus \widetilde{H}_2^{-1+s}(\Omega). \qquad (16.1.9)$$

Here the first subspace on the right-hand side consists of functionals on $\widetilde{H}^{1-s}(\Omega)$ continued by zero to $H_2(-s, \widetilde{L})$, is isomorphic to $H^{-1+s}(\Omega)$ and contains no functionals supported on Γ. The second subspace consists of functionals supported on Γ and is isomorphic to $H^{-1/2+s}(\Gamma)$.

A two-sided a priori estimate for solutions of the Dirichlet problem in the spaces which we now consider is obtained by using the Weyl decomposition. The Neumann problem reduces to the Dirichlet problem (11.2.5) for an inhomogeneous system in $\widetilde{H}^{1+s}(\Omega)$ and to the Neumann problem (11.2.6) for a homogeneous system in $H_2(s, L)$. For a solution $u = u_1 + u_2$, we again obtain a two-sided estimate.

If the form Φ_Ω is strongly coercive on $\widetilde{H}^1(\Omega)$ and the operator $\mathcal{L}_D$ is invertible for $|s| < \varepsilon$, then the Dirichlet-to-Neumann operator D can be extended to a bounded operator from $H^{1/2+s}(\Gamma)$ to $H^{-1/2+s}(\Gamma)$ for the same s. Similarly, if the form Φ_Ω is strongly coercive on $H^1(\Omega)$ and the operator $\mathcal{L}_N$ is invertible for $|s| < \varepsilon$, then the Neumann-to-Dirichlet operator N can be extended to a bounded operator from

$H^{-1/2+s}(\Gamma)$ to $H^{1/2+s}(\Gamma)$ for the same s. Obviously, if the latter assumption holds, then the operators D and N are invertible, and they are mutually inverse.

For the spectral problems for the operators D and N on Γ, statements concerning smoothness, completeness, and the basis property mentioned in Section 11 can be generalized.

The above discussion can be extended to the mixed problems in which $|s| < \varepsilon$ for sufficiently small ε.

The Weyl decomposition of $H^1(\Omega;\Gamma_1)$ specified in Proposition 11.4.4 can be generalized to $H^{1+s}(\Omega;\Gamma_1)$ with s close to zero. *This space is the direct sum of the subspace $\widetilde{H}^{1+s}(\Omega)$ and the subspace of solutions of the Dirichlet problem for a homogeneous system with Dirichlet data vanishing on Γ_1. The latter subspace is isomorphic to $\widetilde{H}^{1/2+s}(\Gamma_2)$.* Shneiberg's theorem can be applied either to the operator which corresponds to a problem with homogeneous boundary conditions or to the operators arising in the reduction of the mixed problem to equations on the boundary.

16.2 The Operators A and H in More General H^s Spaces

We have already mentioned in Section 12.2 that the operators

$$\mathcal{A} = L^{-1}\gamma^* \quad \text{and} \quad A = \gamma L^{-1}\gamma^* \tag{16.2.1}$$

can be regarded as bounded operators from $H^{-1/2+s}(\Gamma)$ to $H^{1+s}(\Omega^\pm)$ and to $H^{1/2+s}(\Gamma)$, respectively, for $|s| < 1/2$.

Recall that, considering these operators, we assume that the domain $\Omega = \Omega^+$, together with its complement Ω^-, lies on the torus $\mathbb{T}^n$ and the form $\Phi_{\mathbb{T}}$ is strongly coercive on $H^1(\mathbb{T})$. Hence the forms $\Phi_{\Omega^\pm}$ are strongly coercive on $\widetilde{H}^1(\Omega^\pm)$. According to the first part of Theorem 16.1.3, the operators $\mathcal{L}_D^\pm$ corresponding to Dirichlet problems with homogeneous boundary conditions in $\Omega^\pm$ are invertible as operators from $\widetilde{H}^{1+s}(\Omega^\pm)$ to $H^{-1+s}(\Omega^\pm)$ for $|s| < \varepsilon(L)$, where $\varepsilon(L)$ is a number in $(0, 1/2]$. Let us look over the considerations of Section 12.3 and of following subsections for the purpose of generalizing the results obtained there to s with $|s| < \varepsilon(L)$.

First, to the solutions of the system $Lu = 0$ in $\Omega^\pm$ that belong to $H^{1+s}(\Omega^\pm)$ relations (12.3.8) and (12.3.9) can be generalized. The operator $\mathcal{B}$ extends to a bounded operator from $H^{1/2+s}(\Gamma)$ to $H^{1+s}(\Omega^\pm)$.

Next, the jump relations (12.3.10) can be generalized as well, because their proof requires only the unique solvability of the Dirichlet problems. The functions φ and ψ in these relations now belong to $H^{1/2+s}(\Gamma)$ and $H^{-1/2+s}(\Gamma)$, respectively. To functions from these spaces Theorem 12.3.7 on problem (12.3.12) extends, and the solution belongs to $H^{1+s}(\Omega^\pm)$.

The operators B and $\widehat{B}$ (see (12.4.1)) are now bounded in $H^{1/2+s}(\Gamma)$ and $H^{-1/2+s}(\Gamma)$, respectively. Relations (12.4.2)–(12.4.4) are preserved. The operators

(12.4.5) are now the projections of $H^{1/2+s}(\Gamma) \times H^{-1/2+s}(\Gamma)$ onto the subspaces of Cauchy data for a homogeneous system in $\Omega^{\pm}$. Relations (12.4.7) remain valid.

Relations (12.6.2) and (12.6.3) remain valid too; they give the invertibility of A and the first relation in (12.6.1).

Now suppose that the forms $\Phi_{\Omega^{\pm}}$ are strongly coercive on $H^1(\Omega^{\pm})$. Then, according to the second part of Theorem 16.1.3, the operators $\mathcal{L}_N^{\pm}$ corresponding to Neumann problems with homogeneous boundary conditions are invertible as operators from $H^{1+s}(\Omega^{\pm})$ to $\widetilde{H}^{-1+s}(\Omega^{\pm})$ for $|s| < \varepsilon(L)$, where $\varepsilon(L)$ is the same. Now we can use the operators $N^{\pm}$ for these s, and we again obtain (12.6.4) and (12.6.5). We see that the operator H is invertible and the second relation in (12.6.1) holds.

The statements on the invertibility of the operators $\frac{1}{2}I \pm B$ and $\frac{1}{2}I \pm \widehat{B}$ can be generalized as well: these operators are invertible in $H^{1/2+s}(\Gamma)$ and $H^{-1/2+s}(\Gamma)$, respectively. As a consequence, the third and the fourth relation in (12.6.1) remain valid.

Below we summarize the main results.

Theorem 16.2.1. *If the form Φ_Ω is strongly coercive on $\widetilde{H}^1(\Omega^{\pm})$ and $|s| < \varepsilon(L)$ for the Dirichlet problem, then A is a bounded invertible operator from $H^{-1/2+s}(\Gamma)$ to $H^{1/2+s}(\Gamma)$, and its inverse is $D^+ + D^-$.*

If the forms $\Phi_{\Omega^{\pm}}$ are strongly coercive on $H^1(\Omega^{\pm})$ and $\varepsilon < \varepsilon(L)$ for the Neumann problem, then H is a bounded invertible operator from $H^{1/2+s}(\Gamma)$ to $H^{-1/2+s}(\Gamma)$, and its inverse is $N^- + N^+$. In this case, the operators $\frac{1}{2}I \pm B$ in $H^{1/2+s}(\Gamma)$ and $\frac{1}{2}I \pm \widehat{B}$ in $H^{-1/2+s}(\Gamma)$ are invertible as well and the third and the fourth relation in (12.6.1) hold.

The following two subsections contain statements according to which $\varepsilon(L) = 1/2$ under additional assumptions about the operator L, the most essential of which is the formal self-adjointness of its leading part.

16.3 The Rellich Identity and Its Consequences

We give no proofs in this section; they can be found in [258] or [286]. The surprising identity discovered by Rellich in the case of the Laplace equation [310] and extended to systems by Payne and Weinberger [295] is as follows. (The authors of [295] considered quite different problems.) The asterisk denotes the passage from a matrix (which may be a column) to its Hermitian conjugate. The symbol $^+$ denotes, as usual, the trace of a function.

Lemma 16.3.1. *Let Ω be a Lipschitz domain, and let $h_1, \ldots, h_n$ be real-valued functions on Ω bounded together with their first derivatives in the sense of distributions. Then any vector-valued functions $u, v \in H^2(\Omega)$ of dimension d satisfy the relation*

$$\sum_{j,k,l} \int_\Gamma \nu_l \{[(h_l a_{j,k} - h_j a_{l,k} - h_k a_{j,l}) \partial_k u]^* \partial_j v\}^+ ds$$

$$= \sum_{j,k} \int_\Omega [(D_{j,k} \partial_k u)^* \partial_j v + (L_0 u)^* (h_j \partial_j v) + (h_k \partial_k u)^* (\widetilde{L}_0 v)]) \, dx, \quad (16.3.1)$$

where

$$D_{j,k} = \sum_l [\partial_l (h_l a_{j,k}) - (\partial_l h_j) a_{l,k} - (\partial_l h_k) a_{j,l}]. \qquad (16.3.2)$$

This lemma is derived from the multidimensional version of the divergence theorem. The lemma is used in the (involved) proof of the following theorem.

Theorem 16.3.2. *Let L be a strongly elliptic operator with formally self-adjoint leading part: $a_{j,k}^* = a_{k,j}$. Suppose that*

$$Lu = f \text{ in } \Omega, \qquad (16.3.3)$$

where $u \in H^1(\Omega)$ but $f \in L_2(\Omega)$. Then the following assertions hold.
   1. If $u^+ \in H^1(\Gamma)$, then $T^+ u \in L_2(\Gamma)$ and

$$\|T^+ u\|_{L_2(\Gamma)} \le C[\|u^+\|_{H^1(\Gamma)} + \|u\|_{H^1(\Omega)} + \|f\|_{L_2(\Omega)}]. \qquad (16.3.4)$$

   2. If L is a scalar operator (i.e., $d = 1$) with real coefficients $a_{j,k} = a_{k,j}$ and $T^+ u \in L_2(\Gamma)$, then $u^+ \in H^1(\Gamma)$ and

$$\|u\|_{H^1(\Gamma)} \le C[\|T^+ u\|_{L_2(\Gamma)} + \|u\|_{H^1(\Omega)} + \|f\|_{L_2(\Omega)}]. \qquad (16.3.5)$$

The first assertion of this theorem implies the following statement.

Theorem 16.3.3. *Suppose that the leading part of L is formally self-adjoint and the form for the Dirichlet problem satisfies the strong coercivity condition. Then the Dirichlet-to-Neumann operator D is a bounded operator from $H^{1/2+s}(\Gamma)$ to $H^{-1/2+s}(\Gamma)$ for $|s| < 1/2$. Moreover, the operator taking Dirichlet data to the solution of the homogeneous equation $Lu = 0$ can be extended to a bounded operator from $L_2(\Gamma)$ to $L_2(\Omega)$. The same is true for the formally adjoint system.*

If the right-hand sides of a system and a Dirichlet condition are smooth (to some extent) and a solution of the corresponding Dirichlet problem has smooth conormal derivative, then this solution is smooth, because the solutions admit a representation of type (12.3.8) in terms of their Cauchy data (with the discussions in Section 16.2 taken into account). Then, interpolation can be applied. Thus, Theorem 16.3.3 leads to a result about the regularity of solutions of the Dirichlet problem for a system with formally self-adjoint leading part. Such regularity results can also be obtained for the solutions of the Neumann problem, but in much lesser generality—in the scalar case and for the Lamé system (the Rellich identity for the latter was obtained in [115]).

16.4 Savaré's Generalized Theorem

In this section, we consider the Dirichlet and Neumann problems under the assumptions of Proposition 16.1.1 and the additional assumptions that the leading part of the operator L is formally self-adjoint and the $a_{j,k}(x)$ are Lipschitz. Moreover, in the case of the matrix Neumann problem, we also assume that the following *additional condition* holds:

$$\sum a_{j,k}^{r,s}(x)\zeta_k^s\overline{\zeta}_j^r \geq 0 \qquad (16.4.1)$$

at all points x near the boundary. This condition essentially means the nonnegativity of the integrand in the form $\Phi_\Omega(u,u)$ at these points. In the scalar case, it holds automatically, and in the general case, it follows from our condition (11.1.17), which is sufficient for the strong coercivity of Φ_Ω on $H^1(\Omega)$; thus, the additional condition is not very restrictive.

Our goal is to prove the following theorem.

Theorem 16.4.1. *Under the assumptions specified above, $\varepsilon(L) = 1/2$ for both the Dirichlet and Neumann problems.*

This generalizes a result of [325] for the scalar equation

$$-\operatorname{div} a(x)\operatorname{grad} u(x) = f(x)$$

with a symmetric real matrix $a(x)$. In [325], other equations were also considered, including abstract and nonlinear ones. Savaré called his method a modification of Nirenberg's method of difference quotients. We touch on this method in the next subsection.

First, consider the formally self-adjoint system

$$Lu := -\sum \partial_j a_{j,k}(x)\partial_k u(x) + \tau u(x) = f(x) \qquad (16.4.2)$$

with homogeneous Dirichlet or Neumann condition in a bounded Lipschitz domain Ω. In the former case, we assume that the system is strongly elliptic and the positive number τ is so large that the corresponding form Φ_Ω is strongly coercive on $\widetilde{H}^1(\Omega)$. In the latter case, we assume that this form is strongly coercive on $H^1(\Omega)$. In the case of the Neumann problem, we impose the additional condition (16.4.1) near the boundary.

The following theorem is the main achievement of this section. We give the involved proof of this important result. It can be skipped if desired.

Theorem 16.4.2. *Suppose that the assumptions specified above hold and $s \in (0, 1/2)$.*

1. If $f \in H^{-1+s}(\Omega)$, then the solution of the Dirichlet problem belongs to $\widetilde{H}^{1+s}(\Omega)$.

2. If $f \in \widetilde{H}^{-1+s}(\Omega)$, then the solution of the Neumann problem belongs to $H^{1+s}(\Omega)$.

4 More general spaces and their applications

This implies the unique solvability of the Dirichlet and Neumann problems for such systems in these spaces.

First, we consider in detail the Neumann problem. Then, we briefly describe the changes required in the case of the Dirichlet problem.

As in [325], we need an information about the Besov spaces with three indices and, later, about interpolation relations for these spaces. Below we give a straightforward definition of the space $B_{2,\infty}^s(\Omega)$ with noninteger $s > 0$, which is convenient for studying the Neumann problem. Given a small positive number q, let Ω^q denote the set of those points in the domain Ω whose distance to the boundary is larger than q. Let Λ be a set of vectors spanning $\mathbb{R}^n$ and star-shaped with respect to the origin. For example, this may be a convex cone with vertex at the origin. Let $\sigma \in (0,1)$. For functions v on Ω, we define a seminorm $[v]_{\sigma,\Omega}$ by

$$[v]_{\sigma,\Omega}^2 = \sup_{h \in \Lambda \setminus 0, |h| < q} |h|^{-2\sigma} \int_{\Omega^{|h|}} |v(x+h) - v(x)|^2 \, dx. \tag{16.4.3}$$

A norm on $B_{2,\infty}^s(\Omega)$ with noninteger $s > 0$ can be defined by

$$\|v\|_{B_{2,\infty}^s(\Omega)} = \|v\|_{H^{[s]}(\Omega)} + \sum_{|\alpha|=[s]} [\partial^\alpha v]_{s-[s],\Omega}. \tag{16.4.4}$$

The space $B_{2,\infty}^s(\Omega)$ is defined as the completion of the linear manifold $C^\infty(\overline{\Omega})$ with respect to this norm. In particular, $\|u\|_{L_2(\Omega)} + [u]_{1/2,\Omega}$ can be taken for a norm on $B_{2,\infty}^{1/2}(\Omega)$ and $\|u\|_{H^1(\Omega)} + [\nabla u]_{1/2,\Omega}$, for a norm on $B_{2,\infty}^{3/2}(\Omega)$; cf. (14.7.6).

In these spaces localization is possible (and plays an important role in what follows). Suppose that the closure $\overline{\Omega}$ is covered by finitely many open balls O_j ($j = 0, \ldots, N$) and $\sum_j \phi_j \equiv 1$ is a partition of unity in a neighborhood of $\overline{\Omega}$, where the ϕ_j are smooth functions supported in O_j. Then

$$\|u\|_{B_{2,\infty}^s(\Omega)} \leq C' \sum \|\phi_j u\|_{B_{2,\infty}^s(O_j \cap \Omega)} \leq C'' \|u\|_{B_{2,\infty}^s(\Omega)}, \tag{16.4.5}$$

where C' and C'' are constants. This allows us to choose different sets Λ for different balls O_j intersecting the boundary.

Let $u \in H^1(\Omega)$ be a solution of the Neumann problem under the above assumptions. The following proposition is key.

Proposition 16.4.3. *If the right-hand side f of system* (16.4.2) *belongs to $L_2(\Omega)$, then the solution $u(x)$ belongs to the Besov space $B_{2,\infty}^{3/2}(\Omega)$ and*

$$\|u\|_{B_{2,\infty}^{3/2}(\Omega)}^2 \leq C_1 \|f\|_{L_2(\Omega)} \|u\|_{H^1(\Omega)}, \tag{16.4.6}$$

where C_1 is a constant not depending on $u(x)$.

The theorem will be derived from this proposition by interpolation. We begin with a preparation for the proof of the proposition.

We set

$$\Phi_\tau(u,v) = \Phi_\Omega(u,v) + \tau(u,v)_\Omega \quad \text{and} \quad \Phi_\tau(v) = \Phi_\tau(v,v). \qquad (16.4.7)$$

By virtue of the assumption that the form Φ_Ω is strongly coercive on $H^1(\Omega)$, we have

$$\|v\|^2_{H^1(\Omega)} \le C_2 \Phi_\tau(v), \qquad (16.4.8)$$

where C_2 is a positive constant.

As mentioned in Section 9.1, Lipschitz surfaces satisfy the uniform cone condition. Each point x_0 of Γ is the vertex of two right circular cones $\Upsilon_+(x_0)$ and $\Upsilon_-(x_0)$ congruent to a fixed cone and contained entirely, except for the vertex x_0, inside Ω and inside the complement to $\overline{\Omega}$, respectively. Locally, the axes of these cones can be chosen parallel. To study the Neumann problem, we need the cones Υ_+. Let ρ_0 be the height of these cones.

By $\Lambda = \Lambda(x_0)$, $x_0 \in \Gamma$, we denote the set of all vectors $h \in \mathbb{R}^n$ for which $x_0 + h \in \Upsilon_+(x_0)$, and by $\Omega_\rho = \Omega_\rho(x_0)$ we denote the intersection of Ω with the ball $O_\rho(x_0)$ of radius ρ centered at x_0. In what follows, we also need $\Omega_{2\rho}$ and $\Omega_{3\rho}$. We fix the direction of the axis of the cone $\Upsilon_+(x_0)$ and the numbers $\rho \in (0, \rho_0)$ and $q(x_0) \in (0, \rho)$ so that

$$x \in \Omega_{2\rho}(x_0), \quad h \in \Lambda(x_0), \quad |h| \le q(x_0) \;\Rightarrow\; x + h \in \Omega_{3\rho}(x_0). \qquad (16.4.9)$$

These numbers ρ and q can be assumed to be independent of x_0.

Given $x_0 \in \Gamma$, let $\varphi(x)$ be a C^∞ real-valued function with values in $[0,1]$ equal to 1 on $O_\rho(x_0)$ and 0 outside $O_{2\rho}(x_0)$. We can assume that these $\varphi(x)$ are the shifts $\psi(x - x_0)$ of a fixed function $\psi(x)$.

We set

$$v_h(x) = \varphi(x)[u(x+h) - u(x)]. \qquad (16.4.10)$$

By virtue of (16.4.8), we have

$$\|v_h\|^2_{H^1(\Omega)} \le C_2 \Phi_\tau(v_h). \qquad (16.4.11)$$

We want to derive the estimate

$$\Phi_\tau(v_h) \le C_3 |h| \|u\|_{H^1(\Omega_{3\rho})} [\|f\|_{L_2(\Omega_{2\rho})} + \|u\|_{H^1(\Omega_{3\rho})}]. \qquad (16.4.12)$$

This will enable us to prove inequality (16.4.6).

Here and in what follows, $h \in \Lambda(x_0)$ and $|h| \le q(x_0) < \rho$.

We use dots to denote summands which are dominated by the right-hand side in (16.4.12).

We need Hadamard's formula

$$g(x+h) - g(x) = \sum_1^n h_j \int_0^1 (\partial_j g)(x+th)\, dt \qquad (16.4.13)$$

for C^1 functions. Using the Schwarz inequality and transposing the integrals, we obtain the following inequality for norms with respect to x:

$$\|g(x+h)-g(x)\|^2_{L_2(G)} \leq C_4|h|^2\|g\|^2_{H^1(\widetilde{G})}. \qquad (16.4.14)$$

Here $G \subset \widetilde{G}$ are subdomains of Ω such that $G+th \subset \widetilde{G}$ for $0 \leq t \leq 1$. This inequality extends to functions in $H^1(G)$.

In particular, inequality (16.4.14) makes it possible to "take the factor φ out of the differentiation" in

$$\partial_k[\varphi(x)(u(x+h)-u(x))] \quad \text{or} \quad \partial_j[\varphi(x)(\overline{u(x+h)}-\overline{u(x)})]$$

in the integral to be estimated, because when differentiating $\varphi(x)$, we can apply (16.4.14) to the difference $u(x+h)-u(x)$ or $\overline{u(x+h)}-\overline{u(x)}$, respectively. Conversely, in, e.g., the expression

$$\varphi^2(x)\partial_j[\overline{u(x+h)}-\overline{u(x)}]$$

we can "take $\varphi^2(x)$ under the differentiation sign."

Proof of Proposition 16.4.3. Obviously,

$$\tau(v_h,v_h)_\Omega = \dots . \qquad (16.4.15)$$

Thus, we shall prove estimate (16.4.12) for $\Phi_0(v_h)$. Taking φ out of the differentiation, we obtain

$$\Phi_0(v_h)$$
$$= \int_\Omega \varphi^2(x) \sum a_{j,k}(x)\partial_k[u(x+h)-u(x)] \cdot \partial_j[\overline{u(x+h)}-\overline{u(x)}]\,dx + \dots$$
$$= I_1 - I_2 - I_3 - I_4 + \dots, \qquad (16.4.16)$$

where

$$I_1 = \int_\Omega \varphi^2(x) \sum a_{j,k}(x)\partial_k u(x+h) \cdot \partial_j \overline{u(x+h)}\,dx,$$

$$I_2 = \int_\Omega \varphi^2(x) \sum a_{j,k}(x)\partial_k u(x) \cdot \partial_j \overline{u(x)}\,dx,$$

$$I_3 = \int_\Omega \varphi^2(x) \sum a_{j,k}(x)\partial_k u(x) \cdot \partial_j [\overline{u(x+h)} - \overline{u(x)}]\,dx,$$

$$I_4 = \int_\Omega \varphi^2(x) \sum a_{j,k}(x)\partial_k [u(x+h)-u(x)] \cdot \partial_j \overline{u(x)}\,dx.$$

In the expression for I_3, we bring φ^2 under the sign ∂_j:

$$I_3 = \Phi_0(u, \varphi v_h) + \dots .$$

Similarly,

$$I_4 = \Phi_0(\varphi v_h, u) + \dots.$$

Since u is a solution of the Neumann problem, it follows that

$$\Phi_0(u, \varphi v_h) = -\tau(u, \varphi v_h)_\Omega + (f, \varphi v_h)_\Omega = \dots. \qquad (16.4.17)$$

We have no such a relation for I_4. But, taking into account the formal self-adjointness of L, which is essentially used here, we obtain

$$I_4 = \overline{\Phi_0(u, \varphi v_h)} + \dots = \overline{I_3} + \dots = \dots. \qquad (16.4.18)$$

It remains to estimate $I_1 - I_2$. These integrals over Ω are equal to integrals over $\Omega_{2\rho}$, because $\varphi(x)$ vanishes outside $O_{2\rho}$. Using the Lipschitz continuity of the functions $\varphi^2(x) a_{j,k}(x)$, we rewrite I_1 in the form

$$I_1 = \int_{\Omega_{2\rho}} \varphi^2(x+h) \sum a_{j,k}(x+h)\partial_k u(x+h)\partial_j \overline{u(x+h)}\,dx + \dots$$

and replace $x + h$ by x:

$$I_1 = \int_{\Omega_{2\rho}+h} \varphi^2(x) \sum a_{j,k}(x)\partial_k u(x)\partial_j \overline{u(x)}\,dx + \dots. \qquad (16.4.19)$$

Now, denoting $\varphi^2(x) \sum a_{j,k}(x)\partial_k u(x)\partial_j \overline{u(x)}$ by $U(x)$, we obtain

$$I_1 - I_2 = \int_{[\Omega_{2\rho}+h]\setminus\Omega_{2\rho}} U(x)\,dx - \int_{\Omega_{2\rho}\setminus[\Omega_{2\rho}+h]} U(x)\,dx + \dots. \qquad (16.4.20)$$

The first of these two integrals vanishes, because $\varphi(x) = 0$ on the domain of integration. The second can be omitted by virtue of the nonnegativity assumption (16.4.1) on the integrand.

We have obtained (16.4.12).

The construction and the result extend to the interior points of the domain. But at these points, we do not need the additional condition; it suffices to replace the integral

$$\int_{\Omega_{2\rho}\setminus[\Omega_{2\rho}+h]} U(x)\,dx$$

by a similar integral with the function $\varphi(x - h)$ (instead of $\varphi(x)$), which vanishes outside $\Omega_{2\rho} + h$.

In the cover of the closure $\overline{\Omega}$ by the neighborhoods O_ρ we choose a finite subcover. Using inequalities (16.4.11) and (16.4.12) and the estimates

$$\|u\|_{H^1(\Omega)} \le C_5\|f\|_{\widetilde{H}^{-1}(\Omega)} \le C_6\|f\|_{L_2(\Omega)}, \qquad (16.4.21)$$

we finally arrive at inequality (16.4.6).   □

Now we give the required interpolation relations. We mean real interpolation. The first six relations can be taken for the definitions of B spaces on their right-hand sides. In these relations, $0 < s < 1$ and $1 \leq q \leq \infty$ (actually, only $q = 1$ and ∞ are needed). For $\mathbb{R}^n$, they can be found in the books [60, Sec. 6.4] and [377, 378]. These relations are

$$(L_2(\Omega), H^1(\Omega))_{s,q} = B_{2,q}^s(\Omega), \tag{16.4.22}$$

$$(L_2(\Omega), H^{-1}(\Omega))_{s,q} = B_{2,q}^{-s}(\Omega), \tag{16.4.23}$$

$$(H^1(\Omega), H^2(\Omega))_{s,q} = B_{2,q}^{1+s}(\Omega), \tag{16.4.24}$$

$$(L_2(\Omega), \widetilde{H}^1(\Omega))_{s,q} = \widetilde{B}_{2,q}^s(\Omega), \tag{16.4.25}$$

$$(L_2(\Omega), \widetilde{H}^{-1}(\Omega))_{s,q} = \widetilde{B}_{2,q}^{-s}(\Omega), \tag{16.4.26}$$

$$(\widetilde{H}^1(\Omega), \widetilde{H}^2(\Omega))_{s,q} = \widetilde{B}_{2,q}^{1+s}(\Omega). \tag{16.4.27}$$

For $t \in (0,1)$ and $1 \leq q \leq \infty$, the relations given below also hold. Recall that $H^s = B_{2,2}^s$, so that the interpolation is with respect to the superscript and the right subscript. The spaces on the right-hand sides do not depend on q.

$$(H^1(\Omega), B_{2,q}^{1+s}(\Omega))_{t,2} = H^{1+ts}(\Omega), \tag{16.4.28}$$

$$(H^{-1}(\Omega), B_{2,q}^{-1+s}(\Omega))_{t,2} = H^{-1+ts}(\Omega), \tag{16.4.29}$$

$$(\widetilde{H}^1(\Omega), \widetilde{B}_{2,q}^{1+s}(\Omega))_{t,2} = \widetilde{H}^{1+ts}(\Omega), \tag{16.4.30}$$

$$(\widetilde{H}^{-1}(\Omega), \widetilde{B}_{2,q}^{-1+s}(\Omega))_{t,2} = \widetilde{H}^{-1+ts}(\Omega). \tag{16.4.31}$$

Proof of Theorem 16.4.2. Let $\mathcal{L}_N$ be the operator $u \mapsto f$ corresponding to the Neumann problem. It follows from (16.4.6) and the left inequality in (16.4.21) that

$$\|u\|_{B_{2,\infty}^{3/2}(\Omega)}^2 \leq C_{12}\|f\|_{L_2(\Omega)}\|f\|_{\widetilde{H}^{-1}(\Omega)}. \tag{16.4.32}$$

By virtue of Theorem 13.8.5, this implies the boundedness of $\mathcal{L}_N^{-1}$ as an operator

$$(\widetilde{H}^{-1}(\Omega), L_2(\Omega))_{1/2,1} \to B_{2,\infty}^{3/2}(\Omega). \tag{16.4.33}$$

Here the spaces on the left-hand side can be permuted by virtue of the first of the five real interpolation properties given in Section 13.3 applied at $\theta = 1 - \theta = 1/2$. Furthermore, $\mathcal{L}_N^{-1}$ is a bounded operator from $\widetilde{H}^{-1}(\Omega)$ to $H^1(\Omega)$. Therefore, it is bounded as an operator

$$(\widetilde{H}^{-1}(\Omega), (\widetilde{H}^{-1}(\Omega), L_2(\Omega))_{1/2,1})_{t,2} \to (H^1(\Omega), B_{2,\infty}^{3/2}(\Omega))_{t,2} \tag{16.4.34}$$

for $t \in (0,1)$. By virtue of (16.4.26) (with the same θ; the spaces can again be permuted) and (16.4.28), it is also bounded as an operator

$$(\widetilde{H}^{-1}(\Omega), \widetilde{B}_{2,1}^{-1/2}(\Omega))_{t,2} \to H^{1+t/2}(\Omega),$$

and by virtue of (16.4.31), it is bounded as an operator

$$\widetilde{H}^{-1+t/2}(\Omega) \to H^{1+t/2}(\Omega).$$

Replacing $t/2$ by s, we obtain the required assertion for the Neumann problem.

In the case of the Dirichlet problem, we make the obvious change of the space $H^{1+s}(\Omega)$ for $\widetilde{H}^{1+s}(\Omega)$ and of $\widetilde{H}^{-1+s}(\Omega)$ for $H^{-1+s}(\Omega)$. It is convenient to assume that the function $u(x)$ vanishes outside Ω and that the shifts h are into the complement of the domain Ω, i.e., to the cones Υ_-. Then the function $v_h(x)$ belongs to the spaces $\widetilde{H}^{1+s}(\Omega)$ together with u; cf. [325]. This allows us to avoid the use of the additional condition (16.4.1) near the boundary as well, replacing $\varphi(x)$ in the integral I_2 by $\varphi(x-h)$.

Accordingly, for the operator $\mathcal{L}_D$ corresponding to the Dirichlet problem, $\mathcal{L}_D^{-1}$ is bounded as an operator

$$(H^{-1}(\Omega), (H^{-1}(\Omega), L_2(\Omega))_{1/2,1})_{t,2} \to (\widetilde{H}^1(\Omega), \widetilde{B}_{2,\infty}^{3/2}(\Omega))_{t,2}, \qquad (16.4.35)$$

as an operator

$$(H^{-1}(\Omega), B_{2,1}^{-1/2}(\Omega))_{t,2} \to \widetilde{H}^{1+t/2}(\Omega)$$

by virtue of (16.4.23) and (16.4.30), and as an operator

$$H^{-1+t/2}(\Omega) \to \widetilde{H}^{1+t/2}(\Omega)$$

by virtue of (16.4.29). Again, it remains to replace $t/2$ by s.

The results obtained above can be generalized to systems with formally self-adjoint leading part and any lower-order terms. This is easy: it suffices to take the lower-order terms to the right-hand side of the equation. □

Theorem 16.4.4. *Under the same assumptions as in Theorem* 16.4.2, *for problems with homogeneous boundary conditions, the following assertions are valid.*

1. For $|s| < 1/2$, the Dirichlet problem with right-hand side $f \in H^{-1+s}(\Omega)$ has a unique solution in $\widetilde{H}^{1+s}(\Omega)$.

2. For $|s| < 1/2$, the Neumann problem with right-hand side $f \in \widetilde{H}^{-1+s}(\Omega)$ has a unique solution in $H^{1+s}(\Omega)$.

For $s \in (0, 1/2)$, these assertions follow at once from Theorem 16.4.2, and for $s \in (-1/2, 0)$, they are proved by using the easy-to-verify adjointness of the operators $\mathcal{L}_D$ and $\widetilde{\mathcal{L}}_D$ and the operators $\mathcal{L}_N$ and $\widetilde{\mathcal{L}}_N$ corresponding to the numbers s and $-s$, respectively:

$$(\mathcal{L}_D u, v)_\Omega = (u, \widetilde{\mathcal{L}}_D v)_\Omega \qquad (u \in \widetilde{H}^{1+s}(\Omega), \quad v \in \widetilde{H}^{1-s}(\Omega)),$$

$$(\mathcal{L}_N u, v)_\Omega = (u, \widetilde{\mathcal{L}}_N v)_\Omega \qquad (u \in H^{1+s}(\Omega), \quad v \in H^{1-s}(\Omega)).$$

The Lipschitz continuity of the coefficients b_j is needed only if $s \in (-1/2, 0)$. For $s \in (0, 1/2)$, we can assume these coefficients to be bounded and measurable.

Note that the correspondence between f and the solution u in Theorem 16.4.4 is an isomorphism.

We repeat that if the system under consideration is only coercive rather than strongly coercive, then it is Fredholm with index zero but not necessarily uniquely solvable; the dimensions of the kernel and the cokernel do not depend on s. The statement about the smoothness of its solutions remains valid.

The results can be extended to problems with inhomogeneous boundary conditions.

16.5 Nirenberg's Method and the Regularity of Solutions inside a Domain

In this subsection we supplement the material of the preceding two by a brief explanation of Nirenberg's method of difference quotients [290] and outline the proof of the following theorem, a close statement of which is contained in [290].

Theorem 16.5.1. *Let u be a solution of the Dirichlet problem for a strongly elliptic system $Lu = f$ whose coefficients $a_{j,k}$ and b_j are Lipschitz continuous in $\overline{\Omega}$, and let u belong to $\widetilde{H}^1(\Omega)$. Suppose that Ω' is an interior subdomain of Ω, i.e., $\overline{\Omega'} \subset \Omega$, and $f \in L_2(\Omega)$. Then $u \in H^2(\Omega')$ and*

$$\|u\|_{H^2(\Omega')} \leq C_1[\|f\|_{L_2(\Omega)} + \|u\|_{H^1(\Omega)}]. \tag{16.5.1}$$

The formal self-adjointness of the leading part of the system is not needed here. The global smoothness of the solution when the right-hand side belongs to $L_2(\Omega)$ in Savaré's theorem is much lower; we now see that this is connected with the behavior of the solution near the boundary.

The main tool in the proof is the difference quotient

$$w_h(x) = \frac{w(x+h) - w(x)}{h}. \tag{16.5.2}$$

In the denominator, h is a small real number, and in the numerator, this is the point in $\mathbb{R}^n$ whose ith coordinate equals h and the other coordinates are zero. The number i is fixed; to simplify the notation, we do not indicate it. We must assume that $|h|$ is less than the distance from an interior point x to the boundary. It is convenient to assume also that all functions in $\widetilde{H}^1(\Omega)$ vanish outside Ω.

Below we state fairly obvious properties of difference quotients. By $\widehat{\Omega}$ we denote any interior subdomain of Ω.

Lemma 16.5.2. *For any $v, w \in H^1(\Omega)$ supported in $\widehat{\Omega}$,*

$$(\partial_k v)_h = \partial_k(v_h) \quad and \quad (v_h, w)_\Omega = -(v, w_{-h})_\Omega \tag{16.5.3}$$

for sufficiently small $|h|$.

Lemma 16.5.3. *For any $v \in H^1(\Omega)$ and sufficiently small h,*

$$\|v_h\|_{L_2(\widehat{\Omega})} \le \|v\|_{H^1(\Omega)} \quad and \quad \lim_{h \to 0} \|v_h - \partial_i v\|_{L_2(\widehat{\Omega})} = 0. \qquad (16.5.4)$$

Here the first inequality is obtained by using Hadamard's formula (16.4.13).

Lemma 16.5.4. *If $v \in \widetilde{H}^1(\Omega)$ and*

$$\|(\partial_k v)_h\|_{L_2(\widehat{\Omega})} \le C_2, \qquad (16.5.5)$$

where C_2 does not depend on h, then $\partial_i \partial_k v \in L_2(\widehat{\Omega})$ exists and has norm not exceeding C_2.

This inequality can be obtained because any bounded set in $L_2(\widehat{\Omega})$ is weakly compact, so that we can choose a sequence of values of h approaching zero along which the $(\partial_k v)_h$ converge to the second derivative of v in the sense of distributions.

We proceed to the proof of the theorem. Without loss of generality, we assume that the form $\Phi_\Omega(u, v)$ contains only leading terms. It is coercive on $\widetilde{H}^1(\Omega)$:

$$\|v\|^2_{\widetilde{H}^1(\Omega)} \le C_3 \operatorname{Re} \Phi_\Omega(v, v) + C_4 \|v\|^2_{L_2(\Omega)}. \qquad (16.5.6)$$

For the solution u and a test function $v \in \widetilde{H}^1(\Omega)$, we have

$$\Phi_\Omega(u, v) = (f, v)_\Omega, \qquad (16.5.7)$$

which implies the key estimate

$$|\Phi_\Omega(u, v)| \le \|f\|_{L_2(\Omega)} \|v\|_{L_2(\Omega)}. \qquad (16.5.8)$$

Let us fix intermediate subdomains Ω'' and Ω''':

$$\overline{\Omega'} \subset \Omega'', \quad \overline{\Omega''} \subset \Omega''', \quad \overline{\Omega'''} \subset \Omega.$$

Let $\zeta(x)$ be a function in $C_0^\infty(\Omega)$ equal to 1 inside Ω'' and 0 outside Ω''' and such that $0 \le \zeta \le 1$. We set $\zeta u = v$ and take $w \in \widetilde{H}^1(\Omega)$. By K_j we denote constants not depending on w and h.

By using the inequality in (16.5.4) and Hadamard's formula, we obtain

$$|\Phi_\Omega(v_h, w)| = \Big| \sum (a_{j,k} \partial_k(\zeta u)_h, \partial_j w)_\Omega \Big|$$
$$\le \Big| \sum (a_{j,k}(\zeta \partial_k u)_h, \partial_j w)_\Omega \Big| + K_1 \|w\|_{H^1(\Omega)} \|u\|_{H^1(\Omega)}. \qquad (16.5.9)$$

The first term on the right-hand side equals

$$\Big| \sum ((a_{j,k} \zeta \partial_k u)_h - (a_{j,k})_h \zeta(x+h) \partial_k u(x+h), \partial_j w)_\Omega \Big|.$$

Applying Lemma 16.5.2, we see that the right-hand side of (16.5.9) is dominated by

$$\left|\sum(a_{j,k}\partial_k u, \zeta\partial_j w_{-h})_\Omega\right| + K_2\|w\|_{H^1(\Omega)}\|u\|_{H^1(\Omega)}.$$

Therefore,

$$|\Phi_\Omega(v_h, w)| \le \left|\sum(a_{j,k}\partial_k u, \partial_j(\zeta w_{-h}))_\Omega\right| + K_3\|w\|_{H^1(\Omega)}\|u\|_{H^1(\Omega)}$$
$$= |\Phi_\Omega(u, \zeta w_{-h})| + K_3\|w\|_{H^1(\Omega)}\|u\|_{H^1(\Omega)}.$$

Now we use inequality (16.5.8) with ζw_{-h} instead of v, which gives

$$|\Phi_\Omega(v_h, w)| \le \|f\|_{L_2(\Omega)}\|\zeta w_{-h}\|_{L_2(\Omega)} + K_3\|w\|_{H^1(\Omega)}\|u\|_{H^1(\Omega)}.$$

Setting $w = v_h$, we obtain

$$|\Phi_\Omega(v_h, v_h)| \le K_4\|v_h\|_{H^1(\Omega)}[\|f\|_{L_2(\Omega)} + \|u\|_{H^1(\Omega)}];$$

this, together with (16.5.6), implies

$$\|v_h\|^2_{H^1(\Omega)} \le K_5\|v_h\|_{H^1(\Omega)}[\|f\|_{L_2(\Omega)} + \|u\|_{H^1(\Omega)}] + C_4\|v_h\|^2_{L_2(\Omega)}.$$

Here the last summand is dominated by $\|v_h\|_{H^1(\Omega)}\|v\|_{H^1(\Omega)}$ in view of the inequality in (16.5.4). After reduction by $\|v_h\|_{H^1(\Omega)}$, we obtain

$$\|v_h\|_{H^1(\Omega)} \le K_6[\|f\|_{L_2(\Omega)} + \|u\|_{H^1(\Omega)}].$$

Now we can apply Lemma 16.5.4, which readily leads to the goal.   □

Nirenberg also proved that *if the boundary is C^2, then the solution belongs to $H^2(\Omega)$.* For this purpose, locally rectifying the boundary, he used a similar argument applied to the tangential derivatives of the first derivatives of the solution and then calculated the "pure" second normal derivative from the remaining derivatives in the system. This result extends to the case of an inhomogeneous Dirichlet condition whose right-hand side belongs to $H^{3/2}(\Gamma)$.

It follows, in particular, that, *for a strongly elliptic system in a smooth domain with right-hand side in $L_2(\Omega)$, the variational setting of the Dirichlet problem is equivalent to its usual setting. Therefore, the strong ellipticity of a system implies the ellipticity of the Dirichlet problem for this system.*

16.6 Fractional Powers of the Operators Corresponding to Problems in Lipschitz Domains and the Kato Problem

We return to the operators considered in Section 11.9 under the same assumptions about spaces and operators and shall use the same notation. We also make two additional assumptions, which will be specified later on.

As we saw in Section 13.9, the domains of the fractional powers S_1^α, $0 \leq \alpha \leq 1$, of a self-adjoint operator S_1 form a scale between the spaces H_{-1} and H_1 for the complex interpolation method. We denote these spaces by H_τ, $\tau = -1 + 2\alpha$, $\alpha \in [0,1]$. Thus, if $\{e_j\}_1^\infty$ is an orthonormal basis of eigenvectors of S_1 in H_{-1} and λ_j are the corresponding eigenvalues, then H_τ consists of the vectors

$$u = \sum \lambda_j^\alpha \langle u, e_j \rangle_{-1} e_j \text{ with } \|u\|_\tau^2 = \sum_1^\infty \lambda_j^{2\alpha} |\langle u, e_j \rangle_{-1}|^2 < \infty. \qquad (16.6.1)$$

The operator S_1^α is bounded and invertible as an operator from H_{τ_1} to H_{τ_2} for $\tau_2 = \tau_1 - 2\alpha$.

First, we make the following natural assumption: *for $\alpha = 1/2$, this space coincides with H_0, and the inner product coincides with $(u, v)_0$:*

$$\sum_1^\infty \lambda_j \langle u, e_j \rangle \overline{\langle v, e_j \rangle} = (u, v)_0. \qquad (16.6.2)$$

The spaces H_τ and $H_{-\tau}$ for $|\tau| \leq 1$ are dual with respect to its extension to their direct product.

Using powers of S_1, we can extend this scale beyond the points ± 1. In our particular problems, these spaces are given a priori. Thus, the operator S_1 remains bounded and invertible as an operator from $H_{1+\gamma}$ to $H_{-1+\gamma}$ for any $\gamma \in \mathbb{R}$. But we shall use the spaces H_τ only for $\tau \in (-3/2, 3/2)$, in order to keep the notation H_2 for $D(A_2)$. In the smooth case, this notation causes no conflict and we can consider H_τ for any τ.

Secondly, *we assume that A_1 and A_1^* remain bounded as operators from $H_{1+\gamma}$ to $H_{-1+\gamma}$ if $|\gamma| < \varepsilon$ for some small $\varepsilon > 0$.*

As we saw in Section 16.1, this assumption causes no trouble in our particular problems. The required boundedness is ensured by the weak smoothness assumption on the coefficients.

Moreover, by Shneiberg's theorem (see Section 13.7), *these operators are invertible for $|\gamma| < \varepsilon_1$,* where ε_1 is a sufficiently small positive number, $\varepsilon_1 \leq \varepsilon$. In our particular problems, these are simultaneously regularity theorems.

In this subsection we explain the way to important results on fractional powers of the operators A_1 and A_2. We start with A_2.

As already mentioned, in the general case, we have no information about $H_2 = D(A_2)$. However, Kato proved that [199, Theorem 1.1]

$$D(A_2^\alpha) = D(A_2^{*\alpha}) = D(S_2^\alpha) \text{ for } 0 \leq \alpha < 1/2. \qquad (16.6.3)$$

Kato also pointed out that these relations do not hold in the general case for $1/2 < \alpha < 1$. This gave rise to the problem of when

$$D(A_2^{1/2}) = D(A_2^{*1/2}) = H_1. \qquad (16.6.4)$$

This problem has become known as the *Kato square root problem* and stimulated a stream of informative studies. We shall give more detailed references in Section 19. In particular, for the problems which we consider, it was shown by various involved methods that the Kato problem has a positive solution.

Kato's observations, the problem, and the obtained results are very interesting, in particular, because, as already mentioned, the domain $D(A_2)$ is generally unknown and may differ from $D(A_2^*)$ (e.g., for a "smooth" non–self-adjoint Neumann problem).

Note that Theorem 13.9.3 [401] implies

$$D(A_2^\alpha) = [H_0, D(A_2)]_\alpha, \quad D(A_2^{*\alpha}) = [H_0, D(A_2^*)]_\alpha \quad (0 \le \alpha \le 1). \quad (16.6.5)$$

But since $D(A_2)$ and $D(A_2^*)$ are generally unknown, these relations do not directly solve the Kato problem.

The operator S_2 is self-adjoint in H_0 by virtue of (11.9.8) and the coincidence of S_2 with S_1 on $D(S_2)$. In such cases, the problem is easy to solve in the affirmative: as shown by Kato [201] (see also [37, Sec. 5.5.1]), if any two spaces in (16.6.4) coincide, then all three spaces automatically coincide.

For smooth elliptic problems, the spaces on the right-hand sides in (16.6.5) were described by Grisvard [172] (see also Seeley [339]), and it is seen from this description that the problem has a positive solution in this case, too.

Below we show that the solution of the Kato problem is positive in the generality of this section.

Theorem 16.6.1. *Under our assumptions, relations* (16.6.4) *hold for the operators* A_2 *and* A_2^*.

Proof. We set

$$B = S_1^{-1/2} A_1 S_1^{-1/2}. \quad (16.6.6)$$

Under our assumptions, this is a bounded invertible operator in H_γ with small $|\gamma|$, and *the operator* A_1 *admits the factorization*

$$A_1 = S_1^{1/2} B S_1^{1/2}. \quad (16.6.7)$$

Remark 16.6.2. The operator B is not arbitrary. Since $S_1 = \frac{1}{2}[A_1 + A_1^*]$, it follows that

$$A_1^* = 2S_1 - A_1 = 2S_1 - S_1^{1/2} B S_1^{1/2} = S_1^{1/2}(2I - B)S_1^{1/2}.$$

Hence $B^* = 2I - B$, i.e.,

$$\frac{1}{2}[B + B^*] = I. \quad (16.6.8)$$

Now, as in [147], we prove the following lemma.

Lemma 16.6.3. *The following inclusions hold:*

$$H_1 \subset D(A_2^{1/2}), \quad H_1 \subset D(A_2^{*1/2}). \quad (16.6.9)$$

This implies Theorem 16.6.1 by virtue of Kato's result in [201, p. 243].

Let γ be a small positive number. Then $D(A_2) \subset H_{1+\gamma}$, because the operator A_1 continuously and bijectively maps these spaces to $H_0 \subset H_{-1+\gamma}$. The second inclusion is dense; therefore, $D(A_2)$ is densely embedded in $H_{1+\gamma}$.

The proof of Lemma 16.6.3 uses the following lemma.

Lemma 16.6.4. *The space $H_{1+\gamma}$ is continuously embedded in $D(A_2^{(1+\gamma)/2})$.*

Proof of Lemma 16.6.4. First, we take u and v in $D(A_2)$ and $D(A_2^*)$, respectively. Let $\delta = (1 - \gamma)/2$. Then $\delta \in (0, 1/2)$.

The operator $A_2^{*-\delta}$ acts continuously from H_0 to $D(A_2^{*\delta})$. Using (16.6.3) and taking into account Triebel's theorem 13.9.2, we obtain

$$D(A_2^{*\delta}) = D(S_2^\delta) = [H_0, D(S_2^{1/2})]_{2\delta} = [H_0, H_1]_{2\delta} = H_{2\delta}.$$

Therefore, $A_2^{*-\delta}$ is a continuous operator from H_0 to $H_{2\delta}$. Since $2\delta - 1 = -\gamma$, it follows that $S_1^{1/2} A_2^{*-\delta}$ is a continuous operator from H_0 to $H_{-\gamma}$; indeed, we have

$$\|S_1^{1/2} A_2^{*-\delta} v\|_{-\gamma} \le C_1 \|v\|_0 \qquad (16.6.10)$$

even for $v \in H_0$.

Moreover, $BS_1^{1/2}$ is a bounded operator from $H_{1+\gamma}$ to H_γ:

$$\|BS_1^{1/2} u\|_\gamma \le C_2 \|u\|_{1+\gamma}. \qquad (16.6.11)$$

Next, we have
$$A_2^{*(1+\gamma)/2} v = A_2^* A_2^{*-\delta} v = A_1^* A_2^{*-\delta} v.$$

Therefore,
$$(u, A_2^{*(1+\gamma)/2} v)_0 = (A_1 u, A_2^{*-\delta} v)_0,$$

or (see (16.6.7))
$$(A_2^{(1+\gamma)/2} u, v)_0 = (BS_1^{1/2} u, S_1^{1/2} A_2^{*-\delta} v)_0.$$

This, together with the generalized Schwarz inequality and inequalities (16.6.10) and (16.6.11), implies

$$|(A_2^{(1+\gamma)/2} u, v)_0| \le C_3 \|BS_1^{1/2} u\|_\gamma \|S_1^{1/2} A_2^{*-\delta} v\|_{-\gamma} \le C_4 \|u\|_{1+\gamma} \|v\|_0.$$

Since $D(A_2^*)$ is dense in H_0, it follows that

$$\|A_2^{(1+\gamma)/2} u\|_0 \le C_4 \|u\|_{1+\gamma}.$$

This inequality is extended to $u \in H_{1+\gamma}$ by the passage to the limit. $\square$

Proof of Lemma 16.6.3. Using the first relation in (16.6.5) and Theorem 13.6.1 on reiteration, we obtain

$$H_1 = [H_0, H_{1+\gamma}]_{1/(1+\gamma)} \subset [H_0, D(A_2^{(1+\gamma)/2})]_{1/(1+\gamma)} = D(A_2^{1/2}).$$

Similarly,

$$H_1 \subset D(A_2^{*1/2}).$$

This proves Lemma 16.6.3 and Theorem 16.6.1.   □
                                                                                □

Now we pass to the operator A_1.

Theorem 16.6.5. *Under our assumptions, the pure imaginary powers of the operator A_1 are locally bounded. For $0 < \alpha < 1$,*

$$D(A_1^\alpha) = [H_{-1}, H_1]_\alpha. \tag{16.6.12}$$

This very important result follows from its equivalence to Theorem 16.6.1, which is explained in [37, Secs. 5.2.2 and 4.4.10].

The explanation is simple. If $D(A_2^{1/2}) = H_1$, then

$$D(A_1^{1/2}) = A_1^{-1/2} H_{-1} = A_1^{1/2} A_1^{-1} H_{-1} = A_1^{1/2} H_1$$
$$= A_1^{1/2} A_2^{-1/2} A_2^{1/2} H_1 = A_1^{1/2} A_2^{-1/2} H_0 = H_0.$$

This is relation (16.6.12) with $\alpha = 1/2$. To the remaining values of α it is extended by using Theorem 16.6.8 of [176]. This theorem implies that the validity of (16.6.12) in the general case follows from its validity for one value of α. (According to Haase, this theorem is essentially due to Komatsu.)

The explanation of the reverse implication ((16.6.12) with $\alpha = 1/2$ implies $D(A_2^{1/2}) = H_1$) given by Arendt [37] is also very short:

$$D(A_2^{1/2}) = A_2^{-1/2} H_0 = A_2^{-1/2} A_1^{-1/2} H_{-1} = H_1.$$

By virtue of this equivalence, relations (16.6.4) are equivalent to

$$D(A_1^{1/2}) = D(A_1^{*1/2}) = H_0, \tag{16.6.13}$$

and Seeley's result [338] mentioned in Section 13.9 is yet another approach to obtain the positive answer to Kato's question for smooth problems.

We leave the specification of these results in the cases of Dirichlet, Neumann, and mixed problems for second-order strongly elliptic systems in Lipschitz domains to the reader. The approach also applies to the Dirichlet and Neumann problems for higher-order systems.

We also add that the interpolation scales $\{H_{-1+2\alpha}\}$, $0 \le \alpha \le 1$, and $\{[H_0, D(A_2)]_{1-\theta}\}$, $0 \le \theta \le 1$, are "glued together" into a single scale by Wolf's Theorem 13.8.4 thanks to the positive solution of the Kato problem (because these scales coincide on the spaces H_0 and H_1). Therefore, the operators A_1 and A_2 turn out to be "compatible" on the spaces between H_{-1} and $D(A_2)$. Any result on the smoothness properties of solutions of the equation $A_1 u = f$ gives an information about $D(A_2)$. See Sections 16.1, 16.3, and 16.4. And, conversely, any information about $D(A_2)$ may provide a result on the smoothness of solutions of $A_1 u = f$.

Now we mention the following important fact.

Remark 16.6.6. *The results obtained above* (*Theorems 16.6.1 and 16.6.5*) *apply to operators on the Lipschitz boundary* Γ, namely, *to the Dirichlet-to-Neumann operator* D, *the operator* A^{-1} (*where* $A\psi$ *is the restriction to the boundary of the single-layer potential* $\mathcal{A}\psi$), *and the hypersingular operator* H (*see Sections 11 and 12*).

In all of these cases, we have $H_1 = H^{1/2}(\Gamma)$, $H_0 = H^0(\Gamma)$, and $H_{-1} = H^{-1/2}(\Gamma)$. In "smooth" problems, we also have $H_2 = H^1(\Gamma)$.

Although the Sobolev spaces $H^t(\Gamma)$ are defined on a Lipschitz surface only for $|t| \leq 1$, their scale can be extended beyond the points ± 1 by using powers of the corresponding operator S_1.

In particular, relation (16.6.13) holds for the Dirichlet-to-Neumann operator $D = A_1$ as an operator from $H^{1/2}(\Gamma)$ to $H^{-1/2}(\Gamma)$. In smooth problems, we have to assume that the domain $D(A_2)$ of the corresponding operator A_2 is $H^1(\Gamma)$ (rather than $H^{3/2}(\Gamma)$). In general nonsmooth problems, this domain remains unknown, although some information about it can be extracted from theorems of Sections 16.1–16.4. In particular, this is $H^1(\Gamma)$ in the case of a scalar operator L with self-adjoint leading part (see Theorem 16.3.2). But in this case, the Dirichlet boundary condition has to be understood more generally than here, in the sense of nontangential convergence; see Section 19.

In all cases of boundary operators, the operator rather than the form is primary, and it determines the form:

$$\Phi(u, v) = (A_1 u, v)_0.$$

The results stated above can be carried over to the Dirichlet-to-Neumann operator for higher-order systems and to operators on a part of a Lipschitz boundary, which we considered in Sections 11–12. Fredholm situations can also be covered.

17 Applications to Boundary Value Problems in Lipschitz Domains. 2

17.1 Further Generalizations of the Settings of the Dirichlet and Neumann Problems

In this section we use classes of spaces larger than those in Section 16. Throughout the section,

$$|s| < 1/2, \quad 1 < p, p' < \infty, \quad \frac{1}{p} + \frac{1}{p'} = 1, \quad \text{and} \quad r = 1/p. \tag{17.1.1}$$

For simplicity, we assume the coefficients $a_{j,k}$ and b_j to be Lipschitz continuous (although, in particular situations, more economical assumptions can be made).

As previously, solutions in Ω of problems with homogeneous boundary conditions are determined by the Green identity

$$(f,v)_\Omega = \Phi_\Omega(u,v). \qquad (17.1.2)$$

But now, in the case of the Dirichlet problem, we assume that

$$u \in \widetilde{H}_p^{1/2+s+1/p}(\Omega), \quad f \in H_p^{-1/2+s-1/p'}(\Omega), \qquad (17.1.3)$$

$$v \in \widetilde{H}_{p'}^{1/2-s+1/p'}(\Omega),$$

and in the case of the Neumann problem, that

$$u \in H_p^{1/2+s+1/p}(\Omega), \quad f \in \widetilde{H}_p^{-1/2+s-1/p'}(\Omega), \qquad (17.1.4)$$

$$v \in H_{p'}^{1/2-s+1/p'}(\Omega).$$

Note that $1/2 + s + 1/p$ lies strictly between $1/p$ and $1 + 1/p$. By virtue of the statement concerning duality mentioned in Section 14.5, in both cases, the right-hand side f and the test function v belong to dual spaces. The difference of the superscripts of the spaces containing u and f equals 2. The inequality for s in (17.1.1) is again determined by the trace theorem. The admissible points (s,r) form the square

$$Q = \{(s,r) : |s| < 1/2, \, 0 < r < 1\}. \qquad (17.1.5)$$

The initial settings of the problems correspond to the center of this square. Results on unique solvability can be obtained at far from all of its points; see Section 17.3 below. In this and the next subsection, we explain what can be obtained.

The form $\Phi_\Omega(u,v)$ is bounded on the direct product of the spaces containing u and v. For simplicity, we demonstrate this in the absence of lower-order terms. For example, in the case of the Neumann problem, we have

$$|\Phi_\Omega(u,v)| \le C \sum |(\partial_k u, \partial_j v)_\Omega|$$

$$\le C' \sum \|\partial_k u\|_{\widetilde{H}_p^{-1/2+s+1/p}(\Omega)} \|\partial_j v\|_{H_{p'}^{1/2-s-1/p}(\Omega)} \qquad (17.1.6)$$

$$\le C'' \|u\|_{H_p^{1/2+s+1/p}(\Omega)} \|v\|_{H_{p'}^{1/2-s+1/p'}(\Omega)}.$$

Here

$$\widetilde{H}_p^{-1/2+s+1/p}(\Omega) = H_p^{-1/2+s+1/p}(\Omega) \quad \text{for } |s| < 1/2,$$

because the last inequality is equivalent to the inclusion

$$-1/2 + s + 1/p \in (-1/p', 1/p)$$

(see Theorem 14.3.3 and the relevant remark in Section 14.5) and $1/2 - s - 1/p = -1/2 - s + 1/p'$.

Furthermore, the operators $\mathcal{L}_D$ and $\mathcal{L}_N$ corresponding to the Dirichlet and Neumann problems with homogeneous boundary conditions are bounded as operators from $\widetilde{H}_p^{1/2+s+1/p}(\Omega)$ to $H_p^{-1/2+s-1/p'}(\Omega)$ and from $H_p^{1/2+s+1/p}(\Omega)$ to $\widetilde{H}_p^{-1/2+s-1/p'}(\Omega)$, respectively. This follows from the Green identity and the duality statements. For example, in the case of the Neumann problem (we again assume for simplicity that there are no lower-order terms), we have

$$\|Lu\|_{\widetilde{H}_p^{-1/2+s-1/p'}(\Omega)} \leq C_1 \sup_{v \neq 0} \frac{|(Lu,v)_\Omega|}{\|v\|_{H_{p'}^{1/2-s+1/p'}(\Omega)}}$$

$$= \sup_{v \neq 0} \frac{|\Phi_\Omega(u,v)|}{\|v\|_{H_{p'}^{1/2-s+1/p'}(\Omega)}} \leq C_2 \|u\|_{H_p^{1/2+s+1/p}(\Omega)}$$

by virtue of (17.1.6).

In the case of the Dirichlet problem with inhomogeneous boundary condition

$$Lu = f \text{ in } \Omega, \quad u^+ = g \tag{17.1.7}$$

we assume that $u \in H_p^{1/2+s+1/p}(\Omega)$ and $g \in B_p^{1/2+s}(\Gamma)$. As to f, we let $f \in H_p^{-1/2+s-1/p'}(\Omega)$ and act as in Section 11.1. Given any function $u_0 \in H_p^{1/2+s+1/p}(\Omega)$ with boundary value g, we define $f_0 = Lu_0 \in H_p^{-1/2+s-1/p'}(\Omega)$ by

$$(f_0,v)_\Omega = \Phi_\Omega(u_0,v), \quad v \in \widetilde{H}_{p'}^{1/2-s+1/p'}(\Omega). \tag{17.1.8}$$

By a solution of problem (17.1.7) we *mean* a function $u = u_0 + u_1$, where u_1 is a solution of the problem

$$Lu_1 = f - f_0 \text{ in } \Omega, \quad u_1^+ = 0 \tag{17.1.9}$$

(provided, of course, that it exists), i.e., u_1 is determined by

$$(f - f_0,v)_\Omega = \Phi_\Omega(u_1,v), \quad v \in \widetilde{H}_{p'}^{1/2-s+1/p'}(\Omega). \tag{17.1.10}$$

As in Section 11.1, it can be shown that if uniqueness holds, then u does not depend on the choice of u_0. Thus, for given s and p, the Dirichlet problem with inhomogeneous boundary condition is uniquely solvable if so is the Dirichlet problem with homogeneous boundary condition.

In the Neumann problem, we define the conormal derivative by the full Green identity

$$(Lu,v)_\Omega = \Phi_\Omega(u,v) - (T^+u,v)_\Gamma. \tag{17.1.11}$$

It is easy to verify that the conormal derivative belongs to $B_p^{-1/2+s}(\Gamma)$, and the operator $(u,f) \mapsto T^+u$ is bounded. Thus, in the inhomogeneous Neumann problem

$$Lu = f \text{ in } \Omega, \quad T^+u = h \tag{17.1.12}$$

we assume that

$$u \in H_p^{1/2+s+1/p}(\Omega), \quad h = T^+u \in B_p^{-1/2+s}(\Gamma), \tag{17.1.13}$$
$$f \in \widetilde{H}_p^{-1/2+s-1/p'}(\Omega).$$

The inhomogeneous Neumann boundary condition is reduced to the homogeneous one by changing f.

17.2 New Corollaries of Shneiberg's Theorem

In this section we again assume the form Φ_Ω to be strongly coercive on $\widetilde{H}^1(\Omega)$ in the case of the Dirichlet problem and on $H^1(\Omega)$ in the case of the Neumann problem.

As we know, it follows from Shneiberg's theorem (see Section 13.7) that there exists an $\varepsilon \in (0, 1/2]$ such that, when $p = 2$, the Dirichlet and Neumann problems are uniquely solvable for $|s| < \varepsilon$. We also know that, under certain additional assumptions, ε equals $1/2$. As in Section 16.1, we denote this ε by $\varepsilon(L)$ and assume that it is the same for the Dirichlet and Neumann problems if the strong coercivity condition holds on $H^1(\Omega)$.

For $0 < \varepsilon \le 1/2$ and $0 < \delta \le 1/2$, we introduce the notation

$$Q_{\varepsilon,\delta} = \{(s, r): |s| < \varepsilon, |r - 1/2| < \delta\}. \tag{17.2.1}$$

Theorem 17.2.1. *There exists a $\delta = \delta(L) > 0$ such that the Dirichlet problem with homogeneous boundary condition remains uniquely solvable for $(s, r) \in Q_{\varepsilon,\delta}$, where $\varepsilon = \varepsilon(L)$. The same is true for the Neumann problem with homogeneous boundary condition.*

Let us explain how this theorem is proved for the Dirichlet problem. The operator $\mathcal{L}_D$ is bounded and invertible as an operator from $\widetilde{H}^{1+s}(\Omega)$ with $|s| < \varepsilon(L)$ to $H^{-1+s}(\Omega)$. We fix s with $|s| < \varepsilon(L)$ and consider $\mathcal{L}_D$ as an operator from $\widetilde{H}_p^{1/2+s+1/p}(\Omega)$ to $H_p^{-1/2+s-1/p'}(\Omega)$. These are interpolation scales, $\mathcal{L}_D$ acts boundedly from the spaces of the former to the spaces of the latter and is invertible for $p = 2$. The required number δ exists by Shneiberg's theorem. It is seen from his estimates given in Section 13.7 that we can assume δ to be independent of s.

For the Neumann problem, the theorem is proved by the same argument applied to the corresponding spaces.   □

Remark 17.2.2. Again, these results are extended to the Dirichlet and Neumann problems with inhomogeneous boundary conditions.

Theorem 16.1.4 carries over to the spaces $H_p^{1/2+s+1/p}(\Omega)$ as follows.

Theorem 17.2.3. *If the Dirichlet problem is uniquely solvable in the space* $H_p^{1/2+s+1/p}(\Omega)$, *then this space is the direct sum of the subspace* $\widetilde{H}_p^{1/2+s+1/p}(\Omega)$ *of functions with zero boundary values and the subspace of solutions of the homogeneous system* $Lu = 0$ *in* Ω. *The latter space is parametrized by the elements of* $B_p^{1/2+s}(\Gamma)$.

As previously, we refer to the corresponding decomposition $u = u_1 + u_2$ of an element of the space $H_p^{1/2+s+1/p}(\Omega)$ as the *Weyl decomposition* corresponding to the operator L in this space. We can suppose that the Dirichlet problem is also uniquely solvable for the formally adjoint system in $H_{p'}^{1/2-s+1/p'}(\Omega)$. We write the first decomposition for $u \in H_p^{1/2+s+1/p}(\Omega)$ and the second decomposition, $v = v_1 + v_2$, for $v \in H_{p'}^{1/2-s+1/p'}(\Omega)$ and the operator $\widetilde{L}$. Setting $Lu_2 = 0$ and $\widetilde{L}v_2 = 0$, we obtain the following generalization of relation (16.1.8):

$$\Phi_\Omega(u,v) = (Lu_1,v_1)_\Omega + (T^+u_2,v_2^+)_\Gamma. \tag{17.2.2}$$

This gives the following analogue of (16.1.9):

$$\widetilde{H}_p^{-1/2+s-1/p'}(\Omega) = \widetilde{H}_{p,1}^{-1/2+s-1/p'}(\Omega) + \widetilde{H}_{p,2}^{-1/2+s-1/p'}(\Omega). \tag{17.2.3}$$

Here the first subspace on the right-hand side is obtained by extending functionals in $\widetilde{H}_p^{-1/2+s-1/p'}(\Omega)$ by zero to the functions v_2. It is isomorphic to $H_p^{-1/2+s-1/p'}(\Omega)$, and these functionals contain no summands supported on Γ. The second subspace consists of functionals supported on Γ and is isomorphic to $B_p^{-1/2+s}(\Gamma)$.

In this way, we obtain a two-sided a priori estimate for solutions of the Dirichlet problem; for the Neumann problem, such an estimate is obtained by reducing this problem to problems (11.2.5)–(11.2.6).

If the form Φ_Ω is strongly coercive on $\widetilde{H}^1(\Omega)$, then we now have the extended *bounded Dirichlet-to-Neumann operator*

$$D: B_p^{1/2+s}(\Gamma) \to B_p^{-1/2+s}(\Gamma) \tag{17.2.4}$$

for s and p specified above, and if this form is strongly coercive on $H^1(\Omega)$, then we also have the extended *bounded Neumann-to-Dirichlet operator*

$$N: B_p^{-1/2+s}(\Gamma) \to B_p^{1/2+s}(\Gamma); \tag{17.2.5}$$

moreover, these are mutually inverse operators if the values of δ corresponding to the two problems coincide.

The operators D_1 and N_1 (first defined in Section 11.4) now act as follows:

$$D_1: \widetilde{B}_p^{1/2+s}(\Gamma_1) \to B_p^{-1/2+s}(\Gamma_1), \tag{17.2.6}$$

$$N_1: \widetilde{B}_p^{-1/2+s}(\Gamma_1) \to B_p^{1/2+s}(\Gamma_1). \tag{17.2.7}$$

As previously, they act on functions in their domains as D and N, and the result is restricted to Γ_1. It is seen from the considerations of Section 14 that these operators act in interpolation scales. At the point $(s,r) = (0, 1/2)$ they are invertible. Therefore, we can (again) apply Shneiberg's theorem and conclude that they are invertible if $|s| < \varepsilon$ and $|r - 1/2| < \delta$ with new numbers ε and δ. Generally, under the second application of Shneiberg's theorem, these numbers decrease. Operators N_2 and D_2 are defined in a similar way, and for them, the same result is obtained.

In the mixed problem (11.4.1), we now have

$$u \in H_p^{1/2+s+1/p}(\Omega), \quad g \in B_p^{1/2+s}(\Gamma_1), \quad h \in B_p^{-1/2+s}(\Gamma_2). \qquad (17.2.8)$$

The test functions belong to $H_{p'}^{1/2-s+1/p'}(\Omega; \Gamma_1)$ (in particular, they vanish on Γ_1), and the right-hand sides f belong to the dual space $[H_{p'}^{1/2-s+1/p'}(\Omega; \Gamma_1)]^*$ (with respect to the extension of the inner product in $L_2(\Omega)$).

Suppose again that the form Φ_Ω is strongly coercive on $H^1(\Omega)$. The mixed problem is reduced to the Dirichlet problem by using the operator D_2 and to the Neumann problem by using the operator N_1 as in Section 11.4. We obtain the following theorem.

Theorem 17.2.4. *The mixed problem is uniquely solvable for* $(s,r) \in Q_{\varepsilon,\delta}$ *with sufficiently small* ε *and* δ.

We pass to potential-type operators. Now we assume that the surface Γ lies on the torus $\mathbb{T}$ and divides it into domains $\Omega^\pm$. We also assume the strong coercivity of the form $\Phi_{\mathbb{T}}$ on $H^1(\mathbb{T})$ and, when necessary, of the corresponding forms $\Phi_{\Omega^\pm}$ on $H^1(\Omega^\pm)$.

First, we must obtain a general statement about the unique solvability of the equation

$$Lu = f \quad (u \in H_p^{1+\sigma}(\mathbb{T}), \quad f \in H_p^{-1+\sigma}(\mathbb{T})) \qquad (17.2.9)$$

on the torus. Under the smoothness assumptions of Section 12.1, the operator L is defined and bounded for all $p > 1$ and $|\sigma| \leq 1$. In Section 12.1 we proved the invertibility of this operator for $p = 2$ and $\sigma = 1$. This result extends to $p = 2$ and $\sigma = -1$ by duality. Using interpolation arguments (see, in particular, Theorem 13.7.1), we obtain invertibility for $p = 2$ and $|\sigma| < 1$, which proves Theorem 12.1.1.

In [274] the invertibility of L was proved by a different method for $p > 1$ and $0 \leq \sigma < 1$ under the assumption that the coefficients belong to $C^1(\overline{\Omega})$.

But for our purposes, it is, in principle, sufficient to apply Shneiberg's theorem, relying on Theorem 12.1.1.

Proposition 17.2.5. *There exists a* $\delta > 0$ *such that Eq.* (17.2.9) *remains uniquely solvable for* $|\sigma| \leq 1$ *and* $|r - 1/2| < \delta$.

Now let us ascertain in what spaces the operators of interest to us act.

As we know, the trace operator γ acts boundedly from $H_{p'}^{1/2-s+1/p'}(\Omega^\pm)$ to $B_{p'}^{1/2-s}(\Gamma)$. Therefore, the adjoint operator γ^* acts boundedly from $B_p^{-1/2+s}(\Gamma)$ to $H_p^{-1/2+s-1/p'}(\Omega^\pm)$. Here (s,r) is any point of the square Q.

Hence, the operator $\mathcal{A} = L^{-1}\gamma^*$ acts boundedly from $B_p^{-1/2+s}(\Gamma)$ to $H_p^{1/2+s+1/p}(\Omega^{\pm})$ for $(s,r) \in Q$. Consequently, for these (s,r), $A = \gamma\mathcal{A}$ is bounded an an operator from $B_p^{-1/2+s}(\Gamma)$ to $B_p^{1/2+s}(\Gamma)$, while $T^{\pm}\mathcal{A}$ and $\widehat{B}$ are bounded operators in $B_p^{-1/2+s}(\Gamma)$.

Since the forms $\Phi_{\Omega^{\pm}}$ are strongly coercive on $\widetilde{H}^1(\Omega^{\pm})$, it follows that the operator A is invertible at $(s,r) = (0,1/2)$. But it acts in interpolation scales. Applying Shneiberg's theorem, we see that it is invertible at $(s,r) \in Q_{\varepsilon,\delta}$ for sufficiently small ε and δ. In what follows, we shall show that $\varepsilon = 1/2$ *under the assumptions of Section* 16.4.

Now we generalize (12.3.4) to $u \in H_p^{1/2+s+1/p}(\Omega^{\pm})$ with (s,r) for which the Dirichlet problem is uniquely solvable. For these (s,r), $\mathcal{B}$ is bounded as an operator from $B_p^{1/2+s}(\Gamma)$ to $H_p^{1/2+s+1/p}(\Omega^{\pm})$ and, as a consequence, $\gamma^{\pm}\mathcal{B}$ and B are bounded as operators in $B_p^{1/2+s}(\Gamma)$ and $H = -T^{\pm}\mathcal{B}$ is bounded as an operator from $B_p^{1/2+s}(\Gamma)$ to $B_p^{-1/2+s}(\Gamma)$.

Now suppose that the forms $\Phi_{\Omega^{\pm}}$ are strongly coercive on $H^1(\Omega^{\pm})$. Then, starting at $(s,r) = (0,1/2)$ and again applying Shneiberg's theorem, we obtain the invertibility of the operators H, $\frac{1}{2}I \pm B$, and $\frac{1}{2}I \pm \widehat{B}$ at $(s,r) \in Q_{\varepsilon,\delta}$ for sufficiently small ε and δ; cf. Section 12.5. In this case, we shall also show that $\varepsilon = 1/2$ *under the assumptions of Section* 16.4.

The point is that if the Dirichlet problem is uniquely solvable at (s,r), then, in addition to expression (12.3.4), the jump relations (12.3.10) hold at this point. It follows that, at (s,r), the operator A is invertible and $A^{-1} = D^+ + D^-$. If the Neumann problem is uniquely solvable too at (s,r), then H is invertible and $H^{-1} = N^+ + N^-$, and, moreover, $\frac{1}{2}I \pm B$ and $\frac{1}{2}I \pm \widehat{B}$ are invertible.

All other relations in Sections 12.5 and 16.2 can be generalized as well, including the expressions for the Calderón projections (which now act in the direct product of $B_p^{1/2+s}(\Gamma)$ and $B_p^{-1/2+s}(\Gamma)$), the expressions for the solutions of the Dirichlet and Neumann problems, and the relations between operators.

Let us summarize.

Theorem 17.2.6. *Suppose that numbers* $\varepsilon = \varepsilon(L)$ *and* $\delta = \delta(L)$ *are such that for* $(s,r) \in Q_{\varepsilon,\delta}$ *the Dirichlet problem is uniquely solvable. Then at* $(s,r) \in Q_{\varepsilon,\delta}$ *the operator* A *is invertible and* $A^{-1} = D^+ + D^-$. *If, in addition, the Neumann problem is uniquely solvable at* $(s,r) \in Q_{\varepsilon,\delta}$, *then the operator* H *is invertible at this* (s,r), *and* $H^{-1} = N^+ + N^-$. *Moreover, all other relations derived in Sections 12.5 and 16.2 are valid.*

The boundedness of the operators (see Section 12.8)

$$A_1 \colon \widetilde{B}_p^{-1/2+s}(\Gamma_1) \to B_p^{1/2+s}(\Gamma_1) \quad \text{and} \quad H_1 \colon \widetilde{B}_p^{1/2+s}(\Gamma_1) \to B_p^{-1/2+s}(\Gamma_1)$$

is derived from the boundedness of A and H, respectively. After this, the invertibility of A_1 and H_1, but only in a neighborhood of the center of Q, is obtained by again applying Shneiberg's theorem.

This makes it possible to generalize results concerning the problems with conditions on a nonclosed surface Γ_1 considered in Section 12.8. We define $H_p^{1/2+s+1/p}(\Omega_0)$ as the space of those functions from $L_p(\Omega_0)$ whose restrictions to $\Omega^\pm$ belong to $H_p^{1/2+s+1/p}(\Omega^\pm)$ and have zero jumps on Γ_2. For example, in the Dirichlet problem, $u^\pm = g^\pm \in B_p^{1/2+s}(\Gamma_1)$ and $[u]_{\Gamma_1} \in \widetilde{B}_p^{1/2+s}(\Gamma_1)$.

Theorem 17.2.7. *The results of Section 12.8 on the operators A_1 and H_1, as well as the results on problems with conditions on a nonclosed surface, remain valid for points in a neighborhood of the center of the square Q.*

The generalization of the result concerning the unique solvability of the transmission problems considered in Section 12.10 to points in a neighborhood of the center of Q is left to the reader.

Now we discuss the smoothness and completeness properties of the generalized eigenfunctions of our operators. We consider them on the example of the operator $L = L_N$, which corresponds to the Neumann problem. We assume the corresponding form to be strongly coercive. Let $Q_{\varepsilon,\delta}$ be the corresponding rectangle in the plane (s,r), $r = 1/p$. It is convenient to assume that $\varepsilon \geq \delta$. For the points of this rectangle, we have

$$u \in H_p^{1/2+s+1/p}(\Omega) \text{ and } Lu = f \in \widetilde{H}_p^{-3/2+s+1/p}(\Omega).$$

Let I_p denote the interval

$$(-3/2 - \varepsilon + 1/p, \ 1/2 + \varepsilon + 1/p). \tag{17.2.10}$$

Here $|1/p - 1/2| < \delta$.

We know that the spaces $H_p^s(\Omega)$ and $\widetilde{H}_p^s(\Omega)$ with fixed p are glued together into a single scale for small $|s|$. In this scale, the superscript ranges, in fact, over I_p. Smaller superscripts correspond to larger spaces. The embedding is dense, because smooth functions are dense in all spaces. (In the Dirichlet problem, these are smooth functions with zero boundary values.) The generalized eigenfunctions belong to the solution space and to the space of right-hand sides simultaneously. If they are complete in one of these spaces, then the same is true in the other, because L determines an isomorphism between these spaces.

Particularly convenient are the spaces corresponding to the points of the interval I_2, because they are Hilbert spaces. If L is a self-adjoint operator for $s = 0$, then, under our inner products, it remains self-adjoint for all s, $|s| < \varepsilon$, and its eigenfunctions form orthonormal bases in these spaces. Of course, the basis property implies completeness. Next, if our operator is a weak perturbation of a self-adjoint operator, then the system of its generalized eigenfunctions is complete for $s = 0$. A more general property sufficient for completeness is given by the Dunford–Schwartz theorem (see Theorem 18.3.2).

Proposition 17.2.8. *The generalized eigenfunctions belong to all spaces corresponding to the points of the interval I_p if they belong to one of these spaces, and if they are complete in one of these spaces, then they are complete in all spaces.*

Proof. Let u be an eigenfunction (for simplicity), and let $Lu = \lambda u$, where $\lambda \neq 0$. Suppose that u belongs to the solution space corresponding to a point A of the interval

$$(1/2 - \varepsilon + 1/p, \, 1/2 + \varepsilon + 1/p). \tag{17.2.11}$$

Then u belongs to the spaces corresponding to all points of the interval I_p on the left of A, in particular, to all points of the interval

$$(-3/2 - \varepsilon + 1/p, \, -3/2 + \varepsilon + 1/p). \tag{17.2.12}$$

But this implies that u belongs to all spaces corresponding to the points of the interval (17.2.11) and, therefore, to all points of the interval I_p.

For similar reasons, completeness in the space corresponding to a point of I_p implies completeness in the spaces corresponding to all points of I_p. $\square$

Theorem 17.2.9. *The generalized eigenfunctions belong to all spaces corresponding to points of the union of intervals I_p with $|1/p - 1/2| < \delta$, and if they are complete for $p = 2$, then the same is true for all p.*

Proof. For simplicity, we shall assume completeness for $p = 2$. In addition to the isomorphisms determined by the operator L, we use embeddings, but in different ways for $p < 2$ and $p > 2$.

For the L_p spaces, Hölder's inequality gives continuous embeddings

$$L_q(\Omega) \subset L_p(\Omega) \text{ for } p < q.$$

They are dense, because smooth functions are dense in all $L_p(\Omega)$.

We fix s and decrease p (i.e., increase r), starting from $p = 2$. Let us compare the spaces

$$H_2^1(\Omega) \text{ and } H_p^{1/2+s+1/p}(\Omega).$$

For $s = 1/2 - 1/p$ (here we use the inequality $\varepsilon \geq \delta$), these are spaces with equal superscripts, and the former is embedded in the latter. This implies our assertion for the spaces corresponding to the points with $p < 2$.

Now suppose that $2 \leq p < q$. Let us compare the solution space $H_p^{1/2+s+1/p}(\Omega)$ with the space $\widetilde{H}_q^{-3/2+s+1/q}(\Omega)$ of right-hand sides. If there were no tilde, the former space would be embedded in the latter for

$$\left(\frac{1}{2} + s + \frac{1}{p}\right) - \frac{n}{p} \geq \left(-\frac{3}{2} + s + \frac{1}{q}\right) - \frac{n}{q}$$

by a known theorem (see Theorem 14.1.5 and its corollaries for spaces on a Lipschitz domain), i.e., for

$$\frac{2}{n-1} \geq \frac{1}{p} - \frac{1}{q}.$$

(We can assume that $s = 0$.) This would make it possible to obtain the result for all needed q in finitely many steps, starting from $p = 2$. But the tilde does not interfere: embeddings hold both for spaces without tilde on a domain and for spaces with tilde; it suffices to increase the number of steps, using the coincidence of spaces with and without tilde which have a common subscript and superscripts small in absolute value. The embeddings are again dense. We have proved our theorem for $p > 2$.   □

If the problem is only coercive, the result remains valid; it is obtained by shifting the spectral parameter.

Similar considerations apply to our other operators. We do not dwell on this here. Certainly, instead of the rectangle $Q_{\varepsilon,\delta}$, more general domains in the (s,r) plane can be considered.

17.3 Examples

Here we give examples of Dirichlet problems with strongly coercive form which are not solvable in the spaces corresponding to a significant part of the points of Q. For this purpose, we use examples well known in the theory of problems in domains with corner and conical points [220, 284].

Let Ω_0 denote the intersection of the sector determined in polar coordinates (r, ω) by the inequalities $0 < \omega < \alpha$ $(\alpha < 2\pi)$ with the unit disk, and let Ω be the domain obtained from Ω_0 by smoothing the boundary near the points $(1, 0)$ and $(1, \alpha)$. For Φ_Ω we take the simplest form

$$\int_\Omega (\nabla u \cdot \nabla \overline{v} + \tau u \overline{v}) \, dx.$$

It is known and easy to verify that, in the sector specified above, the scalar function

$$u(r, \omega) = r^{\pi/\alpha} \sin(\omega \pi / \alpha)$$

is a solution of the homogeneous Dirichlet problem for the Laplace equation. Therefore, in Ω, this is a solution of the Dirichlet problem

$$-\Delta u + \tau u = f, \quad u^+ = g,$$

where τ is an arbitrarily large positive number, $f = \tau u$, and g is a C^∞ function on the boundary (vanishing near the origin). Obviously, f is continuous on $\overline{\Omega}$; therefore, it belongs to all spaces $H_p^{-1/2+s-1/p'}(\Omega)$, and g belongs to all spaces $B_p^{1/2+s}(\Gamma)$ corresponding to the points of the square. On the other hand, u is the product of the positive homogeneous function r^h in (x_1, x_2) of degree $h = \pi/\alpha$ and a C^∞ function bounded away from zero for $0 < \varepsilon \le \omega \le \alpha - \varepsilon$ with arbitrarily small $\varepsilon > 0$. The number h is larger than $1/2$ and arbitrarily close to $1/2$ if α is close to 2π. We write h in the form $h = 1/2 + \eta$, $\eta > 0$.

The first derivatives of u with respect to x_1 and x_2 have summands containing a homogeneous factor of degree $h - 1 = -1/2 + \eta$. They belong to $L_p(\Omega)$ only if

$$\left(-\frac{1}{2}+\eta\right)p+1 > -1.$$

Therefore, u does not belong to $H_p^1(\Omega)$ if this condition is violated. Thus, if $p > 4$, then u does not belong to $H_p^1(\Omega)$ for α sufficiently close to 2π. The corresponding value of s is determined from the condition

$$\frac{1}{2}+s+\frac{1}{p} = 1$$

on the superscripts of the spaces under consideration; it is between $1/4$ and $1/2$. We have specified points (s,r) of the square Q at which there are Dirichlet problems with strongly coercive form which are not uniquely solvable.

Similar examples of Neumann problems are obtained by replacing sine by cosine.

An interesting feature of these examples is that the presence of one corner point on the boundary may substantially affect the smoothness of solutions.

17.4 The Optimal Resolvent Estimate

In this subsection, we first thoroughly consider the Neumann problem in a bounded Lipschitz domain Ω with homogeneous boundary condition and then briefly touch on the similar Dirichlet problem. The corresponding forms are assumed to be strongly coercive.

We saw in Section 11.9 that, for the resolvent of the corresponding operator from $H^1(\Omega)$ to $\widetilde{H}^{-1}(\Omega)$ (we now denote it by L) regarded as an unbounded operator on $\widetilde{H}^{-1}(\Omega)$, under natural assumptions, the optimal estimate

$$\|(L-\lambda I)^{-1}\| \le C(1+|\lambda|)^{-1} \qquad (17.4.1)$$

holds for λ belonging to a sector in the complex plane of opening larger than π with bisector $\mathbb{R}_-$; the constant C does not depend on λ. Such an estimate is important, in particular, because it implies the existence of the analytic semigroup e^{-tL} and the unique solvability of the corresponding "parabolic" problem in the Lipschitz cylinder $\Omega \times \mathbb{R}_+$; we discuss this in the next two subsections. It is interesting to enlarge as much as possible the class of spaces in which such results can be obtained. In [174] and some subsequent papers, among which we mention [171], this was done for the general mixed problem in the Sobolev spaces $W_p^1(\Omega) = H_p^1(\Omega)$ with p close to 2 instead of $H^1(\Omega)$. In our square Q, the points corresponding to these spaces lie on the diagonal $s+r = 1/2$, $r = 1/p$, close to the center $(0, 1/2)$ of the square. We want to show that this result extends to points in a neighborhood of the center.

The approach of the papers mentioned above was based on an idea of Agmon, who used it to study general elliptic problems with parameter [7]. The idea is to introduce an additional variable t and apply known elliptic a priori estimates in a cylindrical domain of dimension higher by 1 to functions containing the exponential

$e^{i\mu t}$ with parameter $\mu > 0$. We proceed as follows: first we apply this idea to the spaces $H^{1+s}(\Omega)$ with s close to 0, and then, having the desired estimates on the two lines $s + r = 1/2$ and $r = 1/2$ passing through the center of Q (to be more precise, on two intervals of these lines centered at the center of the square), we employ complex interpolation. We mean an analogue of the first relation in (14.1.14) for Lipschitz domains and Theorem 13.7.1. We explain details before Theorem 17.4.4, which gives a desired estimate of the form (17.4.1) in a (quadrangular) neighborhood of the center of the square.

First, we derive such an estimate in a sector of opening somewhat smaller than π. We extend it to a sector of opening somewhat larger than π by using the fact that, roughly speaking, the strong ellipticity of the operator L is preserved under multiplication by $e^{i\theta}$ with sufficiently small $|\theta|$. To be more precise, this is so in the case of the Dirichlet problem, and in the case of the Neumann problem, we need strengthened coercivity assumptions, which we specify in Theorem 17.4.3.

If we take a compact part of the required sector, then, for λ from this part, the presence of an s_0 such that the resolvent $(L - \lambda I)^{-1}$ exists in $\widetilde{H}^{-1+s}(\Omega)$ with $|s| \leq s_0$ and satisfies the required estimate follows from Shneiberg's theorem and estimates given in Section 13.7. Moreover, even the uniform estimate

$$\|(L - \lambda I)^{-1} f\|_{H^{1+s}(\Omega)} \leq C(1 + |\lambda|)^{-1} \|f\|_{\widetilde{H}^{-1+s}(\Omega)}$$

holds. But we must handle an infinite sector.

Let us introduce the cylindrical domain $\Theta = \Omega \times I$, $I = [-1, 1]$. It is Lipschitz together with Ω; this is easy to verify by using the cone condition mentioned in Section 9.1, which is necessary and sufficient for the Lipschitz property, at the boundary points of the end-walls of the cylinder. On this domain we introduce the operator

$$L_\Theta = L - \gamma \partial_t^2. \tag{17.4.2}$$

Here $|\gamma| = 1$ and $|\arg \gamma| \leq \pi/2 - \delta$ for arbitrarily small $\delta > 0$; we fix δ. The operator L_Θ is strongly elliptic together with L. We shall consider its action on functions satisfying the homogeneous Dirichlet condition on the end-walls of the cylinder (and the Neumann condition on its lateral surface; this is the standard mixed problem). We permit us to denote the subspace of functions in $H^{1+s}(\Theta)$ satisfying the homogeneous Dirichlet condition on the end-walls by $H^{1+s}(\Theta')$; by $\widetilde{H}^{-1+s}(\Theta')$ we denote its image under L_Θ. We assume the form of this operator to be strongly coercive on $H^1(\Theta')$. (Of course, its coercivity on $H^1(\Theta)$ would suffice.) Using Shneiberg's results, we can assume that, for functions $W = W(x, t) \in H^\sigma(\Theta')$ with $\sigma = 1 + s$ and sufficiently small $|s|$, we have the estimate

$$\|W\|_{H^{1+s}(\Theta')} \leq C_1 \|L_\Theta W\|_{\widetilde{H}^{-1+s}(\Theta')}. \tag{17.4.3}$$

Let $\varphi_\mu(t) = \psi(t) e^{i\mu t}$, where $\mu > 0$ and $\psi(t)$ is a fixed nonnegative function in $C_0^\infty(I)$ taking the value 1 in the interval $[-1/2, 1/2]$. We shall apply estimate (17.4.3) to functions $W(x, t) = u(x)\varphi_\mu(t)$, setting $\lambda = -\gamma\mu^2$.

We need two lemmas. They refer to the same spaces $H^\sigma(\Theta')$.

Recall that there are norms of two kinds on $H^\sigma(\mathbb{R}^{n+1})$: norms defined by means of the Fourier transform and by Slobodetskii's formula. We know that these norms are equivalent.

On $H^\sigma(\Theta)$, there are also norms of two kinds: norms defined as the greatest lower bounds of norms of extensions to $\mathbb{R}^{n+1}$ (e.g., of norms defined by means of the Fourier transform) and those defined by the Slobodetskii's formula. They are equivalent as well and can be used on $H^\sigma(\Theta')$.

We can also consider the anisotropic spaces $H^{\sigma,0}(\Theta')$ and $H^{\sigma,0}(\mathbb{R}^{n+1})$ of smoothness σ with respect to x and 0 with respect to t. Again, there are two kinds of norms of obvious structure, and these norms are equivalent.

Lemma 17.4.1. *The norm on $H^\sigma(\mathbb{R}^{n+1})$ dominates that on $H^{\sigma,0}(\mathbb{R}^{n+1})$. The same is true for the norms on $H^\sigma(\Theta')$ and $H^{\sigma,0}(\Theta')$.*

Proof. The first assertion is verified in an obvious way by using the Fourier transform. The second follows from the first. Indeed, if $w \in H^\sigma(\Theta')$, then there exists an extension w_1 of w to $\mathbb{R}^{n+1}$ such that $\|w_1\|_{H^\sigma(\mathbb{R}^{n+1})}$ is dominated by $\|w\|_{H^\sigma(\Theta')}$. The norm $\|w_1\|_{H^\sigma(\mathbb{R}^{n+1})}$ dominates the norm $\|w_1\|_{H^{\sigma,0}(\mathbb{R}^{n+1})}$, which, in turn, dominates $\|w\|_{H^{\sigma,0}(\Theta')}$. $\qquad\square$

We might consider more general anisotropic norms, but there is no reason for this.

Note that, by virtue of this lemma, we have

$$\|u\|_{H^\sigma(\Omega)} \leq C_2 \|W\|_{H^\sigma(\Theta')} \tag{17.4.4}$$

for $W(x,t) = u(x)\varphi_\mu(t)$, because the left-hand side does not exceed the product $\|u\|_{H^\sigma(\Omega)}\|\varphi_\mu\|_{L_2(I)}$, and this is the anisotropic norm of W in $H^{\sigma,0}(\Theta')$.

Here and in what follows, all constants are independent of the functions involved and of the parameters μ and γ.

Lemma 17.4.2. *Let $w(x,t)$ be any function in $H^\sigma(\Theta')$, and let*

$$v(x) = \int_I w(x,t)\overline{\varphi_\mu(t)}\,dt. \tag{17.4.5}$$

Then there exists a constant C_3 such that

$$\|v\|_{H^\sigma(\Omega)} \leq C_3 \|w\|_{H^\sigma(\Theta')}. \tag{17.4.6}$$

Proof. Obviously, the L_2-norm of v is dominated by the L_2-norm of w. The same is true for the H^1-norm (this is needed if $\sigma > 1$). Applying the Schwarz inequality, we obtain

$$\frac{|v(x) - v(y)|^2}{|x-y|^{n+2\sigma}} \leq \int_I \frac{|w(x,t) - w(y,t)|^2}{|x-y|^{n+2\sigma}}\,dt \int_I |\varphi_\mu(t)|^2 dt;$$

this implies that the Slobodetskii norms satisfy the inequality

$$\|v\|_{H^\sigma(\Omega)} \leq C_4 \|w\|_{H^{\sigma,0}(\Theta')} \tag{17.4.7}$$

for $\sigma < 1$. This inequality holds also for $\sigma > 1$ by virtue of similar estimates for the derivatives $\partial_j v$. It remains to refer to Lemma 17.4.1.   □

We return to estimate (17.4.3) and substitute the function $W(x,t) = u(x)\varphi_\mu(t)$, $u \in H^{1+s}(\Omega)$, into it. We must first assume that $s < 0$. Clearly,

$$L_\Theta W = \varphi_\mu Lu - u\gamma\varphi_\mu''$$

(the prime denotes differentiation with respect to t). It follows from (17.4.3) and (17.4.4) that

$$\|u\|_{H^{1+s}(\Omega)} \leq C_5 \|\varphi_\mu Lu - u\gamma\varphi_\mu''\|_{\widetilde{H}^{-1+s}(\Theta')}. \tag{17.4.8}$$

To calculate and estimate the right-hand side, we take an arbitrary (but smooth to some extent) function $w \in H^{1-s}(\Theta')$ and consider the Green identity

$$(\varphi_\mu Lu - u\gamma\varphi_\mu'', w)_\Theta = \int_I \Phi_\Omega(u\varphi_\mu, w)dt + \gamma \int_\Omega u\,dx \int_I \varphi_\mu' \overline{w}'\,dt.$$

Using the function (17.4.5), we rewrite the right-hand side in the form

$$\Phi_\Omega(u,v) + \gamma \int_\Omega u\,dx \int_I (\psi(t)e^{i\mu t})' \overline{w}'\,dt.$$

We transform the inner integral (over I) applying integration by parts but avoiding the occurrence of μ to the first power, as in [174]. This integral equals

$$2\int_I e^{i\mu t}\psi'\overline{w}'\,dt + \int_I e^{i\mu t}[\mu^2\psi + \psi'']\overline{w}\,dt.$$

Thus,

$$(\varphi_\mu Lu - u\gamma\varphi_\mu'', w)_\Theta = \Phi_\Omega(u,v) - \lambda(u,v)_\Omega + \gamma \int_\Omega u\,dx \int_I e^{i\mu t}[2\psi'\overline{w}' + \psi''\overline{w}]\,dt. \tag{17.4.9}$$

Recall that $|e^{i\mu t}| = 1$ and $\psi \in C_0^\infty(I)$. Hence,

$$|(\varphi_\mu Lu - u\gamma\varphi_\mu'', w)_\Theta| \leq C_6\|Lu - \lambda u\|_{\widetilde{H}^{-1+s}(\Omega)}\|v\|_{H^{1-s}(\Omega)} + C_7\|u\|_{L_2(\Omega)}\|w\|_{H^1(\Theta')}.$$

Here the norm $\|w\|_{H^1(\Theta')}$ does not exceed $\|w\|_{H^{1-s}(\Theta')}$ (because $s < 0$). The norm $\|v\|_{H^{1-s}(\Omega)}$ is dominated by $\|w\|_{H^{1-s}(\Theta')}$ according to Lemma 17.4.2. Dividing both sides of the last inequality by $\|w\|_{H^{1-s}(\Theta')}$ and passing to the least upper bound in this norm, we obtain

$$\|\varphi_\mu Lu - u\gamma\varphi_\mu''\|_{\widetilde{H}^{-1+s}(\Theta')} \leq C_8[\|Lu - \lambda u\|_{\widetilde{H}^{-1+s}(\Omega)} + \|u\|_{L_2(\Omega)}].$$

By virtue of (17.4.8), the left-hand side dominates $\|u\|_{H^{1+s}(\Omega)}$, so that

$$\|u\|_{H^{1+s}(\Omega)} \le C_9[\|Lu - \lambda u\|_{\widetilde{H}^{-1+s}(\Omega)} + \|u\|_{L_2(\Omega)}]. \tag{17.4.10}$$

The second summand can be estimated as

$$\|u\|_{L_2(\Omega)} \le \varepsilon\|u\|_{H^{1+s}(\Omega)} + C_\varepsilon\|u\|_{H^{-1+s}(\Omega)}$$

for any $\varepsilon > 0$ (we discussed such estimates in the first chapter), and here the last norm can be replaced by $\|u\|_{\widetilde{H}^{-1+s}(\Omega)}$. Therefore, it follows from (17.4.10) that

$$\|u\|_{H^{1+s}(\Omega)} \le C_{10}[\|Lu - \lambda u\|_{\widetilde{H}^{-1+s}(\Omega)} + \|u\|_{\widetilde{H}^{-1+s}(\Omega)}]. \tag{17.4.11}$$

The left-hand side dominates $\|Lu\|_{\widetilde{H}^{-1+s}(\Omega)}$. Since $\lambda u = (\lambda u - Lu) + Lu$, we obtain

$$|\lambda|\|u\|_{\widetilde{H}^{-1+s}(\Omega)} \le C_{11}[\|Lu - \lambda u\|_{\widetilde{H}^{-1+s}(\Omega)} + \|u\|_{\widetilde{H}^{-1+s}(\Omega)}]. \tag{17.4.12}$$

For sufficiently large $|\lambda|$, this implies the required estimate

$$(1 + |\lambda|)\|u\|_{\widetilde{H}^{-1+s}(\Omega)} \le C_{12}\|Lu - \lambda u\|_{\widetilde{H}^{-1+s}(\Omega)}. \tag{17.4.13}$$

There remains a compact set of values of λ; we have already explained that on such a set the same estimate is valid (possibly, for smaller s_0).

The estimate can be strengthened: since

$$\|u\|_{H^{1+s}(\Omega)} \le C_{13}\|Lu\|_{\widetilde{H}^{-1+s}(\Omega)} \le C_{13}[\|Lu - \lambda u\|_{\widetilde{H}^{-1+s}(\Omega)} + |\lambda|\|u\|_{\widetilde{H}^{-1+s}(\Omega)}],$$

it follows that

$$\|u\|_{H^{1+s}(\Omega)} + (1 + |\lambda|)\|u\|_{\widetilde{H}^{-1+s}(\Omega)} \le C_{14}\|Lu - \lambda u\|_{\widetilde{H}^{-1+s}(\Omega)}. \tag{17.4.14}$$

The same estimate holds for the formally adjoint operator $\widetilde{L}$. This allows us to carry over the result to small $s > 0$. For this purpose, we note that

$$(u, g)_\Omega = (u, \widetilde{L}v - \overline{\lambda}v)_\Omega = (Lu - \lambda u, v)_\Omega$$

for $u \in H^{1+s}(\Omega)$, $g \in \widetilde{H}^{-1-s}(\Omega)$, and $v = (\widetilde{L} - \overline{\lambda})^{-1}g$ (here $s > 0$ and $v \in \widetilde{H}^{1-s}(\Omega)$), so that

$$|(u, g)_\Omega| \le C_{15}\|Lu - \lambda u\|_{\widetilde{H}^{-1+s}(\Omega)}\|v\|_{H^{1-s}(\Omega)} \le C_{16}\|Lu - \lambda u\|_{\widetilde{H}^{-1+s}(\Omega)}\|g\|_{\widetilde{H}^{-1-s}(\Omega)},$$

because $\|v\|_{H^{1-s}(\Omega)}$ is dominated by $\|g\|_{\widetilde{H}^{-1-s}(\Omega)}$. Therefore,

$$\|u\|_{H^{1+s}(\Omega)} \le C_{17}\|Lu - \lambda u\|_{\widetilde{H}^{-1+s}(\Omega)},$$

where the constant does not depend on u and λ. We already know how to derive that $(1 + |\lambda|)\|u\|_{\widetilde{H}^{-1+s}(\Omega)}$ is dominated by the right-hand side.

This proves (17.4.14) but in a sector of opening smaller than π.

Now suppose that the form of the operator $e^{\pm i\theta}L$ is strongly coercive on $H^1(\Omega)$ for some $\theta > 0$ and that, for this θ, the form of $e^{\pm i\theta}L - \gamma\partial_t^2$ is strongly coercive on

$H^1(\Theta')$ for some $\delta \in (0, \theta)$. Then, as we have already explained, estimate (17.4.14) is obtained in a sector of opening somewhat larger than π.

Theorem 17.4.3. *Suppose that the form of the operator $e^{\pm i\theta} L$ is strongly coercive on $H^1(\Omega)$ for some $\theta > 0$ and that, for this θ, there is a $\delta \in (0, \theta)$ such that the form of $e^{\pm i\theta} L - \gamma \partial_t^2$ is strongly coercive on $H^1(\Theta')$. Then, for the solutions of the Neumann problem with $|s| \leq s_0$ for sufficiently small s_0, the optimal estimate (17.4.14) with constant not depending on u and λ is valid in a sector of opening larger than π with bisector $\mathbb{R}_-$.*

On the line $s + r = 1/2$ the superscript $1/2 + s + r$ of the spaces under consideration equals 1, and the estimate on this line has the form

$$\|u\|_{H_p^1(\Omega)} + (1 + |\lambda|)\|u\|_{\widetilde{H}_p^{-1}(\Omega)} \leq C'\|Lu - \lambda u\|_{\widetilde{H}_p^{-1}(\Omega)} \tag{17.4.15}$$

for small $|p - 2|$ and λ belonging to a sector of the same form. We shall not repeat the work in this case. All ideas are contained in [174], they remain working under our assumptions. In particular, there are no problems with the corresponding lemmas in these spaces.

(In [174], the result was obtained for a scalar equation and $p > 2$. However, the same arguments work for systems under our assumptions, and for $p < 2$ it is possible to use the duality.)

Now a neighborhood of the center of the square is covered by segments with endpoints on the lines $s + r = 1/2$ and $r = 1/2$. Along these segments we apply Theorem 13.7.1 to $L - \lambda I$ treated as an operator from $H_p^{1/2+s+1/p}(\Omega)$ to $\widetilde{H}_p^{-1/2+s-1/p'}(\Omega)$ uniformly bounded in λ. As mentioned above, we use the first relation in (14.1.14). The inverse operator acts in interpolation scales, and we obtain (see (13.2.6)) an estimate, uniform in λ, for the first term on the left-hand side of inequality (17.4.16) below, which, in turn, implies an estimate for the second term.

Theorem 17.4.4. *If the assumptions of Theorem 17.4.3 hold, then, for the solutions of the Neumann problem with homogeneous boundary condition, at points (s, r) of the square Q sufficiently close to the center $(0, 1/2)$, the optimal estimate*

$$\|u\|_{H_p^{1/2+s+1/p}(\Omega)} + (1 + |\lambda|)\|u\|_{\widetilde{H}_p^{-1/2+s-1/p'}(\Omega)} \leq C''\|Lu - \lambda u\|_{\widetilde{H}_p^{-1/2+s-1/p'}(\Omega)} \tag{17.4.16}$$

holds. Here λ belongs to a sector of opening larger than π with bisector $\mathbb{R}_-$ and the constant does not depend on u and λ.

We pass to the Dirichlet problem. Of course, in this case other (simpler) spaces are needed. To obtain a counterpart of Theorem 17.4.3, we must assume that the function u belongs to $\widetilde{H}^{1+s}(\Omega)$ and the right-hand side of the system, to $H^{-1+s}(\Omega)$. Functions on the cylinder are subject to the homogeneous Dirichlet condition on the entire surface of the cylinder, so that instead of $H^{1+s}(\Theta')$ and $\widetilde{H}^{-1+s}(\Theta')$ we must consider $\widetilde{H}^{1+s}(\Theta)$ and $H^{-1+s}(\Theta)$. Recall that the space $\widetilde{H}_p^{1+s}$ is identified with the subspace $\mathring{H}_p^{1+s}$ in H_p^{1+s} for small $|s|$. All considerations are similar to those performed above, but no special coercivity assumptions are needed, it suffices to assume that L is strongly elliptic and its zero order term has large real part. Thus, the statement of the final result is simpler.

Theorem 17.4.5. *Let L be a strongly elliptic operator on Ω whose zero order term has sufficiently large real part. Then for the solutions of the Dirichlet problem with homogeneous boundary condition, the optimal estimate*

$$\|u\|_{\widetilde{H}_p^{1/2+s+1/p}(\Omega)} + (1+|\lambda|)\|u\|_{H_p^{-1/2+s-1/p'}(\Omega)} \le C'''\|Lu - \lambda u\|_{H_p^{-1/2+s-1/p'}(\Omega)}$$
$$(17.4.17)$$

holds at points (s,r) of the square Q sufficiently close to its center $(0,1/2)$. Here λ belongs to a sector of opening larger than π with bisector $\mathbb{R}_-$ and the constant does not depend on u and λ.

A similar theorem is also valid for the mixed problem, but we do not dwell on this. Apparently, similar results can be obtained for Besov spaces.

17.5 Elementary Facts about Semigroups

We begin with definitions (see, e.g., [292, 296, 372]; there are many books on semigroups and nonstationary problems). Let X be an (infinite-dimensional) complex Banach space with norm $\|\cdot\|$. A *strongly continuous semigroup*, or a C_0-*semigroup*, $T(t)$ on X is a mapping of the closed semiaxis $\overline{\mathbb{R}}_+$ to the space $\mathcal{L}(X)$ of bounded linear operators on X with the following properties:

(1) $T(t)x$ is a function on $\mathbb{R}_+$ with values in X continuous at each $x \in X$;
(2) $T(0) = I$ (the identity operator in X);
(3) $T(s+t) = T(s)T(t)$ for all $s,t \in \mathbb{R}_+$.

The *generator* B of such a semigroup is an unbounded operator on X defined by

$$Bx = \lim_{t \to 0} \frac{T(t)x - x}{t};\qquad(17.5.1)$$

its domain $D(B)$ consists of all x for which this limit exists. The domain is dense in X, and the operator B is linear and closed. Instead of $T(t)$, the notation e^{tB} may be used.

For any C_0-semigroup, there exists an $M \ge 1$ and a $\beta \in \mathbb{R}$ such that

$$\|T(t)\| \le Me^{\beta t}.\qquad(17.5.2)$$

If this inequality holds for $M = 1$ and $\beta = 0$, then the semigroup is said to be *contractive*.

It can be derived from (17.5.2) that the resolvent set $\rho(B)$ of the generator B contains the half-plane $\{\lambda : \operatorname{Re}\lambda > \beta\}$, and on this half-plane the resolvent

$$(B - \lambda I)^{-1} = -\int_0^\infty e^{-\lambda t} T(t)\, dt$$

exists; this is the Laplace transform of $T(t)$ with the opposite sign, and if $\operatorname{Re}\lambda > \beta$, then

$$\|(B - \lambda I)^{-n}\| \le M(\operatorname{Re}\lambda - \beta)^{-n} \qquad (17.5.3)$$

for all positive integers n. In particular, this is so for all real $\lambda > \beta$. The converse is also true: if B is a densely defined closed operator on X whose resolvent exists for all real $\lambda > \beta$ and satisfies the relations

$$\|(B - \lambda I)^{-n}\| \le M(\lambda - \beta)^{-n} \qquad (17.5.4)$$

with any positive integers n for these λ, then B is the generator of a C_0-semigroup $T(t)$ satisfying (17.5.2) on X. Together these statements are the *Hille–Yosida theorem*.

It is inconvenient for us to use this theorem as a criterion for the existence of a semigroup in the general case, because (17.5.4) involves all powers of the resolvent. But for $M = 1$ and $\beta = 0$, condition (17.5.4) with $n = 1$ is necessary and sufficient for the existence of a contractive semigroup.

For our purposes, the class of analytic semigroups is more convenient. We say that a strongly continuous semigroup $T(t)$ satisfying (17.5.2) is *analytic*, or *holomorphic*, if, for some $\omega \in (0, \pi/2)$, it can be holomorphically (in t) extended to the sector

$$\Delta_\omega = \{\lambda : |\arg\lambda| < \omega\}$$

in the complex plane so that it retains properties (1)–(3) in this sector and satisfies a uniform estimate of the form (17.5.2) (with $\operatorname{Re}t$ instead of t) in any narrower closed sector with the same bisector $\mathbb{R}_+$. Property (2) is understood in the sense of the strong convergence $T(t)x \to x$ as $\Delta_\omega \ni t \to 0$.

In this case, the resolvent set $\rho(B_\beta)$ of the operator $B_\beta = B - \beta I$ contains the sector $\{\lambda : |\arg\lambda| < \pi/2 + \omega\}$, and in any narrower closed sector $\{\lambda : |\arg\lambda| \le \pi/2 + \omega - \varepsilon\}$, the uniform estimate

$$|\lambda| \|(B_\beta - \lambda I)^{-1}\| \le M_\varepsilon \qquad (17.5.5)$$

holds. The converse statement, which we need, is true as well: *The existence of a sector of opening larger than π in the resolvent set of the operator B_β in which the above estimate holds implies that B generates an analytic semigroup satisfying estimate (17.5.2).*

Remark. If B is replaced by $B - \beta I$, then $T(t)$ is replaced by $e^{-\beta t}T(t)$, and β in (17.5.2) is replaced by zero. The semigroup becomes uniformly bounded (and remains analytic if it was analytic). In our problems $B = -L$, and the addition of the constant β (to be more precise, of βI) to L is an inessential change of this operator for us. In Section 17.4 we in fact preferred to consider an operator with $\beta = 0$ and $0 \in \rho(L)$. This is customary in the literature.

To the semigroup the Cauchy problem

$$\partial_t u(t) = Bu(t) \ (t > 0), \quad u(0) = u_0 \qquad (17.5.6)$$

is associated. In the case of an analytic semigroup, this problem with any $u_0 \in X$ has a unique solution $u(t)$ in X, which is continuous at $t \geq 0$, belongs to $D(B)$, and is differentiable at $t > 0$; this is $T(t)u_0$. To the right-hand side of the equation in (17.5.6) a function $g(t)$ can be added; if it takes values in X and is Hölder continuous in the interval $[0, T]$, then in this interval the problem still has a unique solution, which is of the form

$$u(t) = T(t)u_0 + \int_0^t T(t - s)g(s)\,ds. \qquad (17.5.7)$$

17.6 Parabolic Problems in a Lipschitz Cylinder

Let H_{-1} denote any of the spaces $\widetilde{H}_p^{-1/2+s-1/p'}(\Omega)$ in Theorem 17.4.4, and let H_1 be the corresponding space $H_p^{1/2+s+1/p}(\Omega)$. Comparing this theorem with the considerations of the preceding section, we see that, under the assumptions of Theorem 17.4.3, the following assertion is valid.

Theorem 17.6.1. *The operator* $-L$ *generates the analytic semigroup* $T(t) = e^{-tL}$ *on* H_{-1}. *The norm* $\|T(t)\|$ *is uniformly bounded.*

Now consider the infinite cylindrical domain $\Omega \times \mathbb{R}_+$ and the following Cauchy problem in it:

$$[\partial_t + L]u(x,t) = g(x,t) \ (x \in \Omega, \ t > 0), \quad u(x,0) = u_0(x) \ (x \in \Omega). \qquad (17.6.1)$$

In fact, this is a mixed problem with the homogeneous Neumann condition on the lateral surface of this domain and the Cauchy condition on the end-wall $t = 0$. For this problem, again under the assumptions of Theorem 17.4.3, we obtain the following result.

Theorem 17.6.2. *If* $u_0(x) \in H_{-1}$ *and* $g(x,t)$ *is a function of* t *with values in* H_{-1} *Hölder continuous in any finite interval* $[0, T]$, *then problem* (17.6.1) *has a unique solution, which is a function taking values in* H_{-1}, *continuous in* $[0, \infty)$, *uniformly bounded, differentiable, and belonging to* H_1 *for* $t > 0$.

Similar results are valid for the Dirichlet boundary condition. In this case, in Theorem 17.4.5, by H_{-1} we must understand any of the spaces $H^{-1/2+s-1/p'}(\Omega)$ and by H_1, the corresponding space $\widetilde{H}_p^{1/2+s+1/p}(\Omega)$. Then, under the assumptions of Theorem 17.4.5, Theorems 17.6.1 and 17.6.2 without any changes are valid.

We have touched on only one of the possible interpretations of problems of the form (17.6.1); there are other interpretations and results, but they are beyond the scope of our exposition. The numerous generalizations are not touched on either.

18 Appendix: Definitions and Facts from Operator Theory

For the reader's convenience, in Sections 18.1–18.3 we collect some definitions and theorems from the theory of linear operators mentioned in preceding sections. The material of Sections 18.1–18.2 is used to construct the theory of elliptic operators, and the material of Section 18.3, to study spectral problems for such operators. Section 18.4 contains a brief survey of the theory of pseudodifferential operators.

18.1 Fredholm Operators

As a precaution, we remind the reader that a bounded operator between Banach spaces is said to be *compact*, or *completely continuous*, if it takes any bounded set to a precompact set, i.e., to a set in which any sequence contains a convergent subsequence.

An operator is said to be *finite-dimensional* if its range is finite-dimensional. Any finite-dimensional operator is compact.

Let X and Y be infinite-dimensional Banach spaces. A bounded linear operator A from X to Y is said to be *Fredholm* if it has finite-dimensional kernel, closed range, and finite-dimensional cokernel.

The *kernel* $\operatorname{Ker} A$ of A is the subspace in X consisting of the solutions of the homogeneous equation $Au = 0$, and the *cokernel* $\operatorname{Coker} A$ is the quotient space $Y/R(A)$, where $R(A)$ is the (closed) range of A. The dimension of the cokernel is also called the codimension of the range.

Recall (cf. Section 13.4) that a subspace Z_1 of a Banach space Z is said to be *complemented* if it has a direct complement Z_2, i.e., Z is the direct sum $Z_1 + Z_2$. A subspace Z_1 is complemented if and only if there exists a bounded projection P, $P^2 = P$, of Z onto Z_1 such that $Z_2 = (I - P)Z$. The operator P depends on the choice of Z_2 and is called the projection onto Z_1 parallel to Z_2.

Proposition 18.1.1. 1. *Any finite-dimensional subspace Z_1 in a Banach space Z is complemented.*

2. *If Z_2 is a subspace in Z such that the quotient space Z/Z_2 is finite-dimensional, then Z_2 is complemented.*

Proof. The proof of this proposition can be found, e.g., in [317, Sec. 4.21]. We outline it below.

1. Let Z_1 be a finite-dimensional subspace of Z, and let $z_1, \ldots, z_m$ be a basis in this subspace. Then any vector $z \in Z_1$ has a unique representation

$$z = \alpha_1(z)z_1 + \ldots + \alpha_m(z)z_m.$$

It can be shown that the $\alpha_j(z)$ are continuous linear functionals on Z_1 (the norm on Z_1 is induced from Z). By the Hahn–Banach theorem (see, e.g., [247, Chap. IV, Sec. 1]), they can be extended to continuous linear functionals on Z. For the extended functionals we keep the notation $\alpha_j(z)$. Now we set, already for $z \in Z$,

$$Pz = \alpha_1(z)z_1 + \ldots + \alpha_m(z)z_m.$$

It is easy to understand that this is a projection of Z onto Z_1. Here Z_2 is the intersection of the subspaces on which $\alpha_j(z) = 0$ $(j = 1, \ldots, m)$.

2. (We introduce notation anew.) Let π be the canonical mapping $Z \to Z/Z_2$, and let $e_1, \ldots, e_m$ be a basis in the quotient space. Choose preimages z_j of e_j: $\pi z_j = e_j$. Let Z_1 be a subspace of Z spanned by $z_1, \ldots, z_m$. It can be shown that then Z is the direct sum of Z_1 and Z_2. $\square$

By virtue of Proposition 18.1.1, if $A: X \to Y$ is a Fredholm operator, then Ker A has a direct complement in X, and $R(A)$ has a direct complement in Y. We choose and fix direct complements to Ker A and $R(A)$ and denote them by X_1 and Y_1, respectively. The dimension of Y_1 equals the codimension of the range. It is convenient for us to call Y_1 a *cokernel* of the operator A and denote it by Coker A.

If X and Y are (separable) Hilbert spaces, then it is convenient to use orthogonal complements.

The Fredholm property of an operator should be regarded as its "almost invertibility."

Proposition 18.1.2. *A Fredholm operator A establishes a continuous (in both directions) isomorphism between a direct complement X_1 of its kernel in X and its range $R(A)$ in Y.*

Proof. Obviously, the mapping $A: X_1 \to R(A)$ is one-to-one and continuous. The inverse operator is continuous by Banach's inverse mapping theorem (see, e.g., [247, Chap. III, Sec. 5]). $\square$

The *index* $\varkappa(A)$ of a Fredholm operator A is defined as the difference between the dimension of the kernel of this operator and the dimension of its cokernel:

$$\varkappa(A) = \dim \text{Ker} A - \dim \text{Coker} A. \tag{18.1.1}$$

Let us give the definition of a parametrix for an operator A. A bounded operator B from Y to X is called a *left parametrix* for a bounded operator A from X to Y if

$$BA = I_1 + T_1, \tag{18.1.2}$$

where I_1 and T_1 are, respectively, the identity operator and a compact operator in X; B is called a *right parametrix* for A if

$$AB = I_2 + T_2, \tag{18.1.3}$$

where I_2 and T_2 are, respectively, the identity operator and a compact operator in Y; and a *parametrix*, or a *two-sided parametrix*, if B is a left and a right parametrix. The notation of (18.1.2) and (18.1.3) will be used in what follows.

This definition implies that if B is a parametrix for A, then A is a parametrix for B.

In the literature, the term *regularizer* was used before the name *parametrix* became customary. A parametrix, or regularizer, if it exists, is, to a certain extent, a satisfactory substitute for an inverse operator when the latter does not exist.

Proposition 18.1.3. *If an operator A has a left parametrix, then the kernel of A is finite-dimensional; if A has a right parametrix, then the range of A is closed and the cokernel is finite-dimensional.*

Therefore, the existence of a parametrix B for an operator A implies that A is Fredholm. Moreover, B is Fredholm as well.

Proof. It is well known that the sum $I + T$ of the identity operator and a compact operator in a Banach space is a Fredholm operator of index zero; see, e.g., [247, Chap. VI, Sec.].[3] Therefore, it suffices to note that, in the presence of a left parametrix B, we have

$$\operatorname{Ker} A \subset \operatorname{Ker}(BA) = \operatorname{Ker}(I_1 + T_1),$$

and in the presence of a right parametrix B, we have

$$R(A) \supset R(AB) = R(I_2 + T_2), \quad Y = R(I_2 + T_2) \dotplus Q,$$

where Q is a finite-dimensional subspace of Y; thus,

$$Y = R(A) + Y_1,$$

where Y_1 is a (finite-dimensional) subspace of Q.

The last assertion in the statement follows from A being a parametrix for B. □

Proposition 18.1.4. *If A has a left parametrix B_1 and a right parametrix B_2, then both B_1 and B_2 are two-sided parametrices.*

Proof. By virtue of the relation

$$B_1 A B_2 = (I_1 + T_1)B_2 = B_1(I_2 + T_2),$$

the difference $B_1 - B_2$ of the left and the right parametrix turns out to be a compact operator, because the product of a compact and a bounded operator (in any order) is a compact operator. It remains to recall that the sum of a left or right parametrix for A and a compact operator is, respectively, a left or right parametrix for A. This is easy to verify. □

[3] Let us also mention that the kernel and the cokernel of $I + T$ are trivial if -1 is not an eigenvalue of the compact operator T. It is known that the nonzero eigenvalues of a compact operator are isolated and have finite multiplicity; see Section 18.3.

Proposition 18.1.5. *Any Fredholm operator A has a two-sided parametrix B such that $BA - I_1$ and $AB - I_2$ are finite-dimensional operators.*

Proof. The operator $A: X_1 \to R(A)$ has an inverse B_1. We continue B_1 to Y_1 by zero and extend the resulting operator to the entire space Y by linearity. Let us show that the operator B thus obtained is a two-sided parametrix for A.

   Let P be the projection of X onto $\operatorname{Ker} A$ parallel to X_1, and let Q be the projection of Y onto Y_1 parallel to $R(A)$. Then

$$BA = BA(P + I_1 - P) = BA(I_1 - P) = B_1 A(I_1 - P) = I_1 - P,$$
$$AB = AB(Q + I_2 - Q) = AB(I_2 - Q) = AB_1(I_2 - Q) = I_2 - Q.$$

The operators P and Q are finite-dimensional.   □

   Suppose that, in addition, the space X is continuously embedded in a Banach space X_0.

Proposition 18.1.6. *For any Fredholm operator A from X to Y, the a priori estimate*

$$\|x\|_X \le C(\|Ax\|_Y + \|x\|_{X_0}), \quad x \in X, \tag{18.1.4}$$

with a constant not depending on x holds. If the kernel $\operatorname{Ker} A$ is trivial, then the second term in parentheses can be omitted.

Proof. On the direct complement X_1 to the kernel, we have estimate (18.1.4) without the second term in parentheses by virtue of the invertibility of $A: X_1 \to R(A)$. If the kernel $\operatorname{Ker} A$ is nontrivial, then it is isomorphic to itself in X_0, and $\|x\|_X \le C_1\|x\|_{X_0}$ on $\operatorname{Ker} A$. Let us write each element $x \in X$ in the form $x = x_1 + x_2$, where $x_1 = Px$ is the projection of x on $\operatorname{Ker} A$ and $x_2 = (I - P)x$ is its projection on X_1. We have

$$\|x\|_X \le \|x_1\|_X + \|x_2\|_X.$$

Here

$$\|x_1\|_X \le C_1\|x_1\|_{X_0} \le C_1(\|x\|_{X_0} + \|x_2\|_{X_0}).$$

Next, $\|x_2\|_{X_0} \le C_2\|x_2\|_X$ by virtue of the continuity of the embedding of X in X_0, and

$$\|x_2\|_X = \|x_2\|_{X_1} \le C_3\|Ax_2\|_{R(A)} = C_3\|Ax_2\|_Y = C_3\|Ax\|_Y. \qquad □$$

Proposition 18.1.7. *Suppose that the embedding of X in X_0 is compact and A is a bounded operator for which estimate (18.1.4) holds. Then A has finite-dimensional kernel and closed range.*

Proof. 1. Let us verify that the kernel $\operatorname{Ker} A$ is finite-dimensional. Estimate (18.1.4) implies that

$$\|x\|_X \le C\|x\|_{X_0} \tag{18.1.5}$$

on $\operatorname{Ker} A$. The kernel of any bounded operator is closed. It is seen from (18.1.5) that $\operatorname{Ker} A$ is also closed in X_0. Take any bounded set in $\operatorname{Ker} A$ regarded as a subspace of X_0. By virtue of (18.1.5), this set is bounded in X and, hence, compact in X_0. Thus, in $\operatorname{Ker} A$ any bounded set is compact. It follows that $\operatorname{Ker} A$ is finite-dimensional (see, e.g., [247, Chap. V, Sec. 2, Theorem 4]).

2. Let us verify that the range is closed. First, we show that

$$\|x\|_X \le C_1 \|Ax\|_Y \tag{18.1.6}$$

on X_1. Suppose that this is not the case. Then there is a sequence $\{x_k\}$ in X_1 such that $\|x_k\|_X \to \infty$ and the norms $\|Ax_k\|_Y$ are bounded. Let $\widetilde{x}_k = x_k / \|x_k\|_X$. Then $\|\widetilde{x}_k\|_X = 1$ and $y_k = A\widetilde{x}_k \to 0$ in Y. The sequence $\{\widetilde{x}_k\}$ contains a subsequence convergent in X_0. We assume that $\{\widetilde{x}_k\}$ itself converges in X_0. By virtue of (18.1.4), $\{\widetilde{x}_k\}$ converges to some limit x_0 in X_1, and $\|x_0\|_X = 1$. On the other hand, $A\widetilde{x}_k$ converges in Y to zero. Therefore, $x_0 \in \operatorname{Ker} A$. This is a contradiction.

Estimate (18.1.6) on X_1 implies the closedness of the range. Indeed, let $y_k = Ax_k \to y_0$ in Y. We can assume that $x_k \in X_1$, so that $\{x_k\}$ converges by virtue of (18.1.6) to a limit x_0. We have $Ax_0 = y_0$. $\square$

Remark. If estimate (18.1.6) holds on X, then the kernel is trivial and the range is closed without the assumption that X is compactly embedded in some space.

The reader can easily verify this assertion.

Now let X and Y be infinite-dimensional Banach spaces, which we assume to be reflexive for simplicity, and let A be a bounded operator from X to Y. Suppose that the space X^* is dual to X with respect to a form $(x, g)_1$, i.e., this form determines all continuous linear functionals on X as g ranges over X^*. For convenience, we assume that $(x, g)_1$ is sesquilinear rather than bilinear, as well as the similar form with subscript 2 introduced below.

Let Y^* be the space dual to Y with respect to a form $(y, f)_2$. As is known, X^* and Y^* are Banach spaces, and the relation

$$(Ax, f)_2 = (x, A^*f)_1 \tag{18.1.7}$$

uniquely determines a bounded operator A^* from Y^* to X^*; this A^* is called the *operator adjoint to A in the sense of the theory of operators on Banach spaces.*

Proposition 18.1.8. 1. *The operator A^* is or is not Fredholm simultaneously with A.*

2. *The equation $Ax = y$ is solvable if and only if $(y, f)_2 = 0$ for all $f \in \operatorname{Ker} A^*$. Similarly, the equation $A^*f = g$ is solvable if and only if $(x, g)_1 = 0$ for all $x \in \operatorname{Ker} A$.*

3. *Moreover,*

$$\dim \operatorname{Coker} A = \dim \operatorname{Ker} A^*, \quad \dim \operatorname{Ker} A = \dim \operatorname{Coker} A^*, \tag{18.1.8}$$

and

$$\varkappa(A^*) = -\varkappa(A), \tag{18.1.9}$$

so that the definition (18.1.1) *of index is equivalent to*

$$\varkappa(A) = \dim \operatorname{Ker} A - \dim \operatorname{Ker} A^*. \tag{18.1.10}$$

Proof. 1. If B is a parametrix for A, then, as is easy to verify, B^* is a parametrix for A^*. This follows from the relations $(BA)^* = A^*B^*$ and $(AB)^* = B^*A^*$ and the compactness of the adjoint of any compact operator [247, Chap. VI, Sec. 1, Theorem 4]).

2. It is seen from (18.1.7) that $(Ax, f)_2 = 0$ for all x if and only if $(x, A^*f)_1 = 0$ for all x, i.e., $A^*f = 0$. In [137, Vol. I, Chap. VI, Lemma 2.8] if was shown that if $(y, f)_2 = 0$ for all f from the kernel of A^*, then y belongs to the closure of the range of A. But the latter is closed.

3. Let $e_1, \ldots, e_m$ be a basis in $\text{Coker} A$; then the functionals $f_1, \ldots, f_m$ with $(e_j, f_k) = \delta_{j,k}$ (which vanish on $R(A)$) form a basis in the space of functionals $f \in \text{Ker} A^*$. This gives the first relation in (18.1.8). The second is obtained in a similar way.   □

Remark 18.1.9. This result applies, in particular, to the case where A is a bounded operator in a Hilbert space H identified with its dual. If (x, y) is the inner product in H, then the adjoint A^* of A is related to A by

$$(Ax, y) = (x, A^*y) \tag{18.1.11}$$

and called the *Hilbert adjoint* of A. This notion is very important and well known in spectral theory (see the Section 18.3). In particular, it is convenient for studying so-called elliptic singular integral operators on a closed manifold M (see Section 18.4 below), which is of order zero on the scale of spaces $H^s(M)$ and can be considered as bounded operators in the Hilbert space $H^0(M) = L_2(M)$.

Expression (18.1.10) is also useful in the case where A and A^* are compact operators in H, which act from H to a space compactly embedded in H. This is the case for pseudodifferential elliptic operators of negative order on a closed manifold M. And if A and A^* are elliptic partial differential operators of order $m > 0$ on M, then they can be considered as unbounded operators on $H^0(M)$ with domain $H^m(M)$, and A and A^* are adjoint operators in $H^0(M)$ in the sense of the theory of unbounded operators in a Hilbert space.

Of importance is also the case where A and A^* are unbounded operators in a Hilbert space H with different domains compactly embedded in H and these operators are Fredholm as bounded operators from their domains to H. On the elements of their domains relation (18.1.11) holds. In this case, H is again the orthogonal sum of the range of A and the kernel of A^*, as well as of the range of A^* and the kernel of A, and these operators again have opposite indices. We have encountered such a situation in Section 7.3.

All cases mentioned in this remark deal with *adjointness in the sense of the theory of operators in a Hilbert space.*

Proposition 18.1.10. *Let A_1 be a Fredholm operator from X to Y, and let A_2 be a Fredholm operator from Y to Z. Then $A_2 A_1$ is a Fredholm operator from X to Z and*

$$\varkappa(A_2 A_1) = \varkappa(A_1) + \varkappa(A_2). \tag{18.1.12}$$

Proof. It is easy to show that if B_1 and B_2 are parametrices for A_1 and A_2, then $B_1 B_2$ is a parametrix for $A_2 A_1$. Let us prove (18.1.12).

We represent the space Y as the direct sum of the following four subspaces (this is possible because the kernels and cokernels are finite-dimensional):

(1) $R(A_1) \cap A_2^{-1} R(A_2)$;
(2) $Y_1 = R(A_1) \cap \operatorname{Ker} A_2$;
(3) $Y_2 = \operatorname{Ker} A_2 \cap \operatorname{Coker} A_1$;
(4) $Y_3 = A_2^{-1} R(A_2) \cap \operatorname{Coker} A_1$.

The subspaces Y_1, Y_2, and Y_3 are finite-dimensional; we denote their dimensions by d_1, d_2, and d_3.

The kernel of $A_2 A_1$ is the direct sum of the subspaces $\operatorname{Ker} A_1$ and $A_1^{-1} Y_1$. Its dimension equals $\dim \operatorname{Ker} A_1 + d_1$.

Next, $\operatorname{Coker}(A_2 A_1)$ is the direct sum of the subspaces $\operatorname{Coker} A_2$ and $A_2 Y_3$. Its dimension equals $\dim \operatorname{Coker} A_2 + d_3$.

Furthermore, $\dim \operatorname{Ker} A_2 = d_1 + d_2$ and $\dim \operatorname{Coker} A_1 = d_2 + d_3$.

Therefore, the index $\varkappa(A_2 A_1)$ equals

$$\dim \operatorname{Ker} A_1 + d_1 + d_2 - (\dim \operatorname{Coker} A_2 + d_2 + d_3)$$
$$= \dim \operatorname{Ker} A_1 + \dim \operatorname{Ker} A_2 - (\dim \operatorname{Coker} A_2 + \dim \operatorname{Coker} A_1)$$
$$= \varkappa(A_1) + \varkappa(A_2). \qquad \square$$

Corollary 18.1.11. *If B is a parametrix for A, then $\varkappa(B) = -\varkappa(A)$.*

Indeed, the product of these operators has index zero.

Proposition 18.1.12. 1. *The addition of a compact operator to a Fredholm operator yields a Fredholm operator with the same index.*

2. *Let A be a Fredholm operator from X to Y. Then there exists an $\varepsilon > 0$ such that if A_1 is a bounded operator from X to Y with norm less than ε, then $A + A_1$ is a Fredholm operator with the same index.*

3. *Let $A_t: X \to Y$ ($0 \le t \le 1$) be a family of Fredholm operators continuously depending on t with respect to the operator norm. Then the index $\varkappa(A_t)$ does not depend on t.*

Proof. 1. Let A be a Fredholm operator, and let A_1 be a compact operator from X to Y. Suppose that B is a parametrix for A. It is easy to verify that B is also a parametrix for $A + A_1$, so that $A + A_1$ is Fredholm. Moreover,

$$\varkappa(A + A_1) = -\varkappa(B) = \varkappa(A).$$

2. Let A be a Fredholm operator, and let A_1 be a bounded operator from X to Y. Suppose that B is a parametrix for A. Then $B(A + A_1) = I_1 + T + BA_1$. Here T is a compact operator and BA_1 is a bounded operator with norm $\|BA_1\| \le \|B\|\|A_1\|$. If this product is less than 1, then the operator $I_1 + BA_1$ is invertible and $(I_1 + BA_1)^{-1} B$ is a left parametrix for $A + A_1$. Furthermore, in this case, $B(I_2 + A_1 B)^{-1}$ is a right parametrix for $A + A_1$. Since the index of a parametrix does not change

when the parametrix is multiplied by an invertible operator, it follows that A and $A + A_1$ have equal indices.

3. This assertion follows from assertion 2. □

Property 3 of index is called its *homotopy invariance*.

Proposition 18.1.13. *Any Fredholm operator A can be represented in the form $A_0 + T$, where T is a finite-dimensional operator and*
(1) *if $\varkappa(A) = 0$, then A_0 is invertible;*
(2) *if $\varkappa(A) > 0$, then $R(A_0) = Y$ and $\dim \operatorname{Ker} A_0 = \varkappa(A)$;*
(3) *if $\varkappa(A) < 0$, then $\operatorname{Ker} A_0 = \{0\}$ and $\dim \operatorname{Coker} A_0 = -\varkappa(A)$.*

Proof. Let $x_1, \ldots, x_m$ be a basis in $\operatorname{Ker} A$, and let $y_1, \ldots, y_n$ be a basis in $\operatorname{Coker} A$. We construct an operator F vanishing on X_1 as follows. We set $F x_j = y_j$ for all j in the first case, $F x_j = y_j$ for $j \le n$ and $F x_j = 0$ for all other j in the second, and $F x_j = y_j$ for $j \le m$ in the third. In all the three cases, we set $A_0 = A + F$ and $T = -F$. This gives the required decomposition. □

18.2 The Lax–Milgram Theorem

The results given in this subsection play an important role in the theory of strongly elliptic equations. Theorem 18.2.2 appeared in [228] but in lesser generality (close considerations can be found in the earlier paper [388]); our exposition follows [286].

18.2.1

Let H_1 be a Hilbert space with inner product $(u, v)_1$. According to F. Riesz theorem on continuous linear functionals on H_1, this is a general form of a continuous antilinear functional $F(v)$ on H_1, and u is uniquely determined by F. We deal with antilinear, rather than linear, functionals in order to prepare a statement convenient for applications. We denote the corresponding norm by $\|u\|_1$.

Let $\Phi(u, v)$ be a sesquilinear form on H_1. We assume that it is continuous, i.e.,

$$|\Phi(u, v)| \le \alpha \|u\|_1 \|v\|_1, \tag{18.2.1}$$

where α is a positive constant. For fixed u, this form is a continuous antilinear functional on H_1. Suppose that

$$\|u\|_1^2 \le \beta |\Phi(u, u)| \tag{18.2.2}$$

with a positive constant β.

Theorem 18.2.1. *Under conditions (18.2.1) and (18.2.2), any continuous antilinear functional $F(v)$ on H_1 can be represented in the form $\Phi(u, v)$ with uniquely determined $u \in H_1$.*

Proof. Consider the linear self-mapping $Zu = w$ of H_1 defined by

$$\Phi(u,v) = (w,v)_1 \quad (v \in H_1). \tag{18.2.3}$$

This mapping is continuous:

$$\|Zu\|_1 = \sup_{v \neq 0} \frac{|(Zu,v)_1|}{\|v\|_1} = \sup_{v \neq 0} \frac{|\Phi(u,v)|}{\|v\|_1} \leq \alpha\|u\|_1,$$

so that

$$\|Zu\|_1 \leq \alpha\|u\|_1. \tag{18.2.4}$$

Furthermore,

$$\|u\|_1^2 \leq \beta|\Phi(u,u)| = \beta|(w,u)_1| \leq \beta\|w\|_1\|u\|_1,$$

so that

$$\|u\|_1 \leq \beta\|w\|_1. \tag{18.2.5}$$

It follows from this inequality that the preimage u of any element $w = Zu$ is uniquely determined, and the range of Z is closed (see the remark after Proposition 18.1.7). Suppose that $ZH_1 \neq H_1$. Then there exists an element v orthogonal to the range, i.e., such that

$$(Zu,v)_1 = 0$$

for all u. But then

$$\|v\|_1^2 \leq \beta|\Phi(v,v)| = \beta|(Zv,v)_1| = 0,$$

so that $v = 0$. Therefore, $ZH_1 = H_1$. This implies the desired result. $\quad\square$

Remark. Note that

$$\|u\|_1 \leq \beta\|F\|. \tag{18.2.6}$$

Indeed,

$$\|F\| = \sup_{v \neq 0} \frac{|F(v)|}{\|v\|_1} = \sup_{v \neq 0} \frac{|\Phi(u,v)|}{\|v\|_1} \geq \frac{|\Phi(u,u)|}{\|u\|_1} \geq \beta^{-1}\|u\|_1.$$

18.2.2

Now suppose that the space H_1 is continuously and compactly embedded in a Hilbert space H_0 with inner product $(u,v)_0$. For simplicity, we assume that

$$\|u\|_0 \leq \|u\|_1 \quad (u \in H_1). \tag{18.2.7}$$

For each $u \in H_0$, the form $(u,v)_0$ determines a continuous antilinear functional on H_1, because

$$|(u,v)_0| \leq \|u\|_0\|v\|_0 \leq \|u\|_0\|v\|_1.$$

Let H_{-1} be the space dual to H_1 with respect to the extension of this form, for which we keep the notation $(u,v)_0$. This means that the form $(f,v)_0$ is now defined on $H_{-1} \times H_1$ and determines all antilinear functionals on H_1, and for $f \in H_0$, it coincides with the inner product in H_0. The norm of a functional $f \in H_{-1}$ coincides with $\|f\|_{-1}$:

$$\|f\|_{-1} = \sup_{v \neq 0} \frac{|(f,v)_0|}{\|v\|_1}. \tag{18.2.8}$$

We have the embeddings

$$H_1 \subset H_0 \subset H_{-1}. \tag{18.2.9}$$

The first of them is assumed to be compact. The second is continuous, since

$$\|u\|_{-1} = \sup_{v \neq 0} \frac{|(u,v)_0|}{\|v\|_1} \leq \frac{\|u\|_0 \|v\|_0}{\|v\|_1} \leq \|u\|_0.$$

Therefore, the embedding of H_1 in H_{-1} is compact.

Consider the linear mapping $Lu = f$ of H_1 to H_{-1} defined by

$$(f,v)_0 = \Phi(u,v), \quad v \in H_1. \tag{18.2.10}$$

Taking into account (18.2.8), we can easily verify that this mapping is continuous under condition (18.2.1):

$$\|Lu\|_{-1} = \sup_{v \neq 0} \frac{|\Phi(u,v)|}{\|v\|_1} \leq \alpha \|u\|_1. \tag{18.2.11}$$

Now we fix f and consider the equality $Lu = f$ as an equation with respect to u; its solution, if it exists, can be called a weak solution of this equation.

Theorem 18.2.2. *Under conditions* (18.2.1) *and* (18.2.2), *the equation $Lu = f$ has precisely one weak solution $u \in H_1$ for any $f \in H_{-1}$, and the inverse operator L^{-1} is continuous.*

This is an obvious corollary of the preceding theorem and the remark to it: the functional $(f,v)_0$ can be uniquely written in the form $\Phi(u,v)$, and, by virtue of (18.2.6) and (18.2.8), the norm of the solution is estimated in terms of the norm of the right-hand side. Moreover, a two-sided estimate is valid (see (18.2.11)).

Theorem 18.2.3. *Theorem* 18.2.2 *remains valid under the replacement of condition* (18.2.2) *by the more severe condition*

$$\|u\|_1^2 \leq \beta \operatorname{Re}\Phi(u,u), \quad u \in H_1. \tag{18.2.12}$$

It is this statement which is applied to strongly elliptic equations in the cases of strong coercivity. For convenience of references, we shall name it *the Lax–Milgram theorem.*

18.2.3

Now let us state a similar assertion for the case where the form $\Phi(u,v)$ is coercive but not strongly coercive. First, note that the operator L (defined above as a bounded operator from H_1 to H_{-1}) admits a second interpretation: it can be regarded as an unbounded operator in H_{-1} with domain H_1.

Theorem 18.2.4. *If condition* (18.2.1) *holds and*

$$\|u\|_1^2 \leq \beta \operatorname{Re}\Phi(u,u) + \gamma\|u\|_0^2, \quad u \in H_1, \tag{18.2.13}$$

where $\gamma > 0$, *then* $L: H_1 \to H_{-1}$ *is a Fredholm operator with index zero. If L is treated as an operator in H_{-1} with domain H_1, then its resolvent set contains the half-plane* $\{\lambda : \operatorname{Re}\lambda < -\beta^{-1}\gamma\}$.

Proof. Let us replace the form $\Phi(u,v)$ by $\Phi(u,v) + \mu(u,v)_0$. If $\operatorname{Re}\mu > \beta^{-1}\gamma$, then the new form satisfies the assumptions of Theorem 18.2.3, and therefore the corresponding operator from H_1 to H_{-1} has continuous inverse. This operator can naturally be represented as $L + \mu I$. The identity operator is compact as an operator from H_1 to H_{-1}. By virtue of Proposition 18.1.12, L turns out to be a Fredholm operator of index zero. This proves the first assertion of the theorem; the second is obvious. $\square$

In Section 9, H_1, H_0, and H_{-1} are $\widetilde{H}^1(\Omega)$, $L_2(\Omega)$, and $H^{-1}(\Omega)$ in the case of the Dirichlet problem with homogeneous boundary condition and $H^1(\Omega)$, $L_2(\Omega)$, and $\widetilde{H}^{-1}(\Omega)$ in the case of the Neumann problem with homogeneous boundary condition. In the case of, say, the operators D, A^{-1}, and H considered in Section 12, these are $H^{1/2}(\Gamma)$, $L_2(\Gamma)$, and $H^{-1/2}(\Gamma)$. We leave the case of operators on a non-closed surface to the reader.

Now we consider the adjoint operator L^*. We define it by the relation

$$(Lu,v)_0 = \Phi(u,v) = (u,L^*v)_0. \tag{18.2.14}$$

To each element $v \in H_1$ it assigns the element $L^*v \in H_{-1}$, which is a continuous linear functional on H_1. In Sections 8 and 11 we took special care to ensure that the form $\Phi(u,v)$ remains unchanged under the passage to the formally adjoint equation; in these sections, the operator L^* in the case of the Dirichlet or Neumann problem corresponded to the Dirichlet or Neumann problem for the formally adjoint system. It is easy to verify that, in the cases of the operators D, A^{-1}, and H (see Section 12), L^* also corresponds to similar operators for the formally adjoint systems.

If the form $\Phi(u,v)$ satisfies condition (18.2.1) and condition (18.2.12) or (18.2.13), then Theorem 18.2.3 or 18.2.4, respectively, applies also to the operator L^*. In the former case, this operator is invertible and in the latter, Fredholm. In the latter case, the solvability condition for, say, the equation $Lu = f$ is the orthogonality of the right-hand side to the solutions of the equation $L^*v = 0$ with respect to the form $(f,v)_0$.

18.3 A Few Definitions and Facts from Spectral Theory

The material of this section will be discussed in more detail in [26]. As reference books the reader can use, e.g., [202] and [169].

18.3.1

Let A be a closed linear operator in an infinite-dimensional separable Banach space X with dense domain $D(A)$. This operator may be unbounded; its closedness means that its graph $\{(x, Ax)\}$, $x \in D(A)$, is closed in the direct product $X \times X$. If A is bounded, we assume that $D(A) = X$.

The *resolvent set* $\rho(A)$ of the operator A consists of those points λ of the complex plane at which the operator $A - \lambda I$ has bounded inverse $R_A(\lambda) = (A - \lambda I)^{-1}$. This inverse is called the *resolvent* of A at the point λ. The complement $\sigma(A)$ to the resolvent set is called the *spectrum* of the operator A. If $(A - \lambda I)x = 0$ for a nonzero vector x, then λ is called an *eigenvalue* of A, and x is called an *eigenvector* corresponding to this eigenvalue. The eigenvalues are contained in the spectrum. If

$$(A - \lambda I)^k x = 0$$

for a positive integer k and a vector $x \neq 0$, then x is called a *generalized eigenvector*, or a *root vector*, corresponding to the eigenvalue λ. In particular, for $k = 1$, this is an eigenvector.

In the spectral theory of elliptic equations and problems in a bounded domain or on a compact manifold, of special interest are two classes of operators.

1. *Compact operators T.* (In this case, we use the letter T instead of A.) The spectrum of a compact operator in an infinite-dimensional space consists of zero and isolated nonzero eigenvalues of finite multiplicity, which can accumulate only to zero (if there are infinitely many of them). The finiteness of the multiplicity of an eigenvalue $\lambda_0 \neq 0$ means that the corresponding generalized eigenspace, which consists of the generalized eigenvectors corresponding to λ_0 and the zero vector, is finite-dimensional. If T is a compact self-adjoint operator in a Hilbert space H, then all generalized eigenvectors of T are eigenvectors, and there is an orthonormal basis in H formed by eigenvectors.

2. *Operators with compact resolvent.* An operator with compact resolvent is an unbounded operator A such that, for some λ, its *resolvent* $(A - \lambda I)^{-1}$ exists and is compact. The spectrum of such an operator consists of isolated eigenvalues of finite multiplicity which have no finite accumulation points. As previously, the finiteness of multiplicity means that the corresponding generalized eigenspace is finite-dimensional. The resolvent $(A - \lambda I)^{-1}$ is compact for all $\lambda \in \rho(A)$. Such operators A are also called *operators with discrete spectrum*.

Let X be a separable Banach space. A system of vectors $\{f_j\}_1^\infty$ in X is said to be *complete* in this space if their finite linear combinations are dense in X.

18.3.2

Let again A be an unbounded closed operator in X with dense domain. Suppose that the complex plane contains a ray from the origin on which all λ with sufficiently large absolute values belong to the resolvent set and, for each of these λ,

$$\|R_A(\lambda)\|_X \le C(1+|\lambda|)^{-1}. \tag{18.3.1}$$

The norm of the resolvent cannot decrease faster than $|\lambda|^{-1}$ ([8, p. 182]); thus, we refer to such a ray as a *ray of maximal decrease of the resolvent norm*[4]. It is easy to verify that such rays form an open set. If there is an angle Λ in the complex plane with vertex at the origin on which the resolvent exists and satisfies estimate (18.3.1) for sufficiently large $|\lambda|$, then we call it an *angle of maximal decrease of the resolvent norm*.

Now let $X = H$ be a (separable) Hilbert space with inner product (u, v), and let A be an unbounded self-adjoint operator with discrete spectrum on H. Not dwelling on commonly known definitions (see, e.g., [202], [1], or [66]), we describe it as follows. There is an orthonormal basis $\{e_j\}_1^\infty$ in H consisting of eigenvectors of A. Any vector $x \in H$ can be written in the form

$$x = \sum_1^\infty (x, e_j)e_j, \tag{18.3.2}$$

and $\|x\|_H^2 = \sum |(x, e_j)|^2$. The operator A has the representation

$$Ax = \sum_1^\infty \lambda_j(x, e_j)e_j, \quad x \in D(A), \tag{18.3.3}$$

where λ_j are its eigenvalues corresponding to eigenvectors e_j. All generalized eigenvectors are eigenvectors. The domain $D(A)$ is determined by the condition

$$\sum |\lambda_j|^2 |(x, e_j)|^2 < \infty. \tag{18.3.4}$$

If there are infinitely many negative eigenvalues and infinitely many positive ones, then we can assume that all eigenvalues are numbered by integers in nondecreasing order with multiplicities taken into account, so that $\lambda_j \to \pm\infty$ as $j \to \pm\infty$. An operator is said to be *semibounded from below* if there are no or finitely many negative eigenvalues. In this case, the eigenvalues are usually numbered by positive integers in nondecreasing order with multiplicities taken into account, so that $\lambda_j \to +\infty$. If all eigenvalues are nonnegative or positive, then the operator is said to be *nonnegative* or *positive*, respectively.

Any closed angle with vertex at the origin not containing the real semiaxes $\mathbb{R}_\pm$ is in this case an angle of maximal decrease of the resolvent norm. This follows from the fact that the norm of the resolvent of a self-adjoint operator equals $\sup |\lambda - \lambda_j|^{-1}$.

---

[4] Agmon's term *direction of minimal growth of the resolvent* seems to be less appropriate.

If the operator A is semibounded from below, then this is true for any angle not containing $\mathbb{R}_+$.

18.3.3

Given an unbounded positive self-adjoint operator A on a Hilbert space H, we can define its powers A^t, in particular, for real t. If $t > 0$, then it suffices to replace λ by λ^t in the spectral decomposition

$$A = \int \lambda \, dE_\lambda.$$

We consider in detail the case of a positive self-adjoint operator A with discrete spectrum in a (separable) Hilbert space H.

Let $t > 0$. The operator A^t is defined by

$$A^t x = \sum \lambda_j^t (x, e_j) e_j. \tag{18.3.5}$$

Its domain is the subspace $H_t = D(A^t)$ of H determined by the condition

$$\sum \lambda_j^{2t} |(x, e_j)|^2 < \infty. \tag{18.3.6}$$

This is a positive self-adjoint operator with discrete spectrum; it has the same eigenvectors e_j as A and eigenvalues λ_j^t. The space H_t is endowed with the norm $\|x\|_t$ equal to the square root of the left-hand side of (18.3.6); it is easy to define the corresponding inner product and verify the completeness of the system of vectors e_j.

For $t < 0$, the space H_t consists of generally formal sums

$$x = \sum c_j e_j \tag{18.3.7}$$

(which are surely informal if they are finite) satisfying the condition

$$\sum \lambda_j^{2t} |c_j|^2 < \infty. \tag{18.3.8}$$

Each vector x is determined by the set of its Fourier coefficients $c_j = (x, e_j)$. The norm of a vector x in H_t is defined as the square root of the left-hand side of (18.3.8). Again, it is easy to introduce the corresponding inner product and verify the completeness of the system of vectors e_j.

We have obtained the scale of Hilbert spaces $\{H_t\}$, $-\infty < t < \infty$, which narrow with increasing t. For each s, the operator A^s is a continuous isomorphism of H_t onto H_{t-s}, and all such operators A^t form a group: $A^s A^t = A^{s+t}$.

If A is only nonnegative and has zero eigenvalue, then we can consider the *semigroup* of operators A^t with $t \geq 0$ and the corresponding spaces H_t.

We touched on powers of more general positive operators in Section 13.9.

In the case of a self-adjoint operator A with discrete spectrum and eigenvalues λ_j of both signs, the spaces H_t can be defined by using powers of the operator $|A|$ with eigenvalues $|\lambda_j|$ and the same eigenvectors. Suppose for simplicity that zero is not an eigenvalue of A; then the powers of $|A|$ again form a group, and we have the scale of spaces H_t, $-\infty < t < \infty$. The operator $|A|$ remains self-adjoint on these spaces (as well as its powers $|A|^s$) and retains the same basis of eigenvectors.

Powers of compact self-adjoint operators can also be defined.

18.3.4

When a non-self-adjoint operator in a Hilbert space H is given, there arises the question of whether the system of its generalized eigenvectors is complete in this space. Below we give two convenient completeness conditions.

First note that if T is a compact operator, then the operator T^* is compact and TT^* is compact, self-adjoint, and nonnegative. Let $s_j(T)$ be the nonzero eigenvalues of $(TT^*)^{1/2}$ numbered in nonincreasing order with multiplicities taken into account. These are the so-called *s-numbers* of the operator T. If $\sum s_j^p < \infty$ for some $p > 0$, then we say that this operator is of *finite order*. The greatest lower bound of such numbers p can be called the *order* of T.

Theorem 18.3.1. *Let*

$$A = A_0 + A_1, \tag{18.3.9}$$

where A_0 is a positive self-adjoint operator with discrete spectrum such that A_0^{-1} is of finite order and the operator A_1 is compact with respect to A_0 in the sense that $A_1 A_0^{-1}$ is compact. Then the system of generalized eigenvectors of A is complete in H.

An operator A of the form specified above is called a *weak perturbation of the self-adjoint operator A_0*, or an *operator close to the self-adjoint operator A_0*. Theorem 18.3.1 and its analogue for compact operators are given in the book [169, Chap. V]. The case when A_0 has eigenvalues of both signs is also considered there.

The following theorem is proved in the book [137, Vol. II, Chap. XI, Sec. 9] by Dunford and Schwartz. See also Agmon's paper [7] and book [8].

Theorem 18.3.2. *Suppose that the operator A_0^{-1} is of order p and A has rays of maximal decrease of the resolvent norm such that the maximum angle between neighboring rays is smaller than π/p. Then the system of generalized eigenvectors of A is complete in H.*

Moreover, the condition $\|R_A(\lambda)\| = O(|\lambda|^N)$ for some $N \geq -1$ along these rays is sufficient for the completeness.

This theorem can be generalized to operators in a Banach space (see [79] and [19]).

A system $\{e_j\}_1^\infty$ of vectors in a Banach space X is called a *basis* if any vector $x \in X$ can be represented in the form of a norm convergent series

$$x = \sum_{1}^{\infty} c_j e_j \qquad (18.3.10)$$

with uniquely determined coefficients $c_j = c_j(x)$. If $X = H$ is a Hilbert space and A is a self-adjoint operator in H with discrete spectrum (or a self-adjoint compact operator), then such an (orthonormal) basis can be composed of its eigenvectors. Series (18.3.10) is called the *Fourier series of the vector x in eigenvectors of A*, and the $c_j(x)$ are called the *Fourier coefficients* in this series.

In the general case, prerequisites necessary for a system of vectors to form a basis are completeness and minimality. A complete system of vectors is said to be *minimal* if the removal of any vector from this system implies the loss of completeness.

The minimality of a complete system $\{f_j\}_1^{\infty}$ of vectors is equivalent to the existence of a system $\{g_k\}_1^{\infty}$ biorthogonal to it.

Suppose that an operator A with discrete spectrum has a complete system $\{e_j\}_1^{\infty}$ of generalized eigenvectors, and this system is composed of bases of generalized eigenspaces (in each generalized eigenspace, one basis is taken). Suppose also that the adjoint operator A^* has a similar system $\{g_j\}_1^{\infty}$ of generalized eigenvectors. Then these two systems can be subjected to the biorthogonality condition $(e_j, g_k) = \delta_{j,k}$; see [31, Chap. 5]. In this case, $c_j = (x, g_j)$ in (18.3.10).

All this can be generalized to operators in a Banach space.

A question more delicate than that of completeness is the question of when a system $\{e_j\}$ of generalized eigenvectors retains the basis property to some extent. The present book does not contain a discussion of this question, it is postponed to [26].

18.4 Pseudodifferential Operators

Let $a(x, \xi)$ be a C^{∞} function on $\mathbb{R}^n \times \mathbb{R}^n$. The *pseudodifferential operator with symbol $a(x, \xi)$* is defined by

$$Au(x) = a(x, D)u(x) = \frac{1}{(2\pi)^n} \int e^{ix \cdot \xi} a(x, \xi)(Fu)(\xi) \, d\xi, \qquad (18.4.1)$$

first for functions $u \in C_0^{\infty}(\mathbb{R}^n)$. For simplicity, we first assume these functions to be scalar. Here F is the Fourier transform $u(x) \mapsto (Fu)(\xi)$. Expression (18.4.1) is, so to speak, the "inverse Fourier transform" of aFu, but it depends on x "two-fold": x is contained in the exponent and in the symbol $a(x, \xi)$.

For definition (18.4.1) to make sense and be useful, we must impose certain conditions on the symbol. The simplest set of conditions is as follows: there exists a real number m such that

$$|\partial_x^{\alpha} \partial_{\xi}^{\beta} a(x, \xi)| \leq C_{\alpha, \beta} (1 + |\xi|)^{m - |\beta|} \qquad (18.4.2)$$

for all multi-indices α and β, where the constants depend on a and do not depend on x and ξ.[5] Below we assume this condition to be fulfilled.

Under these conditions, operator (18.4.1) can be extended to an operator acting boundedly from $H^s(\mathbb{R}^n)$ to $H^{s-m}(\mathbb{R}^n)$ for all s. The number m is called the *order* of the pseudodifferential operator A. The term *pseudodifferential operator* is abbreviated as ΨDO.

The simplest example is a partial differential operator

$$A = \sum_{|\alpha| \le m} a_\alpha(x) D^\alpha \qquad (18.4.3)$$

of positive integer order m with C^∞ coefficients which have bounded derivatives of all orders. Obviously, the symbol of this operator is

$$a(x, \xi) = \sum_{|\alpha| \le m} a_\alpha(x) \xi^\alpha. \qquad (18.4.4)$$

An example of a ΨDO of order zero is the operator of multiplication by a C_b^∞ function $a(x)$, which is the symbol of this operator. More general ΨDOs of order zero are specified below in (18.4.8).

Now suppose that m is negative. Then, at least for $u \in C_0^\infty(\mathbb{R}^n)$, we can apply Fubini's theorem and rewrite (18.4.1) in the form

$$Au(x) = \int K(x, y) u(y) \, dy,$$
$$\text{where } K(x, y) = \frac{1}{(2\pi)^n} \int e^{i(x-y)\cdot\xi} a(x, \xi) \, d\xi. \qquad (18.4.5)$$

This is a ΨDO in the *x-representation*. The function $K(x, y)$ is called its *kernel*. Thus, the class of ΨDOs contains many integral operators. In fact, a ΨDO of any order admits an x-representation, provided that its kernel is understood in the sense of Schwartz (see [335]).

The most important facts of ΨDO *calculus* are as follows.

1. The product $AB = a(x, D) b(x, D)$ of two ΨDOs of orders m_1 and m_2 is a ΨDO C of order $m_1 + m_2$ with symbol

$$c(x, \xi) = a(x, \xi) b(x, \xi) \qquad (18.4.6)$$

up to the addition of a ΨDO of order at most $m_1 + m_2 - 1$.

2. The operator formally adjoint to a ΨDO $A = a(x, D)$ of order m with respect to the inner product in $L_2(\mathbb{R}^n)$ is a ΨDO A^* of the same order with symbol $a^*(x, \xi)$ Hermitian conjugate to $a(x, \xi)$ up to the addition of a ΨDO of order at most $m-1$. In the scalar case, $a^* = \bar{a}$, but one can also consider the case of matrix operators, in which $u(x)$ is a column vector and $a(x, \xi)$ is a square (for simplicity) matrix. In this

---

[5] A theory of more general pseudodifferential operators is presented, e.g., in [348] and [373]. See also [400].

case, a^* is the matrix Hermitian conjugate to a. Formal adjointness means that

$$(A\varphi, \psi)_{\mathbb{R}^n} = (\varphi, A^*\psi)_{\mathbb{R}^n}$$

for any functions $\varphi, \psi \in C_0^\infty(\mathbb{R}^n)$.

More precisely, for the symbols of the ΨDOs AB and A^*, asymptotic expansions in the symbols of ΨDOs of decreasing order are constructed.

This calculus is particularly rich in content and useful in the case of *polyhomogeneous* ΨDOs. This is the case where the symbol admits an asymptotic expansion of the form (we give its simplest version)

$$a(x, \xi) \sim a_0(x, \xi) + a_1(x, \xi) + \dots, \tag{18.4.7}$$

where the functions $a_j(x, \xi)$ are positive homogeneous in ξ of degree $m - j$. This expansion means that if $\theta(\xi)$ is any C^∞ function vanishing in a neighborhood of the origin and taking the value 1 outside a larger neighborhood, then, for any nonnegative integer N, the function

$$\theta(\xi)\left[a(x, \xi) - \sum_{j=0}^N a_j(x, \xi)\right]$$

is the symbol of a ΨDO of order at most $m - N - 1$. Obviously, differential operators form a subclass of polyhomogeneous ΨDOs.

Polyhomogeneous ΨDOs of order $m = 0$ belong to the class of so-called *singular integral operators*. In x-representation they look as

$$Au(x) = a(x)u(x) + \lim_{\varepsilon \to 0} \int_{|x-y|>\varepsilon} K(x, y)u(y)\, dy. \tag{18.4.8}$$

Here the integral is understood in the sense of the Cauchy principal value. In more detail, the kernel $K(x, y)$ has the form $k(x, x - y)$, where the function $k(x, z)$ (or its principal part) is positive homogeneous in z of critical degree $-n$: $k(x, tz) = t^{-n}k(x, z)$, $t > 0$. The integral of this function over the unit sphere $|z| = 1$ is assumed to vanish; it is this assumption which ensures the existence of integral (18.4.8) in the sense of the Cauchy principal value.

The function $a_0(x, \xi)$ is called the *principal symbol* of the polyhomogeneous ΨDO $A = a(x, D)$.

The product AB of polyhomogeneous ΨDOs A and B is a polyhomogeneous ΨDO. Its principal symbol is the product $a_0 b_0$ of the principal symbols a_0 and b_0 of the ΨDOs A and B. Therefore, if polyhomogeneous ΨDOs A and B of orders m_1 and m_2 have commuting principal symbols (e.g., if these are scalar operators), then the commutator $AB - BA$ is a polyhomogeneous ΨDO of order at most $m_1 + m_2 - 1$. The ΨDO A^* formally adjoint to the polyhomogeneous ΨDO A is again a polyhomogeneous ΨDO, and its principal symbol is Hermitian conjugate to that of A.

The uniform *ellipticity condition* is, in the scalar case, the inequality

$$|a_0(x,\xi)| \geq C|\xi|^m, \tag{18.4.9}$$

where C is a positive constant. In the simplest matrix situation, a similar condition is imposed on the determinant of the matrix a_0.

It can be proved that, under this condition, $a_0^{-1}(x,\xi)$ is the principal symbol of a ΨDO B of order $-m$, which is also polyhomogeneous and elliptic. Clearly, AB and BA differ from the identity operator by a ΨDO of order -1. Therefore, B is a quasi-parametrix for the elliptic ΨDO A (cf. Section 6.2).

It can be verified that, under any diffeomorphism $y = y(x)$ such that all derivatives of all orders of this diffeomorphism and its inverse are bounded, the polyhomogeneous ΨDO A transforms into a polyhomogeneous ΨDO B with principal symbol

$$b_0(y,\eta) = a_0\Big(x(y), \Big(\Big[\frac{\partial x}{\partial y}\Big]'\Big)^{-1}\eta\Big). \tag{18.4.10}$$

This expression is important for the passage to ΨDOs on a manifold. In particular, it shows the invariance of the ellipticity (and strong ellipticity) conditions with respect to smooth nonsingular transformations of coordinates.

Now let M be a smooth closed manifold. A ΨDO A on M is defined as follows. If φ and ψ are C^∞ functions supported in a coordinate neighborhood U, then

$$\varphi A\psi u = \varphi A_U \psi u, \tag{18.4.11}$$

where A_U is a ΨDO on $\mathbb{R}^n$ in the corresponding local coordinates. If the supports of φ and ψ are disjoint, then $\varphi A\psi$ is an operator of order $-\infty$ in the scale $\{H^s(M)\}$.

Expression (18.4.10) makes it possible to define a polyhomogeneous ΨDO of order m on M; by virtue of this expression, the principal symbol $a_0(x,\xi)$ turns out to be invariantly defined as a function on the cotangent bundle $T^*(M)$ of M. Such a ΨDO acts boundedly from $H^s(M)$ to $H^{s-m}(M)$ for all s. The calculus is constructed globally at the level of principal symbols. Elliptic polyhomogeneous ΨDOs on M are defined in a natural way. It can be proved that the ellipticity of a ΨDO A of order m is equivalent to A being Fredholm as an operator from $H^s(M)$ to $H^{s-m}(M)$, and the index of A does not depend on s by virtue of an appropriate smoothness theorem. This generalizes the material of Section 6. All analytical work turns out to have already been done in the course of constructing the calculus of ΨDOs on $\mathbb{R}^n$.

In particular, we can consider singular integral operators on a manifold.

ΨDOs on the standard torus can be defined by using Fourier series instead of the Fourier transform:

$$a(x,D)u(x) = \sum_{\alpha\in\mathbb{Z}^n} a(x,\alpha)e^{-i\alpha\cdot x}c_\alpha(u),$$

where $c_\alpha(u)$ are the Fourier coefficients of the function u; the symbol $a(x,\xi)$ is employed only at the points $\xi = \alpha$ with integer coordinates. We used this in Section 12.1.

The calculus of pseudodifferential operators in its current form emerged in the 1960s. Originally, elliptic ΨDOs were introduced for and have played a crucial role in solving the index problem for elliptic differential operators, namely, in applying homotopies. It is hard to arrange homotopies in the class of polynomials (principal symbols of differential elliptic operators). Although the principal symbols of ΨDOs are infinitely differentiable at $\xi \neq 0$, a homotopy in the class of continuous functions (over "unit" cotangent vectors if M is a Riemannian manifold) can be approximated by a homotopy in the class of infinitely differentiable functions.

It was very soon found that ΨDOs arise and are very necessary and very useful in numerous problems of analysis, especially those on determining various asymptotics.

The strong ellipticity of a polyhomogeneous ΨDO is defined as the uniform positive definiteness of the real part of its principal symbol; see Costabel and Wendland's paper [103] and Hsiao and Wendland's monograph [186].

We have outlined the theory of ΨDOs for "smooth" situations. Actually, ΨDOs can also be defined and studied in situations where the smoothness of the functions under consideration is bounded or very low. However, in "Lipschitz" situations, it is usually impossible to apply ΨDO calculus in full, and put in the forefront are theorems on the boundedness, e.g., of singular integral operators. This is a profound rich-of-content science, which goes back to works of Calderón and Zygmund. It is exposed, e.g., in Christ's survey [93].

19 Additional Remarks and Literature Comments

The reader should bear in mind that some of the questions discussed in this book
have been studied in parallel by many mathematicians, and it is an unrewarding and
often impracticable task to investigate in detail who had first done certain things.
This refers, in particular, to works on Sobolev-type spaces and on elliptic problems.

Significantly more references can be found in the monographs which we cite.

We also warn the reader that the journal articles included in the list of references
may be preceded by brief communications of the same authors published earlier,
and books may be preceded by many articles.

19.1 To Chapter 1

The theory of Sobolev-type spaces and, to a fair degree, distribution theory originate
in the 1930s, in Sobolev's papers reflected in his monograph [354]. The literature on
this topic is very rich; this theory is essentially a vast domain in functional analysis,
still being developed. As already mentioned, Sobolev considered the spaces W_p^s
with nonnegative integer s. The spaces W_p^s of nonnegative fractional order s were
introduced by Slobodetskii in [350,351]. To the case $p = 2$ his detailed paper [350] is
devoted. As it later turned out, spaces of fractional order appeared somewhat earlier
in [39]. The article [43] was one of the first papers concerning H^s spaces on $\mathbb{R}^n$. It
was continued in [3,4,42,44]; see also Calderón's paper [81] on the spaces H_p^s.

To Sections 1.14 and 1.15

In the current literature, discrete norms are used in considering very general
spaces: Triebel–Lizorkin and Besov spaces with three indices. Their definitions are
given in Section 14.7. In Section 1.14 we gave the definitions of the simplest discrete norms on the H^s spaces and in Section 1.15 showed how these norms can be
used to prove one of Sobolev's embedding theorems.

Issues related to extending functions from domains or surfaces of lower dimension were considered by a great number of mathematicians; detailed references can
be found in the paper [288] and the books [63], [197], and [289].

An analogue of the Hestenes operator extending functions defined on a half-
space or on a smooth domain to $\mathbb{R}^n$ for the spaces H^s with $s \in \mathbb{R}$ (and more general
function spaces) can be found in [378, Sec. 4.5.5].

The Seeley operator [336] extending functions defined on the half-space $\mathbb{R}_+^n$
to $\mathbb{R}^n$ is constructed as follows. It can be shown that there exist numerical sequences
$\{a_k\}_0^\infty$ and $\{b_k\}_0^\infty$ such that

(1) $b_k < 0$ and $b_k \to -\infty$,

(2) $\sum_{0}^{\infty} |a_k| \, |b_k|^m < \infty$ for all nonnegative integers m, and

(3) $\sum_{0}^{\infty} a_k (b_k)^m = 1$ for all nonnegative integers m.

Let $\phi(t)$ be a C^∞ function on the line taking the value 1 on $[0,1]$ and 0 at $t \geq 2$. For $x_n < 0$, we set

$$Eu(x', x_n) = \sum_{0}^{\infty} a_k \phi(b_k x_n) u(x', b_k x_n). \qquad (19.1.1)$$

This operator maps $C^\infty(\overline{\mathbb{R}^n_+})$ to $C^\infty(\mathbb{R}^n)$ and, as mentioned in [336], can serve as an extension operator for the Sobolev spaces W_p^s.

For comments on the case of more general spaces, see also our remarks to Section 14 below.

To Section 3.4

One of the proofs of Theorem 3.4.1 is contained in [34]; see also [376]. Here we follow Yakovlev's paper [402].

19.2 To Chapter 2

The general definition of elliptic systems was given by Petrovskii in [302], where the main result was the analyticity of the solutions of elliptic systems with analytic coefficients.

To Section 7

The general result "a problem is elliptic $\Leftrightarrow$ it is Fredholm" was preceded by numerous studies. An impression of the development of the theory of elliptic equations until 1953 can be gained from Miranda's 1955 monograph republished in 1970 [266]. The survey [391] gives a view of the achievements of the Soviet mathematical school in the field of elliptic equations until 1957.

As we already mentioned, the general definition of strong ellipticity was given in 1951 by Vishik [388]. This paper, written before the development of the general theory of elliptic problems, had met with a wide response. The general Gårding inequality appeared in [163] for the scalar Dirichlet problem and, later, in [290] for the general matrix Dirichlet problem. The problem of determining coercivity conditions in other problems for higher-order strongly elliptic equations and systems was studied, e.g., in [6, 40, 328, 352]. But, apparently, it is easier to see the monographs [8, 237, 258, 286], where further references can also be found.

In 1952, Vishik's paper [389] appeared, which proposed an abstract approach to problems for elliptic equations in terms of operator extensions. Extensions with certain properties are described by using boundary conditions. This approach was developed by many authors, particularly in relation to questions of spectral theory.

Of great importance for developing the general theory of elliptic problems was Lopatinskii's paper [242] already mentioned in Section 7. It was also included in the book [243]. This paper considered general boundary value problems for elliptic system in a multidimensional convex domain in which the orders of the boundary operators are lower than that of the system. Actually, just in this paper the "Lopatinskii condition" appeared. Lopatinskii's method of study was to solve explicitly a problem with constant coefficients without lower-order terms in a half-space and then reduce the problem in a bounded domain to Fredholm integral equations on the boundary. The same idea appeared in Shapiro's papers [340] (Lopatinskii used results of [340] as an example) and [341], but in lesser generality.

It should be mentioned that the notion of ellipticity arose much earlier in a different area of analysis under a different name. We mean the theory of one-dimensional singular integral equations (scalar and matrix), where elliptic equations were called equations of normal type (see, e.g., [280]). It was found that problems for elliptic partial differential equations in two-dimensional domains can be reduced to one-dimensional singular equations on the boundary; this not only provided a method for proving these problems being Fredholm but also opened a way to calculate their index, because the index formula for one-dimensional singular integral operators had long been known. A general theory of elliptic problems in two-dimensional domains with index calculation was constructed by Vol'pert [394].

Only later did the understanding come that these notions of ellipticity, for boundary value problems and for singular integral equations, are of the same nature.

As mentioned in Section 6.3, a great influence on the study of elliptic equations and problems has been exerted by Gelfand's paper [164], in which the problem of finding homotopy invariants and the index of an elliptic operator was posed. It stimulated, in the first place, completing the theory of elliptic differential equations and problems and generalizing it to pseudodifferential equations and problems, which was needed for applying homotopies, in particular, to index calculation.

General a priori estimates for solutions of elliptic problems in Schauder and Sobolev L_p-norms were obtained by Agmon, Douglis, and Nirenberg in Part I of [9] for the scalar case and in Part II for the matrix case of Douglis–Nirenberg elliptic systems introduced in [131]. But a problem can be proved to be Fredholm by using solely a priori estimates only if the adjoint problem can be employed.

Our list of references contains a series of papers by Browder [71–74] and Schechter [328–334], which have played a substantial role in the completion of the proof that ellipticity is equivalent to the Fredholm property. Parametrices were introduced to the study of elliptic problems by Browder [72]. Their analogues in the theory of singular integral equations were regularizers. A general transparent construction of a parametrix was proposed by Dynin [139, 140], who has essentially completed the proof of the theorem "a problem is elliptic $\Leftrightarrow$ it is Fredholm" in the scalar case.

A singular integro-differential operator in Dynin's paper [139] was a differential operator whose coefficients are multidimensional singular integral operators. In essence, this was an immediate predecessor of a pseudodifferential operator. General elliptic operators of such form on a closed manifold were considered in [139] and general elliptic boundary value problems with such boundary conditions for scalar elliptic partial differential equations, in [140]. For further generalizations see [11] and [30]. The general theory of pseudodifferential elliptic problems was constructed by Vishik and Eskin in a series of papers (see the monograph [149]) and by Boutet de Monvel [68,69]. See also [311].

See also our references in Section 15.2.

Elliptic problems with parameter were introduced and first considered by Agmon [7]. As mentioned in Section 17.4, he obtained an a priori estimate with parameter by introducing an additional variable t in a cylindrical domain. But he had no right inverse operator, and to obtain the result on unique solvability, he employed the adjoint problem. In [34] the problem with parameter was considered by the same scheme as the problem without parameter; in particular, a right inverse operator was constructed on the pattern of Dynin's right parametrix. A continuation of [7] is Agmon and Nirenberg's paper [10], in which, as well as in [34], parabolic problems were also considered. See also [36] and [29].

The notions of a normal system of boundary conditions, a Dirichlet system, and an adjoint problem, as well as the general Green identity, appeared in Aronszajn and Milgram's paper [41]. This material was expounded anew and used by Schechter in [330, 331]. It was subsequently included in Lions and Magenes' book [237], which was preceded by their articles. Our exposition in Section 7.3 is somewhat simpler, because there we can already use the theory of general elliptic problems with inhomogeneous boundary conditions. Generalizations to systems can be found in the notes [314] of Roitberg and Sheftel and [248] of L'vin and in Duduchava's paper [133], where also surface potentials and Calderón projections are discussed in the general context of the theory of elliptic problems.

The remark on the normality of boundary conditions in the presence of a ray along which the problem is elliptic with parameter was made by Burak [78] and Seeley [338].

The Calderón projections [82] are projections of the corresponding space on the boundary onto the Cauchy data for an elliptic equation in the interior and exterior domains. It was to study general elliptic problems that Calderón introduced these projections; see also [337] and [103]. In Chapter 3 we consider surface potentials and Calderón projections for second-order strongly elliptic systems in Lipschitz domains. In the context of the general theory, it is convenient to use the calculus of pseudodifferential operators when dealing with these projections.

In the book [237] Lions and Magenes constructed scales of spaces with s taking any real values, for which they proved the Fredholm property of elliptic problems. For s not smaller than the order of the equation, these are spaces of ordinary functions of smoothness increasing with s. For negative s, the transposition (the passage to the adjoint problem) was used, and some distribution spaces were introduced; for intermediate s, interpolation was applied. The book [237] is available and well known, and we did not dwell on these results; here we only mention that, in transposition, the spaces containing smooth solutions and right-hand sides of the adjoint problem are constructed anew, in order to obtain spaces of distributions on Ω after transposition. The space $\widetilde{H}^{-s}(\Omega)$ dual to $H^s(\Omega)$ with $s > 1/2$ is not such a space, because it contains functionals supported on the boundary.

Of course, the operator corresponding to an elliptic problem has an adjoint also for all problems in which the boundary conditions are not normal, and adjoint operators may be studied and used as well; see, e.g., [129].

We dwell on a different approach to general elliptic problems for all values of the parameter s. To this approach Roitberg's monograph [312] is devoted, which was preceded by papers of Berezanskii, Krein, and Roitberg (see, e.g., [59]) and by many works of Roitberg and other mathematicians. This topic was also elucidated in Berezanskii's monograph [58].

Consider the elliptic problem (7.1.3), (7.1.4) in a bounded domain Ω with infinitely smooth (for simplicity) boundary Γ, in which the orders r_j of the boundary operators are lower than the order of the system. All coefficients are assumed to be C^∞. Consider the space $H_s(\Omega, \Gamma)$ consisting of $2l$-tuples $U = (u, u_1, \ldots, u_{2l})$. Very roughly speaking, u is a solution of the elliptic equation in Ω and the u_j are, in the general case, the generalized traces of its normal derivatives of orders $0, \ldots, 2l - 1$.

More precisely, u is a function in $H^s(\Omega)$ if $s \geq 0$ and in $\widetilde{H}^s(\Omega)$ if $s < 0$, and u_j belong to $H^{s-j+1/2}(\Gamma)$. For simplicity, we exclude all half-integer s between zero and $2l$. If $s > 2l - 1/2$, then u_j with $j \geq 1$ are the traces of the normal derivatives $\partial_\nu^{j-1} u$ of u on the boundary in $H^{s-j+1/2}(\Gamma)$; they exist in the usual sense in this case. If $s < 1/2$, then u_j are any elements of $H^{s-j+1/2}(\Gamma)$. For intermediate s, these are the traces of normal derivatives of order $j - 1$ in $H^{s-j+1/2}(\Gamma)$ for those j for which they exist in the usual sense and any elements of this space for all other j. On this space $H_s(\Omega, \Gamma)$ the norm

$$\|U\|_{H_s(\Omega,\Gamma)} = \left(\|u\|_{H^s(\Omega)}^2 + \sum_1^{2l} \|u_j\|_{H^{s-j+1/2}(\Gamma)}^2 \right)^{1/2} \qquad (19.2.1)$$

is introduced, where $\|u\|_{H^s(\Omega)}$ is replaced by $\|u\|_{\widetilde{H}^s(\Omega)}$ for negative s. It is verified that this is a complete space and that the linear manifold consisting of "smooth" $2l$-tuples $U = (u, u_1, \ldots, u_{2l})$, where u belongs to $C^\infty(\overline{\Omega})$ and u_j are the traces of its normal derivatives, is dense in it.

The operator $\mathfrak{A}$ corresponding to problem (7.1.3), (7.1.4) is considered as an operator from this space to the direct product of (i) the space $H^{s-2l}(\Omega)$ (if $s \geq 2l$) or $\widetilde{H}^{s-2l}(\Omega)$ (if $s < 2l$) and (ii) the spaces $H^{s-r_j-1/2}(\Gamma)$ ($j = 1, \ldots, l$) with the corresponding norms. The action of the operator is defined as follows: each element U is approximated in $H_s(\Omega, \Gamma)$ by smooth elements U_k, $k \to \infty$, for which the $\mathfrak{A}U_k$ are understood in the usual sense; it is verified that the limit $(f, g_1, \ldots, g_l)$ exists and does not depend on the choice of the approximating sequence. This limit is taken for $\mathfrak{A}U$.

The main result is that ellipticity is equivalent to the Fredholm property and the index does not depend on s. A local theorem on increasing smoothness is also proved. In the case of ellipticity with parameter, the unique solvability for large parameter is obtained. If the problem is normal, then the range of the operator $\mathfrak{A}$ is described in a standard way. The results extend to Douglis–Nirenberg elliptic systems and to Bessel potential spaces and Besov spaces with $p \neq 2$. A number of applications have been developed. For a narrower range of s, the smoothness assumptions can be relaxed.

Thus, the main idea of this theory is to consider sets of generalized solutions together with their generalized normal derivatives and approximate these sets by ordinary smooth solutions with their ordinary normal derivatives.

Our exposition of the theory of general elliptic problems does not touch on certain topics. First, these are general mixed problems, in which different boundary conditions are imposed on the solution on different parts of the boundary. Such problems are often encountered in applications, and the literature on particular mixed problems is vast. In the context of the general theory of elliptic problems, we mention Schechter's paper [333]. Secondly, these are transmission problems, whose solutions must satisfy generally different elliptic equations in, say, two contiguous domains and conditions on their common boundary, which relate the boundary values of the solutions and of their derivatives on both domains. For the general theory of elliptic problems, see, e.g., Schechter's paper [332], Roitberg and Sheftel's paper [313], and Sheftel's paper [342]. In Chapter 3 we consider some classes of mixed problems and transmission problems for second-order strongly elliptic systems in Lipschitz domains.

Problems with parameter become complicated when it is impossible to assign a certain weight to the parameter with respect to differentiation. Spectral problems of this form for Douglis–Nirenberg systems were first considered by Kozhevnikov (see, e.g., [219]). Afterwards, such problems were studies from different points of view by Denk, Mennicken, and Volevich (see, e.g., [126]) and by Denk and Volevich (see, e.g., [127, 128]).

Sternin considered elliptic equations with boundary conditions on submanifolds of any dimension [368].

An extensive literature is devoted to elliptic problems with degeneracy. In the skew derivative problem, which we mentioned in Section 11.8, degeneracy occurs at those boundary points at which the direction of differentiation is tangent to the boundary (if such points exist): the Lopatinskii condition is violated at these points. Such problems in smooth domains were studied by many authors; see Levendorskii and Paneyakh's survey [230]. We also mention Grushin and Vishik's paper [390].

19.3 To Chapter 3

To Section 9

Theorem 9.1.3 is a part of the following more general assertion.

Suppose that a function $\varphi(x)$ is given in an open neighborhood of a measurable set E in $\mathbb{R}^n$. Then it is differentiable at almost all points of E if and only if, for almost all $x^0 \in E$,

$$\varphi(x^0 + y) - \varphi(x^0) = O(|y|) \quad as \ |y| \to 0;$$

the estimate is not assumed to be uniform in x^0.

This assertion was stated in Chapter VIII of Stein's book [359] as "a celebrated theorem of Denjoy, Rademacher, and Stepanov" with the only reference to Saks' book [322, Chap. IX]. It was proved in these books and in Whitney's book [398, Chap. IX, Sec. 11], but the proof there relies on the preceding material in these books. In our self-contained exposition, we mainly follow Stepanov's paper [363]. Stepanov also constructed an example of a continuous function having all first partial derivatives almost everywhere but not differentiable at any point.

We do not touch at all on function spaces on domains and surfaces more general than Lipschitz. Initial information on this subject can be found in Grisvard's book [173] and McLean's book [258].

To Section 10

Before the appearance of the Rychkov operator, the extension operators most generally used in the literature were the operators (of extension beyond the boundary of a Lipschitz domain) constructed by Calderón [81] (see also [258, Appendix A]) and Stein [359] for the H_p^k spaces with nonnegative integer k. The Stein operator, unlike the Calderón operator, does not depend on k and applies to $p = 1$ and $p = \infty$. See also Muramatu [278].

Rychkov constructed his universal extension operator in [320] for general Triebel–Lizorkin and Besov spaces; we mentioned this in Section 14. In his difficult proof, Rychkov uses "maximal Peetre functions," which are not present in the final statement, including the "intrinsic" description of the spaces under consideration on special Lipschitz domains, which is interesting by itself. Trying to simplify the exposition, we restricted ourselves to the H^s spaces. We were able to prove the fairly transparent auxiliary Proposition 10.1.2 for this case, which made it possible to obviate the necessity of using maximal functions. Note also that, unlike in [320], we do not need to iterate the construction of Section 10.3, and that in the representation of functions in $H^s(\Omega)$, we obtain the convergence of series in the norms of these spaces.

To Sections 11 and 12

To Section 11.1

There is an abstract approach to the derivation of Green identities close to those used in our book and applicable to the problems which we are interested in; see [55, 218, 392] and the references therein.

To Section 11.2

In our opinion, here of interest are the Weyl decomposition and its applications, in particular, the reduction of a complete Neumann problem with nonuniquely determined right-hand sides to incomplete Dirichlet and Neumann problems with uniquely determined right-hand sides [24].

To Section 11.4

General mixed problems in Lipschitz domains were considered, e.g., in Pal'tsev's paper [293] and in McLean's book [258].

Note that it is often required to divide the boundary into many parts; see, e.g., Lebedev and Agoshkov's book [229]. The approach using the spaces $H^s(\Omega; \Gamma_1)$ works in this case, too.

To Sections 11.4 and 12.8

With regard to mixed problems, we have already mentioned that they have a vast literature. This refers to domains with smooth and nonsmooth boundary. In the books [157] and [353], complicated explicit expressions for solutions of particular mixed problems can be found. Even if the surfaces Γ_1 and Γ_2 and the "edge" between them are smooth, a solution may have a singularity near the edge, and great attention was paid to the analysis of this singularity. An additional geometric condition at the edge points was often imposed, which, roughly speaking, consists in that the angle between the normals to Γ_1 and Γ_2 at these points is less than π and is uniformly bounded away from π; see, e.g., [75], [270], and [324]. We did not touch on these questions.

Several German and Georgian mathematicians considered mixed problems for the Laplace–Helmholtz equation and elasticity systems (the Lamé system and anisotropic elasticity systems) in the case of smooth Γ_1, Γ_2, and $\partial \Gamma_j$; they determined a variational solution, reduced the problem to equations on the boundary by using potential-type pseudodifferential operators, and effectively found the asymptotics of the solution near the edge. We mention Stephan's papers [364] and [367] and Natroshvili, Chkadua, and Shargorodskii's paper [283]. From these papers we borrowed the approach to obtain a variational solution by using potentials and generalized it to second-order strongly elliptic systems in Lipschitz domains. Under our assumptions, we obtained some results on smoothness (which are weaker than results obtained in the above-mentioned papers under stronger assumptions about the smoothness of the coefficients and the parts of the boundary) but without an asymptotics. They can be found in [23], where spectral problems were also considered.

The papers [364–366] of Stephan, [134] of Duduchava and Natroshvili, [135] of Duduchava, Natroshvili, and Shargorodskii, and [136] of Duduchava and Wendland studied problems with boundary conditions on a smooth surface with smooth boundary for the same equations and systems by using potentials and determined the asymptotics of a solution near this boundary. Again, we borrowed the approach using potentials, generalizing it to second-order strongly elliptic systems in domains with Lipschitz boundary, and obtained some results on regularity. These are results of [22], where spectral problems were considered, too.

To Section 12.9

Problems with a spectral parameter in transmission conditions for the Helmholtz equation were proposed by physicists Katsenelenbaum, Voitovich, and Sivov; see [31] (this is a revision of the 1977 Russian edition). The idea was to obtain problems with a spectral parameter for this equation in $\mathbb{R}^n$ or an unbounded domain which have discrete spectrum. The author studied these problems by means of the theories of non-self-adjoint operators and of elliptic pseudodifferential operators in this book.

To Section 12.10

The topic of this subsection was suggested by questions posed by E. Krypchuk and S. E. Mikhailov in September of 2011.

19.4 To Chapter 4

To Section 13

Interpolation theory in Banach spaces extends to quasi-Banach spaces; see, in particular, [198].

To Section 14

As previously, the definitions with which we begin here are maximally close to those used in works on partial differential equations, including works on equations in Lipschitz domains; see, e.g., [195] and [274]. But it is already impossible to avoid using the "discrete definitions" that have arisen afterwards.

There exist other definitions of spaces with $s > 0$, which involve differences (instead of derivatives) or approximations, e.g., by entire analytic functions of exponential type; see, e.g., [63, 288, 289, 376].

Spaces with three indices were considered, e.g., in [60, 298, 376–378]. As mentioned in [319], the initial definitions are due to

Besov ($B_{p,q}^s$, $s > 0$, $1 \le p, q \le \infty$) [61, 62];
Taibleson ($B_{p,q}^s$, $s \in \mathbb{R}$) [371];
Lizorkin ($F_{p,q}^s$, $s > 0$, $1 < p, q < \infty$) [240, 241];
Triebel ($F_{p,q}^s$, $s \in \mathbb{R}$, $1 < p, q < \infty$) [375];
Peetre ($B_{p,q}^s$ and $F_{p,q}^s$, $0 < p, q \le \infty$) [298, 299]).

To Section 15

As already mentioned, analogues of these results for spaces with three indices can be found in [161] and [376]; see also [318].

The monographs [91] and [167] considered less general linear problems but under minimal smoothness assumptions; these monographs contain very profound results for the scalar case (due to De Giorgi, Nash, and Moser).

To Section 16

To Section 16.4

We largely follow Savaré's paper [325]. We discussed it very fruitfully with T. A. Suslina. She, in particular, suggested the use of relation (16.4.17) and noticed that the nonnegativity assumption (16.4.1) is not required for the Dirichlet problem, whereas for the Neumann problem it is required only near the boundary.

To Section 16.5

The assumption of Theorem 16.5.1 about the Lipschitz continuity of the coefficients and the inequality are borrowed from Gilbarg and Trudinger's book [167, Chap. 8]. But Gilbarg and Trudinger considered a scalar equation with real coefficients and used its specifics. Nirenberg assumed the coefficients $a_{j,k}$ and b_j to be C^1.

Nirenberg also considered higher-order equations and systems; under stronger smoothness assumptions on the coefficients and the boundary, he obtained stronger results on the smoothness of a solution. The variational solutions of the Dirichlet problem turn out to be smooth ordinary solutions, so that, in the general case, the strong ellipticity of a system implies the ellipticity of the Dirichlet problem.

Below we mention several directions of study of problems in Lipschitz domains which have not been reflected in our book.

1. We did not consider problems in domains complementary to bounded domains. Here the choice of a fundamental solution and the corresponding conditions on solutions at infinity are in the forefront. For example, in the case of the Helmholtz equation $\Delta u + k^2 u = 0,\ k > 0$, these are the radiation conditions, which can be found in textbooks on mathematical physics. Solutions, which locally belong to the same spaces as previously, obey similar conditions and, thanks to this, admit representations in terms of jumps on the boundary; see [14, 15, 101, 281, 282, 381].

2. Numerous important books and papers are devoted to problems in polyhedral domains and "curvilinear polyhedra." We have already mentioned, in particular, the monographs [173] by Grisvard, [117] by Dauge, and [253] by Maz'ya and Rossman.

3. *The Stokes system.* This is a linearization of the Navier–Stokes equations of motion of a viscous incompressible fluid. The stationary version of the Stokes system has the form

$$-\nu \Delta u + \mathrm{grad}\, p = f,$$
$$\mathrm{div}\, u = 0; \tag{19.4.1}$$

it can be considered in a bounded Lipschitz domain Ω. Here $u = (u_1, \ldots, u_n)'$ is the velocity of the fluid, p is the pressure, and ν is a positive constant (the kinematic coefficient of viscosity). The simplest boundary condition is the homogeneous Dirichlet condition

$$u^+ = 0. \tag{19.4.2}$$

Of interest for us here is the fact that the Stokes system is "*strongly elliptic in a subspace*" (to be more precise, the Gårding inequality is valid in this subspace), namely, in the subspace V of $\overset{\circ}{H}^1(\Omega)$ formed by all functions with zero divergence. Suppose that the functions u and v belong to V.

Multiplying the first equation by v and integrating by parts, we obtain

$$\nu \Phi_\Omega(u,v) = (f,v)_\Omega, \tag{19.4.3}$$

where

$$\Phi_\Omega(u,v) = \int_\Omega \nabla u \cdot \nabla \bar{v}\,dx, \qquad (19.4.4)$$

because

$$(\operatorname{grad} p, v)_\Omega = (p, \operatorname{div} v)_\Omega = 0.$$

Taking into account the zero boundary condition (and using Remark 8.1.3), we arrive at the inequality

$$\|u\|_V^2 \le C\Phi_\Omega(u,u); \qquad (19.4.5)$$

thus, if f belongs to the space dual to V, then a solution u exists and is unique in V. Knowing u, we can determine $p \in L_2(\Omega)$ from the first equation in (19.4.1).

Some further details can be found, e.g., in Temam's book [374]. The literature on Navier–Stokes and Stokes equations is vast. In particular, it is useful to read the paper [154] and the book [186]; see also the references therein.

4. The stationary *Maxwell system* has the form

$$\begin{aligned}\operatorname{rot} E - ikH &= 0,\\ \operatorname{rot} H + ikE &= J\end{aligned} \qquad (19.4.6)$$

in the simplest case (see, e.g., [203]). Here E and H are vector-valued functions in a three-dimensional domain, the complex amplitudes of the electric and magnetic fields; k is the wave number; and J is the density of extraneous currents. Although exterior problems and problems in $\mathbb{R}^n$ with transmission conditions on a closed compact surface are of great importance, there is an important problem in a bounded domain with boundary condition

$$\nu \times E = 0 \text{ on } \Gamma \qquad (19.4.7)$$

(a resonator with a metal wall; ν is the unit normal vector). The Maxwell system, as well as the Stokes system, is nonelliptic, but there are various approaches to reveal its "hidden" ellipticity or strong ellipticity. We mention that (19.4.6) directly implies (because $\operatorname{div}\operatorname{rot} = 0$) the relation $\operatorname{div} E = J_1$, whose restriction to the boundary can be added to the boundary condition (19.4.7):

$$\operatorname{div} E = J_1 \text{ on } \Gamma. \qquad (19.4.8)$$

Relations (19.4.6) imply also (because $\operatorname{rot}\operatorname{rot} = -\Delta + \operatorname{grad}\operatorname{div}$) the *vector Helmholtz equation*

$$\Delta E + k^2 E = F \text{ in } \Omega. \qquad (19.4.9)$$

As is easy to verify, here we have an elliptic problem for the vector Helmholtz equation with boundary conditions (19.4.7) and (19.4.8); this problem is not equivalent to the initial one, but it is interesting by itself.

The Maxwell system (in particular, in Lipschitz domains) has been studied in a huge number of works. We only mention [65], [100], [267], [271], and [276].

5. The development of the theory of classical problems for strongly elliptic equations in Lipschitz domains, beginning with the Laplace equation, in a direction quite different from our was initiated by Dahlberg, Calderón, Jerison, and Kenig; see, in particular, Calderón's paper [87], the survey [193] by Jerison and Kenig, and Kenig's monograph [208], which contains an extensive bibliography. This theory was not reflected in our book. This direction is known as the *Calderón program*. It has an extensive literature; some, but far from all, related works are included in our list of references. Here we very briefly describe the Calderón program and give several references.

The initial setting of the Dirichlet problem is as follows. The equation is homogeneous: $Lu = 0$. The right-hand side of the Dirichlet boundary condition $u^+ = g$ belongs to $L_2(\Gamma)$ or $H^1(\Gamma)$. There are no corresponding trace theorems in the framework of the theory of Sobolev spaces. *Nontangential convergence* defined below is used. The Dirichlet condition is understood in the sense of the pointwise convergence

$$u(y) \to g(x) \quad \text{for } K(x) \ni y \to x \qquad (19.4.10)$$

at almost all $x \in \Gamma$. Here $\{K(x)\}$ is a regular family of cones in Ω with vertices at all points of Γ (but we do not include the vertices in the cones). This convergence is controlled by the *nontangential maximal function*

$$u^*(x) = \sup |u(y)|, \quad y \in K(x), \qquad (19.4.11)$$

on Γ, for which the a priori estimate

$$\|u^*\|_{L_2(\Gamma)} \le C\|g\|_{L_2(\Gamma)} \qquad (19.4.12)$$

or a similar inequality between L_p-norms is derived in the case of unique solvability. In the second setting, a similar convergence of the first derivatives of the solution is required, and the nontangential maximal function of its gradient is majorized by the H^1-norm of g. A similar meaning is attached to the Neumann boundary condition $T^+u = h$ with right-hand side belonging to $L_2(\Gamma)$ or $L_p(\Gamma)$, for which again the nontangential maximal function of the gradient is estimated. The study is largely aimed at determining those p for which the unique solvability or the Fredholm property of the problems can be proved.

Thus, in this theory, the "starting" points are the boundary points $(\pm 1/2, 1/2)$ of our square Q, while we start at its center $(0, 1/2)$.

The most important, but difficult, technical tools are the single- and double-layer potential-type operators defined by the classical integral formulas. A substantial progress was made in the paper [153] on the Laplace equation in a C^1 domain. It was found out that a theorem on the boundedness of a singular integral operator on a Lipschitz surface was needed. Calderón [85] proved such a theorem for a Lipschitz surface with small Lipschitz constant, which gave rise to the important problem of generalizing it to arbitrary Lipschitz surfaces. This problem was solved by Coifman, McIntosh, and Meyer [96] for the case of equations with constant co-

efficients. To the general case of variable coefficients the result was generalized by
Mitrea and Taylor [273].

See also [86, 94, 95, 120, 121, 138, 156].

Verchota obtained in [382, 383] important results for the Laplace equation in a
Lipschitz domain on the basis of the theorem in [96]. To prove the solvability of
equations on the boundary, he used the Rellich identity combined with an approximation of the Lipschitz boundary by smooth boundaries (see Proposition 9.1.5), for
which the proof of solvability is comparatively simple.

Other authors used the Rellich identity as well; this determined the generality of
possible extensions. For the Dirichlet problem, the case of arbitrary second-order
strongly elliptic systems with formally self-adjoint principal part was successfully
treated; see [269]. For the Neumann problem, the cases of general scalar equations
and of the Lamé system, for which the required identity was specially proved, were
handled [115]. See also [112].

For the Laplace–Poisson equation and, later, the Beltrami–Laplace equation, the
problem of extending the results to as large as possible set of points (s, r) in the
square Q (in our notation) was studied. Problems for the Laplace equation were
considered in [195] and, later, in [155] and [405]; the most general results for the
Beltrami–Laplace equation were obtained by Mitrea and Taylor (see [273–275]).
Very deep results specific to the theory of harmonic functions were used, a difficult
analysis of the problems at the boundary points of Q (which we did not consider)
was implemented, and delicate interpolation theorems were applied. Positive results
were obtained for all points of the hexagon in Q bounded by the lines $s = r - \frac{1}{2} \pm \varepsilon$,
where $\varepsilon > 0$ depends on the domain Ω. It is impossible to obtain results for the
entire square Q in the general case. Relevant examples can be found, e.g., in [155]
and [195]; see also our Section 17.3. On the other hand, in the case of C^1 boundaries, the results are valid for the entire square Q; this can be deduced from results
of [153].

Special attention in related works was paid to the mixed problem [75, 171, 174,
177, 270], the oblique derivative problem [211, 231, 272, 303], the Robin problem [116, 227], problems for the Lamé system [112, 115], the Stokes system [154]
and [115], and the Maxwell system [267, 271, 276], and problems for higher-order
equations [5, 114, 304, 305, 384, 386]. The literature on this topic is vast. Our bibliography contains also [106–111, 113, 148, 150–152, 159, 162, 190–192, 194, 205–
207, 209, 212, 259, 268, 343–346]. But this list is far from complete.

We do not touch on certain spaces popular in modern analysis, namely, the spaces
of functions of bounded mean oscillation and of vanishing mean oscillation [158,
196, 323] and the Hardy spaces (see, e.g., [97, 158, 160, 181, 210, 275]). They play
an important role in approaching the boundary points of our rectangle Q.

Our Sections 11, 12, 16, and 17 essentially present an alternative theory, closer
to the classical setting of variational problems, to which the majority of authors
adhere. This theory goes back, in particular, to Costabel's papers [98, 99] and
McLean's book [258]. The personal contribution of the author to this theory includes the idea to apply the Weyl decompositions of the solution space (see Sections 11.2 and 11.4); theorems on the regularity of solutions (see Sections 16 and

17); relations between operators on the boundary (see Section 12.6); equations on the boundary for mixed problems (see Section 11.4); problems with conditions on a nonclosed surface (see Section 12.8); general transmission problems (see Section 12.10); spectral problems with a spectral parameter in the boundary or transmission conditions [14–17, 19, 20, 22–25]; some new remarks about the smoothness and completeness of generalized eigenfunctions (in Section 17.2), as well as some progress in obtaining optimal resolvent estimates (in Section 17.4). Unlike in papers [98, 99], the major part of the book [258], and the book [186, Chap. 5], we do not assume the coefficients to be infinitely smooth.

To Sections 11.9, 13.9, and 16.6

This material was included in the book the day before the book was sent to press in Moscow. The author thought over it together with A. M. Selitskii, and the new approach was elaborated jointly [33].

As we have already mentioned, the Kato problem has an extensive literature. Final results for the Dirichlet and Neumann problems in Lipschitz domains and some other problems have been obtained long ago; see, e.g., [47, 180, 257] and the references therein. Important results are due to Auscher and Tchamitchian, McIntosh, and other mathematicians. Our bibliography contains only several references to works in this direction; further references can be found therein.

The terminology in these works usually differs from ours, but it is explained, so that this difference should not impede reading.

The main attention there was given to the operator A_2, although the operator A_1 was also known, as well as the equivalences between statements concerning these operators mentioned in Section 16.6.

Our approach extensively uses the possibility of simultaneously considering these operators and certain special features of the solutions of the equation $A_1 u = f$. By using the operator A_1, the Kato problem for A_2 is solved, which, in turn, implies a desired result for A_1. Apparently, this approach deserves attention because of the simplicity and uniformity of considerations. In our opinion, results for the operator A_1 are as interesting as the solution of the Kato problem for the operator A_2.

In particular, we have succeeded in obviating the necessity of preliminary considerations in $\mathbb{R}^n$, a half-space, and unbounded "special" Lipschitz domains. The authors of [51, 53] used this way. Although the problems in $\mathbb{R}^n$ and unbounded domains or with unbounded potentials are also important, we did not consider them. However, results, e.g., in $\mathbb{R}^n$ for systems with coefficients of low smoothness are easy to obtain by replacing the assumption that the embeddings (11.9.1) are compact by an appropriate strong coercivity assumption on the corresponding form on H_1.

The purpose to fully minimize the smoothness assumptions was not pursued. As in [147], we need some low regularity of coefficients. About the higher-order coefficients it suffices to assume that these are multipliers in the spaces $H^s(\Omega)$ with small $|s|$. The "intermediate" coefficients can be assumed to be bounded and measurable (as well as the lower-order coefficients), at least, in the cases of the Dirichlet, Neumann, and mixed problems by virtue of the following argument from [147]. The full operator, e.g., in the Dirichlet problem is a weak perturbation of the operator without intermediate terms. After results for the latter are obtained, we use the fact that the intermediate coefficients do not affect the domain of the corresponding operator A_2 (see Section 11.9). By virtue of Theorem 13.9.3, they do not affect $D(A_2^{1/2})$ either, so that the equality $D(A_2^{1/2}) = H_1$ is preserved under the addition of the intermediate terms. As a consequence, a similar equality for $D(A_2^{*1/2})$ holds.

Next, we had not to restrict consideration to the scalar operator $\mathrm{div}(a(x)\nabla)$ or imitate its structure (cf. [147]). In the matrix case, we had not to construct a complicated factorization by using a Dirac-type operator (cf. [49]): the factorization (16.6.7) essentially sufficed. It was not required to consider separately the simplest mixed problem (cf. [54] and [255]). Our approach applies also to problems for higher-orders systems; cf. [48]. Apparently, boundary operators have not previously been discussed in the literature on the Kato problem.

We did not mention H^∞-bounded functional calculus (see, in particular, [256] and [49]); in fact, its presence for A_1 is equivalent to the validity of the main theorems of Section 16.6 (see Arendt's survey [37]), and this is an important fact.

The Kato problem was also considered in Banach spaces [46, 52, 187].

To Section 17.2

Considering the question about completeness, we essentially followed Agmon's paper [7] on smooth problems.

To Sections 17.4–17.6

We used analytic semigroups. Note that the approach with contractive semigroups is also interesting and fruitful for an important class of scalar equations, see [292]. See also [92]. Our approach covers more general systems but less general spaces.

An additional information on semigroups and their applications to nonstationary equations can be found, e.g., in the classical monograph [179] by Hille and Phillips and in Yosida's textbook [403], as well as in [37, 80, 104, 122, 176].

19.5 To Section 18

To Section 18.1

The operators called now Fredholm operators were previously known as Noetherian operators (after F. Noether, who introduced them in the early 1920s in studying one-dimensional singular integral equations); the term *Fredholm* was used for Noetherian operators of index zero. The reader must have heard about Fredholm's study of integral equations.

To Section 18.3

Important results on the spectral theory of elliptic equations and problems are contained in Edmunds and Triebel's monograph [142] and in the works [123–125] by Davies and [166] by Geymonat and Grisvard.

To Section 18.4

We mention the classical monographs [264] by Mikhlin and [280] by Muskhel-ishvili on singular integral equations.

The way to the ΨDO calculus was paved by Calderón and Zygmund's paper [89], which considered partial differential operators and singular integral operators. The calculus itself appeared, in almost final form, in the paper [213] by Kohn and Niren-berg.

References

1. Achieser, N.I., Glasman, I.M.: Theory of Linear Operators in Hilbert Space. Pitman, London (1981)
2. Adams, R.A.: Sobolev Spaces. Academic, Boston (1975)
3. Adams, R.D., Aronszajn, N., Hanna, M.S.: Theory of Bessel potentials, III: Potentials on regular manifolds. Ann. Inst. Fourier (Grenoble) **19** (2), 279–338 (1969)
4. Adams, R.A., Aronszajn, N., Smith, K.T.: Theory of Bessel potentials, II. Ann. Inst. Fourier (Grenoble) **17**, 1–135 (1967)
5. Adolfsson, V., Pipher, J.: The inhomogeneous Dirichlet problem for Δ^2. J. Funct. Anal. **159**, 137–190 (1998)
6. Agmon, Sh.: The coerciveness problem for integro-differential forms. J. Anal. Math. **6**, 183–223 (1958)
7. Agmon, Sh.: On the eigenfunctions and on the eigenvalues of general elliptic boundary value problems. Commun. Pure Appl. Math. **15**, 119–147 (1962)
8. Agmon, Sh.: Lectures on Elliptic Boundary Value Problems. Van Nostrand, Princeton, N.J. (1965)
9. Agmon, Sh., Douglis, A., Nirenberg, L.: Estimates near the boundary for solutions of elliptic partial differential equations satisfying general boundary conditions, I, II. Commun. Pure Appl. Math. **12**, 623–729 (1959); **17**, 35–92 (1964)
10. Agmon, Sh., Nirenberg, L.: Properties of solutions of ordinary differential equations in Banach space. Commun. Pure Appl. Math. **16** (2), 121–239 (1963)
11. Agranovich, M.S.: Elliptic singular integro-differential operators. Russian Math. Surveys **20** (5), 1–121 (1965)
12. Agranovich, M.S.: Elliptic operators on closed manifolds. In: Encyclopaedia of Mathematical Sciences, vol. 63, pp. 1–130. Springer, Berlin (1994)
13. Agranovich, M.S.: Elliptic boundary problems. In: Encyclopaedia of Mathematical Sciences, vol. 79, pp. 1–144. Springer, Berlin (1997)
14. Agranovich, M.S.: Spectral properties of potential type operators for a certain class of strongly elliptic systems on smooth and Lipschitz surfaces. Trans. Mosc. Math. Soc. **62**, 1–47 (2001)
15. Agranovich, M.S.: Spectral problems for second-order strongly elliptic systems in smooth and non-smooth domains. Russian Math. Surveys **57** (5), 847–920 (2002)
16. Agranovich, M.S.: Regularity of variational solutions to linear boundary-value problems in Lipschitz domains. Funct. Anal. Appl. **40** (4), 313–329 (2006)
17. Agranovich, M.S.: To the theory of the Dirichlet and Neumann problems for strongly elliptic systems in Lipschitz domains. Funct. Anal. Appl. **41** (4), 247–263 (2007)
18. Agranovich, M.S.: Obobshchennye funktsii (Generalized Functions). MTsNMO, Moscow (2008)

© Springer International Publishing Switzerland 2015

M.S. Agranovich, *Sobolev Spaces, Their Generalizations and Elliptic Problems in Smooth and Lipschitz Domains*, Springer Monographs in Mathematics, DOI 10.1007/978-3-319-14648-5

19. Agranovich, M.S.: Spectral problems in Lipschitz domains for strongly elliptic systems in Banach spaces H_p^s and B_p^s. Funct. Anal. Appl. **42** (4), 249–267 (2008)
20. Agranovich, M.S.: Potential type operators and transmission problems for strongly elliptic second-order systems in Lipschitz domains. Funct. Anal. Appl. **43** (3), 165–183 (2008)
21. Agranovich, M.S.: Remarks on potential spaces and Besov spaces in a Lipschitz domain and on Whitney arrays on its boundary. Russian J. Math. Phys. **15** (2), 146–155 (2008)
22. Agranovich, M.S.: Strongly elliptic second-order systems with boundary conditions on a nonclosed Lipschitz surface. Funct. Anal. Appl. **45** (1), 1–12 (2011)
23. Agranovich, M.S.: Mixed problems in a Lipschitz domain for strongly elliptic second-order systems. Funct. Anal. Appl. **45** (2), 81–98 (2011)
24. Agranovich, M.S.: Remarks on strongly elliptic second-order systems in Lipschitz domains. Russian J. Math. Phys. **20** (4), 5–16 (2012)
25. Agranovich, M.S.: Spectral problems in Lipschitz domains. J. Math. Sci. (N. Y.) **190** (1), 8–33 (2013)
26. Agranovich, M.S.: Elliptic Pseudodifferential Operators and Spectral Problems. A project.
27. Agranovich, M.S., Amosov, B.A.: Estimates for s-numbers and spectral asymptotics for integral operators of potential type on nonsmooth surfaces. Funct. Anal. Appl. **30** (2), 75–89 (1996)
28. Agranovich, M.S., Amosov, B.A., Levitin, M.: Spectral problems for the Lamé system with spectral parameter in boundary conditions on smooth or nonsmooth boundary. Russian J. Math. Phys. **6** (3), 247–281 (1999)
29. Agranovich, M., Denk, R., Faierman, M.: Weakly smooth nonselfadjoint spectral elliptic boundary problems. In: Demuth, M., Schrohe, E., Schulze, B.-W., Sjöstrand, J. (eds.) Spectral Theory, Microlocal Analysis, Singular Manifolds, pp. 138–199. Akademie, Berlin (1997)
30. Agranovich, M.S., Dynin, A.S.: General boundary-value problems for elliptic systems in multidimensional domains. Sov. Math., Dokl. **3**, 1323–1327 (1962)
31. Agranovich, M.S., Katsenelenbaum, B.Z., Sivov, A.N., Voitovich, N.N.: Generalized Method of Eigenoscillations in Diffraction Theory. Wiley-VCH, Berlin (1999)
32. Agranovich, M.S., Mennicken, R.: Spectral properties for the Helmholtz equation with spectral parameter in boundary conditions on a nonsmooth surface. Russian Acad. Sci. Sb. Math. **190** (1), 29–69 (1999)
33. Agranovich, M.S., Selitskii, A.M.: Fractional powers of operators corresponding to coercive problems in Lipschitz domains. Funkts. Anal. Appl. **47** (2), 83–95 (2013)
34. Agranovich, M.S., Vishik, M.I.: Elliptic problems with a parameter and parabolic problems of general type. Russian Math. Surveys **19** (3), 53–157 (1964)
35. Agranovich, M.S., Volevich, L.R., Dynin, A.S.: Solvability of general boundary-value problems for elliptic systems in domains of arbitrary dimension. In: Abstracts of the Joint Soviet-American Symposium on Partial Differential Equations, Novosibirsk (1963)
36. Alvino, A., Trombetti, G.: Problemi ellitici dipendenti di un parametro in L_p. Ricerche Mat. **26**, 335–348 (1977)
37. Arendt, W.: Semigroups and Evolution Equations, Functional Calculus, Regularity and Kernel Estimates. In: Handbook of Differential Equations: Evolutionary Differential Equations, vol. 1, pp. 1–85. North-Holland, Amsterdam (2004)
38. Arendt, W., Elst, A.F.M. ter: Gaussian estimates for second order elliptic operators with boundary conditions. J. Operator Theory **38**, 87–130 (1997)
39. Aronszajn, N.: Boundary values of functions with finite Dirichlet integral. In: Studies in eigenvalue problems. Conference on Partial Differential Equations, University of Kansas, Summer 1954
40. Aronszajn, N.: On coercive integro-differential quadratic forms. In.: Studies in eigenvalue problems. Conference on Partial Differential Equations, University of Kansas, Summer 1954
41. Aronszajn, N., Milgram, A.N.: Differential operators on Riemannian manifolds. Rendiconti Circ. Mat. Palermo **2** (3), 266–325 (1953)
42. Aronszajn, N., Mulla, F., Szeptycki, P.: On spaces of potentials connected with L^p classes. Ann. Inst. Fourier (Grenoble) **13**, 211–306 (1963)

43. Aronszajn, N., Smith, K.T.: Theory of Bessel potentials, I. Ann. Inst. Fourier (Grenoble) 11, 385–475 (1961)
44. Aronszajn, N., Szeptycki, P.: Theory of Bessel potentials, IV: Potentials on subcartesian spaces with singularities of polyhedral type. Ann. Inst. Fourier (Grenoble) 25 (3–4), 27–69 (1975)
45. Atiyah, M.F., Singer, I.M.: The index of elliptic operators on compact manifolds. Bull. Amer. Math. Soc. 69, 422–433 (1963)
46. Auscher, P., Badr, N., Haller-Dintelmann, R., Rehberg, J.: The square root problem for second-order, divergence form operators with mixed boundary conditions on L^p. To appear in J. Evol. Eq.
47. Auscher, P., Hofmann, S., Lacey, M., Lewis, J., McIntosh A., Tchamitchian P.: The solution of Kato's conjectures. C.R. Acad. Sci. Paris, Sér. 1 332, 601–606 (2001)
48. Auscher P., Hofmann S., McIntosh, A., Tchamitchian, P.: The Kato square root problem for higher order elliptic operators and systems on $\mathbb{R}^n$. J. Evolution Equations 1 (4), 361–385 (2001)
49. Auscher, P., McIntosh, A., Nahmod, A.: Holomorphic functional calculi of operators, quadratic estimates and interpolation. Indiana Univ. Math. J. 46 (2), 375–403 (1997)
50. Auscher, P., Mourgoglou, M.: Boundary layers, Rellich estimates and extrapolation of solvability for elliptic systems. http://arxiv.org/abs/1305.4115v1 (2007). Accessed 18 Jun 2013.
51. Auscher, P., Tchamitchian, Ph.: Square Root Problem for Divergence Operators and Related Topics. Astérisque 249 (1998)
52. Auscher, P., Tchamitchian, Ph.: Square roots of elliptic second order divergence operators on strongly Lipschitz domains: L^p theory. Math. Ann. 320 (3), 577–623 (2001)
53. Auscher, P., Tchamitchian, Ph.: Square roots of elliptic second order divergence operators on strongly Lipschitz domains: L^2 theory. J. Anal. Math. 90, 1–12 (2003)
54. Axelsson, A., Keith, S., McIntosh, A.: The Kato square root problem for mixed boundary value problems. J. London Math. Soc. 74 (1), 113–130 (2006)
55. Azizov, T.Ya., Kopachevskii, N.D.: Abstraktnaya formula Grina i ee prilozheniya (Green's Abstract Formula and Its Applications), Special Course. Tavricheskii Natsional'nyi Universitet, Simferopol' (2011)
56. Babich, V.I.: On extension of functions. Uspekhi Mat. Nauk 8 (2), 111–113 (1953)
57. Bennett, C., Sharpley, R.: Interpolation of Operators. Academic, Boston (1988)
58. Berezanskii, Yu.M.: Expansions in Eigenfunctions of Self-Adjoint Operators. Translations of Mathematical Monographs, vol. 17. Amer. Math. Soc., Providence, R.I. (1968)
59. Berezanskii, Yu.M., Krein, S.G., Roitberg, Ya.A.: Theorem on homeomorphisms and local increase of smoothness of solutions to elliptic equations up to the boundary. Sov. Math., Dokl. 4 152–155 (1963)
60. Bergh, J., Löfström, J.: Interpolations Spaces: An Introduction. Springer, Berlin (1976)
61. Besov, O.V.: On some families of functional spaces. Imbedding and extension theorems. Dokl. Akad. Nauk SSSR 126 (6), 1163–1165 (1959)
62. Besov, O.V.: Investigation of a class of function spaces in connection with imbedding and extension theorems. Amer. Math. Soc. Transl. 40, 85–126 (1964)
63. Besov, O.V., Il'in, V.P., Nikol'skii, S.M.: Integral Representations of Functions and Imbedding Theorems, vols. 1, 2. Winston, Washington, D.C. (1978, 1979)
64. Birman, M.S., Solomyak, M.Z.: Asymptotics of the spectrum of differential equations. J. Soviet Math. 12, 247–283 (1979)
65. Birman, M.S., Solomyak, M.Z.: L_2-theory of the Maxwell operator in arbitrary domains. Russian Math. Surveys 42 (6), 75–96 (1987)
66. Birman, M.Sh., Solomyak, M.Z.: Spectral Theory of Self-Adjoint Operators in Hilbert Space. Kluwer, Dordrecht (1987)
67. Bitsadze, A.V.: On the uniqueness of the solution of the Dirichlet problem for elliptic partial differential equations. Uspekhi Mat. Nauk 3 (6(28)), 241–242 (1948)
68. Boutet de Monvel, L., Opérateurs pseudo-différentiels analytiques et problèmes aux limites elliptiques. Ann. Inst. Fourier (Grenoble) 19, 169–268 (1969)

69. Boutet de Monvel, L., Boundary problems for pseudodifferential operators. Acta Math. **126**, 11–51 (1971)
70. Boyd, D.V.: The spectrum of a Cesàro operator. Acta Sci. Math. (Szeged) **29**, 31–34 (1968)
71. Browder, F.: On the regularity properties of solutions of elliptic differential equations. Commun. Pure Appl. Math. **9**, 351–361 (1956)
72. Browder, F.: Estimates and existence theorems for elliptic boundary value problems. Proc. Nat. Acad. Sci. U.S.A. **45** (3), 365–372 (1959)
73. Browder, F.: A priori estimates for solutions of elliptic boundary value problems, I, II, III. Koninkl. Nederl. Akad. Wetenschap. **22**, 145–159, 160–169 (1960); Indag. Math. **23**, 404–410 (1961)
74. Browder, F.: A continuity property for adjoints of closed operators in Banach spaces and its applications to elliptic boundary value problems. Duke Math. J. **28** (2), 157–182 (1961)
75. Brown, R.M.: The mixed problem for Laplace's equation in a class of Lipschitz domains. Commun. Partial Differ. Equations **19** (7–8), 1217–1233 (1994)
76. Brudnyi, Yu.A., Krein, S.G., Semenov, E.M.: Interpolation of linear operators. J. Soviet Math. **42** (6), 2009–2112 (1988)
77. Brudnyi, Yu.A., Krugljak, N.A.: Interpolation Functors and Interpolation Spaces, vol. I. North-Holland, Amsterdam (1991)
78. Burak, T.: On spectral projections of elliptic operators. Ann. Scuola Norm. Sup. Pisa (3) **24**, 209–230 (1970)
79. Burgoyne J.: Denseness of the generalized eigenvectors of a discrete operator in Banach space. J. Operator Theory **33**, 279–297 (1995)
80. Butzer, P.L., Berens, H.: Semi-Groups of Operators and Approximation. Springer, Berlin (1967)
81. Calderón, A.: Lebesgue spaces of differentiable functions and distributions. In: Proceedings of Symposia in Pure Mathematics, vol. 4: Partial Differential Equations, pp. 33–49. Amer. Math. Soc., Providence, R.I. (1961)
82. Calderón, A.P.: Boundary-value problems for elliptic equations. In: Outlines of the Joint Soviet-American Symposium on Partial Differential Equations, Novosibirsk, 1963
83. Calderón, A.P.: Algebras of singular integral operators. In: Proceedings of Symposia in Pure Mathematics, vol. 10, pp. 18–55. Amer. Math, Soc., Providence, R.I. (1967)
84. Calderón, A.: Intermediate spaces and interpolation, the complex method. Studia Math. **24**, 113–190 (1964)
85. Calderón, A.: Cauchy integrals on Lipschitz curves and related operators. Proc. Nat. Acad. Sci. USA **74**, 1324–1327 (1977)
86. Calderón, A.P., Calderón, C.P., Fabes, E., Jodeit, M., Riviere, N.M.: Applications of the Cauchy integral on Lipschitz curves. Bull. Amer. Math. Soc. **84** (2), 287–290 (1978)
87. Calderón, A.: Boundary value problems for the Laplace equation in Lipschitz domains. In: Recent Progress in Fourier Analysis, pp. 33–48. North-Holland, Amsterdam, (1983)
88. Calderón, A., Zygmund, A.: On the existence of certain singular integrals. Acta Math. **88**, 85–139 (1952)
89. Calderón, A.P., Zygmund, A.: Singular integral equations and differential operators. Amer. J. Math. **79**, 901–921 (1957)
90. Cao, W., Sager, Y.: Stability of Fredholm properties on interpolation spaces. Arkiv Math. **28** (2), 249–258 (1990)
91. Chen, Ya-Zhe, Wu, Lan-Cheng: Second Order Elliptic Equations and Elliptic Systems. Amer. Math. Soc., Providence, R.I. (1998)
92. Chialdea, A., Maz'ya, V.: Semi-Bounded Partial Differential Operators. Springer, Berlin (in press)
93. Christ, M.: Lectures on Singular Integral Operators. Amer. Math. Soc., Providence, R.I. (1989)
94. Coifman, R.R., David, G., Meyer, Y.: La solution des conjectures de Calderón. Adv. Math. **48** (2), 144–148 (1983)
95. Coifman, R.R., Jones, P.W., Semmes, S.: Two elementary proofs of the L^2 boundedness of Cauchy integrals on Lipschitz curves J. Amer. Math. Soc. **2** (3), 553–564 (1989)

96. Coifman, R., McIntosh, A., Meyer, Y.: L'intégrale de Cauchy définit un opérateur borné sur L^2 pour les courbes Lipschitziennes. Ann. Math. **116**, 361–388 (1982)

97. Coifman, R.R., Weiss, G.: Extensions of Hardy spaces and their use in analysis. Bull. Amer. Math. Soc. **83**, 569–645 (1977)

98. Costabel, M.: Starke Ellipticität von Randintegraloperatoren Erster Art. Habilitation Thesis, Technische Hochschule Darmstadt, Preprint 868 (1984)

99. Costabel, M.: Boundary integral operators on Lipschitz domains: Elementary results. SIAM J. Math. Anal. **19** (3), 613–626 (1988)

100. Costabel, M.: A coercive bilinear form for Maxwell's equations. J. Math. Anal. Appl. **157**, 527–541 (1991)

101. Costabel, M., Dauge, M.: On representation formulas and radiation conditions, Math. Methods Appl. Sci. **20**, 133–150 (1997)

102. Costabel, M., Stephan, E.: A direct boundary integral equation method for transmission problems. J. Math. Anal. Appl. **106**, 367–413 (1985)

103. Costabel, M., Wendland, W.L.: Strong ellipticity of boundary integral operators. J. Reine Angew. Math. **372**, 34–63 (1986)

104. Cowling, M.G.: Harmonic analysis on semigroups. Ann. of Math. **117**, 267–283 (1983)

105. Cwickel, M.: Real and complex interpolation and extrapolation of compact operators. Duke Math. J. **65** (2), 503–538 (1992)

106. Dahlberg, B.E.J.: Estimates of harmonic measure. Arch. Ration. Mech. Anal. **65** (3), 275–288 (1977)

107. Dahlberg, B.E.J.: On the Poisson integral for Lipschitz and C^1-domains. Studia Math. **66**, 13–32 (1979)

108. Dahlberg, B.E.J.: L^q-estimates for Green potentials in Lipschitz domains. Math. Scand. **44**, 149–170 (1979)

109. Dahlberg, B.E.J.: Weighted norm inequalities for the Lusin area integral and the nontangential maximal functions for functions harmonic in a Lipschitz domain. Studia Math. **27**, 297–314 (1980)

110. Dahlberg, B.E.J., Jerison, D., Kenig, C.E.: Area integral estimates for elliptic differential operators with non-smooth coefficients. Ark. Mat. **22**, 97–108 (1984)

111. Dahlberg, B.E.J., Kenig, C.E.: Hardy spaces and the Neumann problem in L^p for Laplace's equation in Lipschitz domains. Ann. Math. **125**, 437–466 (1987)

112. Dahlberg, B.E.J.; Kenig, C.E.: L^p estimates for the three-dimensional systems of elastostatics on Lipschitz domains. In: Analysis and Partial Differential Equations, Lecture Notes in Pure and Applied Mathematics, vol. 122, pp. 621–634. Dekker, New York (1990)

113. Dahlberg, B.E.J., Kenig, C.E., Pipher, J., Verchota, G.C.: Area integral estimates for higher order elliptic equations and systems. Ann. Inst. Fourier (Grenoble) **47** (5), 1425–1461 (1997)

114. Dahlberg, B., Kenig, C., Verchota, G.: The Dirichlet problem for the biharmonic equation in a Lipschitz domain. Ann. Inst. Fourier (Grenoble) **36** (3), 109–135 (1986)

115. Dahlberg, B., Kenig, C., Verchota, G.: Boundary value problems for the system of elastostatics in Lipschitz domains. Duke Math. J. **57**, 795–818 (1988)

116. Daners, D.: Robin boundary value problems on arbitrary domains. Trans. Amer. Math. Soc. **352** (9), 4207–4236 (2000)

117. Dauge, M.: Elliptic Boundary Value Problems on Corner Domains—Smoothness and Asymptotics of Solutions. Lecture Notes in Mathematics, vol. 1341. Springer, Berlin (1988)

118. Dautray, R., Lions, J.-L.: Mathematical Analysis and Numerical Methods for Science and Technology, vol. 2: Functional and Variational Methods. Springer, Berlin (1988)

119. Dautray, R., Lions, J.-L.: Mathematical Analysis and Numerical Methods for Science and Technology, vol. 5: Evolution Problems. Springer, Berlin (1992)

120. David G.: Wavelets and Singular Integrals on Curves and Surfaces. Springer, Berlin (1991)

121. David, G., Journé, J.-L.: A boundedness criterion for generalized Calderón–Zygmund operators. Ann. Math. **120**, 371–397 (1984)

122. Davies, E.B.: One-Parameter Semigroups. Academic, London (1980)

123. Davies, E.B.: Heat Kernels and Spectral Theory. Cambridge Univ. Press, Cambridge (1989)

124. Davies, E.B.: Limits of L^p regularity of self-adjoint elliptic operators. J. Differ. Equations **135**, 83–102 (1997)

125. Davies, E.B. L^p spectral theory of higher order elliptic differential operators. Bull. London Math. Soc. **29** (5), 513–546 (1997)

126. Denk, R., Mennicken, R., Volevich, L.: The Newton polygon and elliptic problems with parameter. Math. Nachr. **192**, 125–157 (1998)

127. Denk, R., Volevich, L.: Elliptic boundary value problems with large parameter for mixed order systems. In: Partial Differential Equations, pp. 29–64. Amer. Math. Soc., Providence, R.I. (2002)

128. Denk, R., Volevich, L.: Parameter-elliptic boundary value problems connected with the Newton polygon. Differ. Integral Equations **15** (3), 289–326 (2002)

129. Dikanskii, A.S.: Conjugate problems of elliptic differential and pseudodifferential boundary value problems in a bounded domain. Math. USSR, Sb. **20** (1), 67–83 (1973)

130. Dmitriev, V.I., Krein, S.G., Ovchinnikov, V.I.: Fundamentals of the theory of interpolation of linear operators. In: Geometriya lineinykh prostranstv i teoriya operatorov (Geometry of Linear Spaces and Operator Theory), pp. 31–74. Yaroslavskii Gosudarstvennyi Universitet, Yaroslavl' (1977)

131. Douglis, A., Nirenberg, L.: Interior estimates for elliptic systems of partial differential equations. Commun. Pure Appl. Math. **8**, 503–538 (1955)

132. Dubrovin, B.A., Novikov, S.P., Fomenko, A.T.: Modern Geometry—Methods and Applications, parts 1, 2, 3. Springer, Berlin (1982, 1985, 1990)

133. Duduchava, R.: The Green formula and layer potentials. Integral Equations Oper. Theory **41** (2), 127–178 (2001)

134. Duduchava, R., Natroshvili, D.: Mixed crack type problem in anisotropic elasticity. Math. Nachr. **191**, 83–107 (1998)

135. Duduchava, R.V., Natroshvili, D.G, Shargorodsky, E.M.: Boundary-value problems of the mathematical theory of cracks. Tr. Inst. Prikl. Mat. im. Vekua Tbilis. Univ. **39**, 68–84 (1990)

136. Duduchava, R., Wendland, W.L.: The Wiener–Hopf method for systems of pseudodifferential equations with an application to crack problems. Integral Equations Oper. Theory **23**, 294–335 (1995)

137. Dunford, N., Schwartz, J.T.: Linear Operators, part 1: General Theory, part 2: Spectral Theory. Interscience, New York (1958, 1963)

138. Duong X.T., McIntosh, A.: Singular integral operators with non-smooth kernels on irregular domains. Rev. Mat. Iberoam. **15**, 233–255 (1999)

139. Dynin, A.S.: Singular operators of arbitrary order on a manifold. Sov. Math., Dokl. **2**, 1375–1377 (1961)

140. Dynin, A.S.: Multidimensional elliptic boundary-value problems with one unknown function. Sov. Math., Dokl. **2**, 1431–1433 (1961)

141. Dynin, A.S.: Fredholm elliptic operators on manifolds. Uspekhi Mat. Nauk **17** (2), 194–195 (1962)

142. Edmunds, D.E., Triebel, H.: Function Spaces, Entropy Numbers and Differential Operators. Cambridge Univ. Press, Cambridge (1996)

143. Egorov, Yu.V., Shubin, M.A.: Linear partial differential equations: Elements of the modern theory. Encyclopaedia of Mathematical Sciences, vol. 31, pp. 1–120. Springer, Berlin (1988)

144. Egorov, Yu.V., Shubin, M.A.: Linear Partial Differential Equations: Foundations of Classical Theory. Encyclopaedia of Mathematical Sciences, vol. 30. Springer, Berlin (1992)

145. Eidel'man, S.D.: Parabolic equations. Encyclopaedia of Mathematical Sciences, vol. 63, pp. 203–316. Springer, Berlin (1991)

146. Eidel'man, S.D.: Parabolic systems. North-Holland, Amsterdam (1999)

147. Elst, A.F.M. ter, Robinson D.W.: On Kato's square root problem. Hokkaido Math. J. **26**, 365–376 (1997)

148. Escauriaza, L., Mitrea, M.: Transmission problems and spectral theory for singular integral operators on Lipschitz domains. J. Funct. Anal. **216** (1), 141–171 (2004)

149. Eskin, G.I.: Boundary Value Problems for Elliptic Pseudodifferential Equations. Amer. Math. Soc., Providence, R.I. (1981)

150. Fabes, E. B.: Layer potential methods for boundary value problems on Lipschitz domains. In: Potential Theory—Surveys and Problems, pp. 55–80. Springer, Berlin (1988)
151. Fabes, E.B., Jerison, D.S., Kenig, C.E.: Necessary and sufficient conditions for absolute continuity of elliptic-harmonic measure. Ann. Math. **119** (1), 121–141 (1984)
152. Fabes, E.B., Jodeit, M., Jr., Lewis, J.E.: Double layer potentials for domains with corners and edges. Indiana Univ. Math. J. **26** (1), 95–114 (1977)
153. Fabes, E., Jodeit, M., Jr., Riviére, N.: Potential techniques for boundary value problems on C^1 domains. Acta Math. **141**, 165–186 (1978)
154. Fabes, E., Kenig, C., Verchota, G.: The Dirichlet problem for the Stokes system on Lipschitz domains. Duke Math. J. **57** (3), 769–793 (1988)
155. Fabes, E., Mendez, O., Mitrea, M.: Boundary layers on Sobolev–Besov spaces and Poisson's equation for the Laplacian in Lipschitz domains. J. Funct. Anal. **159** (2), 323–368 (1998)
156. Fabes, E.B., Mitrea, I., Mitrea, M.: On the boundedness of singular integrals. Pacif. J. Math. **189**, 21–29 (1999)
157. Fabrikant, V.I.: Mixed Boundary Value Problems of Potential Theory and Their Applications in Engineering. Kluwer, Dordrecht (1991)
158. Fefferman, R.: Some applications of Hardy spaces and BMO in harmonic analysis and partial differential equations. In: Harmonic Analysis and Partial Differential Equations: Proceedings of a Conference, Boca Raton, Fl., 4–5 April 1988. Contemporary Mathematics, vol. 107, pp. 61–69. American Math. Soc., Providence, R.I. (1990)
159. Fefferman, R. A., Kenig, C. E., Pipher, J.: The theory of weights and the Dirichlet problem for elliptic equations. Ann. of Math. **134**, 65–124 (1991)
160. Fefferman, C.L., Stein, E.M.: H^p spaces of several variables. Acta Math. **129**, 137–193 (1972)
161. Franke, J., Runst, T.: Regular elliptic boundary value problems in Besov–Triebel–Lizorkin spaces. Math. Nachr. **174**, 113–149 (1995)
162. Gao, W.: Layer potentials and boundary value problems for elliptic systems in Lipschitz domains. J. Funct. Anal. **95** (2), 377–399 (1991)
163. Gårding, L.: Dirichlet's problem for linear elliptic partial differential equations. Math. Scand. **1**, 55–72 (1953)
164. Gel'fand, I.M.: On elliptic equations. Russian Math. Surveys **15** (3), 113–123 (1960)
165. Gel'fand, I.M., Shilov, G.E.: Generalized Functions, vol. 1: Generalized Functions and Operators on Them. Academic, New York (1964)
166. Geymonat, G., Grisvard, P.: Alcuni risultati di teoria spettrale per i problemi al limiti lineari ellittici. Rend. Sem. Mat. Univ. Padova **38**, 121–173 (1967)
167. Gilbarg, D., Trudinger, N.S.: Elliptic Partial Differential Equations of Second Order, 2nd. ed. Springer, Berlin (1983)
168. Gindikin, S.G., Volevich, L.R.: Distributions and Convolution Equations. Gordon & Breach, Philadelphia, Pa. (1992)
169. Gohberg, I.Ts., Kreĭn, M.G.: Introduction to the Theory of Linear Nonselfadjoint Operators in Hilbert Spaces. Amer. Math. Soc., Providence, R.I. (1969)
170. Goluzin, G.M.: Geometricheskaya teoriya funktsii kompleksnogo peremennogo (Geometric Theory of Functions of a Complex Variable), 2nd ed. Nauka, Moscow (1966)
171. Grippentrog, J., Kaiser, H.-C., Rehberg, J.: Heat kernel and resolvent properties for second order elliptic differential operators with general boundary conditions on L_p. Advances Math. Sci. Appl. **11** (1), 87–112 (2001)
172. Grisvard, D.: Caractérisation de quelques espaces d'interpolation. Arc. Rat. Mech. Anal. **25**, 40–63 (1967)
173. Grisvard, P.: Elliptic Problems in Nonsmooth Domains. Pitman, Boston (1985)
174. Gröger, K., Rehberg, J.: Resolvent estimates in $W^{-1,p}$ for second order elliptic differential operators in case of mixed boundary problems. Math. Ann. **285**, 105–113 (1989)
175. Grubb, G.: Functional Calculus of Pseudodifferential Boundary Problems, 2nd ed. Birkhäuser, Boston (1996)
176. Haase M.: The Functional Calculus for Sectorial Operators. Birkhäuser, Basel (2006)

177. Haller-Dintelmann, R., Kaiser, H.-C., Rehberg, J.: Elliptic model problems including mixed boundary conditions and material heterogeneities. J. Math. Pures Appl. **89**, 25–48 (2008)
178. Hestenes, M.R.: Extension of the range of a differentiable function. Duke Math. J. **8**, 183–192 (1941)
179. Hille, E., Phillips, R.: Functional Analysis and Semi-Groups. Colloquium Publications, vol. 31. Amer. Math. Soc., Providence, R.I. (1957)
180. Hofmann S.: A short course on the Kato problem. In: Contemporary Mathematics, vol. 289, pp. 61–67. Amer. Math. Soc., Providence, R.I. (2001)
181. Hoffman, S., Mayboroda, S., McIntosh, A.: Second order elliptic equations with complex bounded measurable coefficients in L^p, Sobolev and Hardy spaces. Ann. Sci. Éc. Norm. Supér. (4) **44** (5), 723–800 (2011)
182. Hörmander, L.: Estimates for Translation Invariant Spaces in L^p spaces. Acta Math. **104**, 93–140 (1960)
183. Hörmander, L.: Linear Partial Differential Operators. Springer, Berlin (1963)
184. Hörmander, L.: The spectral function of an elliptic operator. Acta Math. **121**, 193–218 (1968)
185. Hörmander, L.: The Analysis of Linear Partial Differential Operators, vols. 1–4. Springer, Berlin (1983, 1985)
186. Hsiao, G.C., Wendland, W.L.: Boundary Integral Equations. Springer, Berlin (2008)
187. Hytönen, T., McIntosh, A., Portal, P.: Kato's square root problem in Banach spaces. J. Funct. Anal. **254** (3), 675–726 (2008)
188. Ivrii, V.Ja.: Accurate spectral asymptotics for elliptic operators that act in vector bundles. Funct. Anal. Appl. **16** (2), 101–108 (1982)
189. Janson, S., Nilsson, P., Peetre, J.: Notes on Wolff's note on interpolation spaces. Proc. London Math. Soc. **48**, 283–299 (1984)
190. Jerison, D.S., Kenig, C.E.: An identity with applications to harmonic measure. Bull. Amer. Math. Soc. **2** (3), 447–451 (1980)
191. Jerison, D.S., Kenig, C.E.: The Dirichlet problem in non-smooth domains. Ann. Math. **113** (2), 367–382 (1981)
192. Jerison, D.S., Kenig, C.E.: The Neumann problem on Lipschitz domains. Bull. Amer. Math. Soc. **4** (2), 203–207 (1981)
193. Jerison, D.S., Kenig, C.E.: Boundary value problems on Lipschitz domains. In: Studies in partial differential equations, pp. 1–68. MAA Studies in Mathematics, vol. 23. Mathematical Association of America, Washington, D.C. (1982)
194. Jerison, D.S., Kenig, C.E.: Boundary behavior of harmonic functions in non-tangentially accessible domains. Adv. Math. **46** (1), 80–147 (1982)
195. Jerison, D., Kenig, C.: The inhomogeneous Dirichlet problem in Lipschitz domains. J. Funct. Anal. **130**, 164–219 (1995)
196. John, F., Nirenberg, L.: On functions of bounded mean oscillation. Commun. Pure Appl. Math. **14**, 415–426 (1961)
197. Jonsson, A., Wallin, H.: Function Spaces on Subsets of $\mathbb{R}^n$. Mathematical Reports, vol. 2, part 1. Academic, Harwood (1984)
198. Kalton, N., Mitrea, M.: Stability results on interpolation scales of quasi-Banach spaces and applications. Trans. Amer. Math. Soc. **350**, 3903–3922 (1998)
199. Kato, T.: Fractional powers of dissipative operators. J. Math. Soc. Japan **13** (3), 246–274 (1961)
200. Kato, T.: A generalization of the Heinz inequality. Proc. Japan Acad. **37** (6), 305–308 (1961)
201. Kato, T.: Fractional powers of dissipative operators, II. J. Math. Soc. Japan **14** (2), 242–248 (1962)
202. Kato, T.: Perturbation Theory for Linear Operators. Grundlehren der mathematischen Wissenschaften, vol. 132. Springer, Berlin (1966)
203. Katsenelenbaum, B.Z.: High-Frequency Electrodynamics. Wiley, New York (2006)
204. Kenig, C.E.: Weighted H^p spaces on Lipschitz domains. Amer. J. Math. **102** (1), 129–163 (1980)

205. Kenig, C.E.: Recent progress on boundary value problems on Lipschitz domains. In: Proceedings of Symposia in Pure Mathematics, vol. 43, pp. 175–205. Amer. Math. Soc., Providence, R.I. (1985)
206. Kenig C.E.: Elliptic boundary value problems on Lipschitz domains: Beijing lectures in harmonic analysis. Ann. Math. Studies **112**, 131–183 (1986)
207. Kenig C.E.: Potential theory of non-divergence form elliptic equations. In: Dirichlet Forms. Lecture Notes in Mathematics, vol. 1563, pp. 89–128. Springer, Berlin (1993)
208. Kenig, C.E.: Harmonic Analysis Techniques for Second Order Elliptic Boundary Value Problems. Amer. Math. Soc., Providence, R.I. (1994)
209. Kenig, C.E., Koch, H., Pipher, J., Toro, T.: A new approach to absolute continuity of elliptic measure, with application to non-symmetric equations. Adv. Math. **153**, 231–298 (2000)
210. Kenig, C.E., Pipher, J.: Hardy spaces and the Dirichlet problem on Lipschitz domains. Rev. Mat. Iberoam. **3** (2), 191–248 (1987)
211. Kenig, C.E., Pipher, J.: The oblique derivative problem on Lipschitz domains with L^p data. Amer. J. Math. **110** (4), 715–738 (1988)
212. Kenig, C.E., Pipher, J.: The Neumann problem for elliptic equations with non-smooth coefficients. Invent. Math. **113** (1993), 447–509; part II. Duke Math. J. **81**, 227–250 (1995)
213. Kohn, J.J., Nirenberg, L.: An algebra of pseudodifferential operators. Commun. Pure Appl. Math. **18** (1-2), 269–305 (1965)
214. Kolmogorov, A.N., Fomin, S.V.: Elements of the Theory of Functions and Functional Analysis. Dover, New York (1999)
215. Komatsu H.: Fractional powers of operators. Pacif. J. Math. **19**, 285–346 (1966)
216. Kondrat'ev, V.A.: Boundary problems for elliptic equations in domains with conical or angular points. Trans. Mosc. Math. Soc. **16**, 227–313 (1967)
217. Kondrashov, V.I.: On some properties of functions in the space L_p, Dokl. Akad. Nauk SSSR **48**, 563–566 (1945)
218. Kopachevskii, N.D.: On Green's abstract formula for mixed boundary-value problems and some of its applications. In: Spektral'nye i evolyutsionnye zadachi (Spectral and Evolutionary Problems). Proceedings of the 21st Crimean Autumn Mathematical Workshop-Symposium, Taurida National University, Simferopol', 2011
219. Kozhevnikov, A.: Asymptotics of the spectrum of Douglis–Nirenberg elliptic operators on a compact manifold. Math. Nachr. **182**, 261–293 (1886)
220. Kozlov, V.A., Maz'ya, V.G., Rossman, J.: Elliptic Boundary Value Problems in Domains with Point Singularities. Amer. Math. Soc., Providence, R.I. (1997)
221. Kozlov, V.A., Maz'ya, V.G., Rossman, J.: Spectral Problems Associated with Corner Singularities of Solutions to Elliptic Equations. Amer. Math. Soc., Providence, R.I. (2001)
222. Krasnosel'sky, M.A., Zabreiko, P.P., Pustyl'nik, E.I., Sobolevsky, P.E.: Integral Operators in Spaces of Summable Functions. Nordhoff, Leiden (1976)
223. Krein, S.G., Petunin, Yu.I., Semenov, E.M.: Interpolation of Linear Operators. Amer. Math. Soc., Providence, R.I. (1982)
224. Kupradze, V.D., Gegelia, T.G., Basheleishvili, M.O., Burchuladze, T.V.: Tree-dimensional Problems of the Mathematical Theory of Elasticity and Thermoelasticity. North-Holland, Amsterdam (1979)
225. Ladyzhenskaya, O.A.: The Boundary Value Problems of Mathematical Physics. Applied Mathematical Sciences, vol. 49. Springer, Berlin (1985)
226. Ladyzhenskaya, O.A., Ural'tseva, N.N.: Linear and Quasilinear Elliptic Equations. Academic, New York (1968)
227. Lanzani, L., Shen, Z.: On the Robin boundary condition for Laplace's equation in Lipschitz domains. Commun. Partial Differ. Equations **29** (1–2), 91–109 (2004)
228. Lax, P.D., Milgram, N.: Parabolic equations, Contributions to the Theory of Partial Differential Equations. Annals of Mathematical Studies, vol. 33, pp. 167–190. Princeton Univ. Press, Princeton, N.J. (1954)
229. Lebedev, V.I., Agoshkov, V.I.: Operatory Puankare–Steklova i ikh prilozheniya v analize (Poincaré–Steklov Operators and Their Applications in Analysis). Akad. Nauk SSSR, Vychislitel'nyi Tsentr, Moscow (1983)

230. Levendorskii, S.Z., Paneyakh, B.: Degenerate Elliptic Equations and Boundary Problems. In: Encyclopaedia Mathematical Sciences, vol. 63, pp. 131–201. Springer, Berlin (1994)
231. Lieberman, G.M.: Oblique derivative problems in Lipschitz domains, I: Continuous boundary data. Boll. Un. Mat. Ital. Sez. B **7**, 1185–1210 (1987)
232. Lions, J.L.: Espaces intermédiares entre espaces hilbertiens et applications. Bull. Math. Soc. Sci. Math. Phys. R.P. Roumaine (N.S.) **2** (50), 419–432 (1958)
233. Lions, J.L.: Théorèmes de trace et d'interpolation, I. Ann. Scuola Norm. Sup. Pisa **13** (4), 389–403 (1959)
234. Lions, J.L.: Une construction d'espaces d'interpolation. C. R. Acad. Sci. Paris **251** (3), 1853–1863 (1960)
235. Lions, J.L.: Équations différentielles opérationelles et problèmes aux limites. Springer, Berlin (1961)
236. Lions, J.L.: Espaces d'interpolation et domaines de puissances fractionaires d'opérateurs. J. Math. Soc. Japan **14** (2), 233–241 (1962)
237. Lions, J.-L., Magenes, E.: Non-Homogeneous Boundary Value Problems and Applications, vol. 1. Springer, Berlin (1972)
238. Lions, J.L., Peetre, J.: Sur une classe d'espaces d'interpolation. Publ. Math., Inst. Hautes Étud. Sci. **19**, 5–68 (1964)
239. Lizorkin, P.I.: (L_p, L_q)-multipliers of Fourier integrals. Sov. Math., Dokl. **4**, 1420–1424 (1963)
240. Lizorkin, P.I.: Operators connected with fractional differentiation, and classes of differentiable functions. Proc. Steklov Inst. Math. **117**, 251–286 (1972)
241. Lizorkin, P.I.: Properties of functions in the spaces $\Lambda_{p,\theta}^r$. Proc. Steklov Inst. Math. **131**, 165–188 (1974)
242. Lopatinskii, Ya.B.: A method of reduction of boundary-value problems for systems of differential equations of elliptic type to a system of regular integral equations. Ukr. Mat. Zh. **5** (2), 123–151 (1956). English translation: Amer. Math. Soc. Transl., Ser. II **89**, 149–183 (1970)
243. Lopatinskii, Ya.B. Teoriya obshchikh granichnykh zadach. Izbrannye trudy (Theory of General Boundary Value Problems. Selected Works). Naukova Dumka, Kiev (1984)
244. Lopatinskii, Ya.B. Factorization of a polynomial matrix. Nauchn. Zap. L'vov. Politekh. Inst., Ser. Fiz.-Mat. **38** (2), 3–7 (1957)
245. Lumer, G., Phillips, R.S.: Dissipative Operators in a Banach Spaces. Pacif. J. Math. **11**, 679–698 (1961)
246. Lunardi, A.: Interpolation Theory, 2nd ed. Edizioni della Normale, Piza (2009)
247. Lusternik, L.A., Sobolev, V.J.: Elements of Functional Analysis. Wiley, New York (1974)
248. L'vin, S.Ya.: Green's formula and the solvability of elliptic problems with boundary conditions of arbitrary order. Tr. Nauchn.-Issled. Inst. Mat. Voronezh. Gos. Univ. **17** (2), 49–56 (1975)
249. Markus, A.S.: Introduction to the Spectral Theory of Polynomial Operator Pencils. Amer. Math. Soc., Providence, R.I. (1988)
250. Maz'ya, V.G: Sobolev Spaces: With Applications to Elliptic Partial Differential Equations. Grundlehren der mathematischen Wissenschaften, vol. 342. Springer, Berlin (2011)
251. Maz'ya, V.G.: Boundary integral equations. In: Encyclopaedia Mathematical Sciences, vol. 27, pp. 127–222. Springer, Berlin (1991)
252. Maz'ya, V., Mitrea, M., Shaposhnikova, T.: The Dirichlet problem in Lipschitz domains for higher order elliptic systems with rough coefficients. J. Anal. Math. **110**, 167–239 (2010)
253. Maz'ya, V., Rossmann, J.: Elliptic Equations in Polyhedral Domains. Amer. Math. Soc., Providence, R.I. (2010)
254. Maz'ya, V.G., Shaposhnikova, T.: Theory of Sobolev Multipliers with Applications to Differential and Integral Operators. Grundlehren der mathematischen Wissenschaften, vol. 337. Springer, Berlin (2009)
255. McIntosh, A.: Square roots of elliptic operators. J. Funct. Anal. **61**, 307–327 (1985)
256. McIntosh, A.: Operators which have an H^∞ functional calculus. In: Miniconference on Operator Theory and Partial Differential Equations, North Ryde, 1986. Proceedings of the Centre for Mathematical Analysis, Australian National University, vol. 14, pp. 210–231. Australian National University, Centre for Mathematical Analysis, Canberra (1986)

257. McIntosh, A.: The square root problem for elliptic operators: A survey. In: Functional-Analytic Methods for Partial Differential Equations. Lecture Notes in Mathematics, vol. 1450, pp. 122–140. Springer, Berlin (1990)

258. McLean, W.: Strongly Elliptic Systems and Boundary Integral Equations. Cambridge Univ. Press, Cambridge (2000)

259. Mendez, O., Mitrea, M.: The Banach envelopes of Besov and Triebel–Lizorkin spaces and applications to partial differential equations. J. Fourier Analysis and Appl. **6**, 503–531 (2000)

260. Métivier, G.: Valeurs propres de problèmes aux limites elliptiques irreguliers. Mém. Soc. Math. France **51–52**, 125–219 (1977)

261. Mikhailov, S.E.: Traces, extensions and co-normal derivatives for elliptic systems on Lipschitz domains. J. Math. Anal. Appl. **378**, 324–342 (2011)

262. Mikhailov, S.E.: Solution regularity and co-normal derivatives for elliptic systems with non-smooth coefficients on Lipschitz domains. J. Math. Analysis and Appl. **400** (1), 48–67 (2013)

263. Mikhlin, S.G.: On the multipliers of Fourier integrals. Dokl. Akad. Nauk SSSR **109**, 701-703 (1956)

264. Mikhlin, S.G.: Multidimensional Singular Integrals and Integral Equations. Pergamon, Oxford (1965)

265. Mikhlin, S.G.: Variational Methods in Mathematical Physics. Macmillan, New York (1964)

266. Miranda, C.: Partial Differential Equations of Elliptic Type, 2nd ed. Springer, Berlin (1970)

267. Mitrea, D., Mitrea, M., Pipher, J.: Vector potential theory on nonsmooth domains in $\mathbb{R}^3$ and applications to electromagnetic scattering. J. Fourier Anal. Appl. **3** (2), 131–192 (1997)

268. Mitrea, D., Mitrea, M.: General second order strongly elliptic systems in low dimensional nonsmooth manifolds. Contemp. Math. **277**, 61–86 (2001)

269. Mitrea, D., Mitrea, M., Taylor, M.: Layer Potentials, the Hodge Laplacian, and Global Boundary Value Problems in Nonsmooth Riemannian Manifolds. Memoirs of the American Mathematical Society, vol. 150, no. 713. Amer. Math. Society, Providence, R.I. (2001)

270. Mitrea, I., Mitrea, M.: The Poisson problem with mixed boundary conditions in Sobolev and Besov spaces in non-smooth domains. Trans. Amer. Math. Soc. **359**, 4113–4182 (2007)

271. Mitrea, M.: The method of layer potentials in electromagnetic scattering theory on nonsmooth domains. Duke Math. J. **77** (5), 111–133 (1955)

272. Mitrea, M.: The oblique derivative problem for general elliptic systems in Lipschitz domains. In: Integral Methods in Science and Engineering. Proceedings of the 5th International Conference, Houghton, 1998. Chapman & Hall/CRC Research Notes in Mathematics, vol. 418, pp. 240–245. Chapman & Hall/CRC, Boca Raton (2000)

273. Mitrea, M., Taylor, M.: Boundary layer methods for Lipschitz domains in Riemannian manifolds. J. Funct. Anal. **163**, 181–251 (1999)

274. Mitrea, M., Taylor, M.: Potential theory on Lipschitz domains in Riemannian manifolds: Sobolev–Besov space results and the Poisson problem. J. Funct. Anal. **176** (1), 1–79 (2000)

275. Mitrea, M., Taylor, M.: Potential theory on Lipschitz domains in Riemannian Manifolds: L^p, Hardy, and Hölder space results. Commun. Anal. Geom. **9**, 369–421 (2001)

276. Mitrea, M., Torres, R., Welland, G.: Regularity and approximation results for the Maxwell problem on C^1 and Lipschitz domains. In: Clifford Algebras in Analysis and Related Topics, pp. 297–308. CRC, Boca Raton (1995)

277. Morrey, C.B.: Second order elliptic systems of differential equations, In: Contributions to the Theory of Partial Differential Equations, pp. 101–159. Princeton Univ. Press, Princeton, N.J. (1954)

278. Muramatu, T.: On Besov spaces and Sobolev spaces of generalized functions defined on a general region. Publ. Res. Inst. Math. Sci. Kyoto Univ. **9**, 325–396 (1974)

279. Muramatu, T.: On the dual of Besov spaces. Publ. Res. Inst. Math. Sci. Kyoto Univ. **12**, 123–140 (1976)

280. Muskhelishvili, N.I.: Singular Integral Equations. Noordhoff, Groningen (1953)

281. Natroshvili, D.G.: Investigation of boundary value and initial-boundary value problems of mathematical elasticity and thermoelasticity theory for homogeneous anisotropic media by the method of potentials. Doctoral dissertation, Tbilisskii Matematicheskii Institut im. Razmadze Akademii Nauk Gruzinskoi SSR (1984)

282. Natroshvili, D.: Boundary integral equation method in the steady state oscillation problems for anisotropic bodies. Math. Methods Appl. Sci. **20**, 95–119 (1997)
283. Natroshvili, D.G., Chkadua, G.D., Shargorodsky, E.M: Mixed problems for homogeneous anizotropic elastic media. Tr. Inst. Prikl. Mat. im. Vekua Tbiliss. Univ. **39**, 133–178 (1990)
284. Nazarov, S.A., Plamenevsky, B.A.: Elliptic Problems in Domains with Piecewise Smooth Boundaries. Walter de Gruyter, Berlin (1994)
285. Nečhas, J. On domains of type 𝔑. Czech. Math. J. **12** (87), 274–287 (1962)
286. Nečhas, J.: Direct Methods in the Theory of Elliptic Equations. Springer, Heidelberg (2012)
287. Netrusov, Yu., Safarov, Yu.: Weyl asymptotic formula for the Laplacian on domains with rough boundaries. Commun. Math. Phys. **253**, 481–509 (2005)
288. Nikol'skii, S.M.: On imbedding, continuation and approximation theorems for differentiable functions of several variables. Russian Math. Surveys **16** (5), 55–104 (1961)
289. Nikol'skiĭ, S.M.: Approximation of Functions of Several Variables and Imbedding Theorems. Springer, New York (1975)
290. Nirenberg, L.: Remarks on strongly elliptic partial differential equations. Commun. Pure Appl. Math. **8**, 649–675 (1965)
291. Oleinik, O.A., Shamaev, A.S., Yosifian, G.A.: Mathematical Problems in Elasticity and Homogenization. Studies in Mathematics and Its Applications, vol. 26. North-Holland, Amsterdam (1992)
292. Ouhabaz, E.M.: Analysis of Heat Equations on Domains. London Mathematical Society Monographs. Princeton Univ. Press, Princeton (2005)
293. Pal'tsev, B.V.: Mixed problems with non-homogeneous boundary conditions in Lipschitz domains for second-order elliptic equations with a parameter. Russian Acad. Sci., Sb. Math. **187** (4), 525–580 (1996)
294. Papageorgiou, N.S., Kyritsi-Yiallourou, S.Th.: Handbook of Applied Analysis. Springer, Berlin (2009)
295. Payne, L.E., Weinberger, H.F.: New bounds for solutions of second order elliptic partial differential equations. Pacific J. Math. **8**, 551–573 (1958)
296. Pazy, A.: Semigroups of Linear Operators and Applications to Partial Differential Equations. Springer, Berlin (1983)
297. Peetre, J.: A Theory of Interpolation of Normed Spaces, Notes de Matemática, vol. 39. Instituto de Matemática Pura e Aplicada, Conselho Nacional de Pesquisas, Rio de Janeiro (1968)
298. Peetre, J.: Remarques sur les spaces de Besov. Le cas $0 < p < 1$. C. R. Acad. Sci., Paris, Sér. A–B **277**, 947–950 (1973)
299. Peetre, J.: On spaces of Triebel–Lizorkin type. Ark. Mat. **13**, 123–130 (1975)
300. Peetre, J.: New Thoughts on Besov Spaces. Duke University Mathematics Series, vol. 1. Mathematics Department, Duke University, Durham, N.C. (1976)
301. Petersdorff, T. von, Boundary integral equations for mixed Dirichlet, Neumann and transmission problems. Math. Methods Appl. Sci. **11** (2), 183–213 (1989)
302. Petrovsky, I.G.: On systems of differential equations whose all solutions are analytic. Dokl. Akad. Nauk SSSR **17**, 339–342 (1937)
303. Pipher, J.: Oblique derivative problems for the Laplacian in Lipschitz domains. Rev. Math. Iberoam. **3**, 455–472 (1987)
304. Pipher, J., Verchota, G.C.: The Dirichlet problem in L^p for the biharmonic equation on Lipschitz domains. Amer. J. Math. **114** (5), 923–972 (1992)
305. Pipher, J., Verchota, G.C.: Dilation invariant estimates and the boundary Gårding inequalities for higher order elliptic operators. Ann. Math. **142**, 1–38 (1995)
306. Poincaré, H.: La méthode de Neumann et le probléme de Dirichlet. Acta Math. **20** (1), 59–142 (1897)
307. Rademacher, H.: Über partielle und totale Differenzierbarkeit von Funktionen mehrer Variabeln und über die Transformation der Doppelintegrale. Math. Ann. **79**, 340–359 (1919)
308. Ransford, T.J.: The spectrum of an interpolated operator and analytic multivalued functions. Pacif. J. Math. **121** (2), 445–466 (1986)
309. Rellich, F.: Ein Satz über mittlere Konvegenz. Nachr. Akad. Wiss. Göttingen Math. Phys. Kl. II, 30–35 (1930)

310. Rellich, F.: Darstellung der Eigenverte von $\Delta u + \lambda u = 0$ durch ein Randintegral, Math. Z. **46**, 635–636 (1940)
311. Rempel, S., Schulze, B.-W.: Index Theory of Elliptic Boundary Problems. Akademie, Berlin (1992)
312. Roitberg, Ya.: Elliptic Boundary Value Problems in the Spaces of Distributions. Kluwer, Dordrecht (1996)
313. Roitberg, Ya.A., Sheftel, Z.G.: General boundary-value problems for elliptic equations with discontinuous coefficients. Soviet Math., Dokl. **4**, 231–234 (1963)
314. Roitberg, Ya.A., Sheftel', Z.G.: Green's formula and a theorem on homeomorphisms for elliptic systems. Uspekhi Mat. Nauk **22** (5), 181–182 (1967)
315. Rozenblum, G.V., Shubin, M.A., Solomyak, M.Z.: Partial Differential Equations, VII: Spectral Theory of Differential Operators. Encyclopaedia of Mathematical Sciences, vol. 64. Springer, Berlin (1994)
316. Rozenblum, G., Tashchiyan, G.: Eigenvalue asymptotics for potential type operators on Lipschitz surfaces. Russian J. Math. Phys. **13** (3), 326–339 (2006)
317. Rudin, W.: Functional Analysis. McGraw-Hill, New York (1973)
318. Runst, T., Sickel, W.: Sobolev Spaces of Fractional Order, Nemytskij Operators, and Nonlinear Partial Differential Equations. Walter de Gruyter, Berlin (1996)
319. Rychkov, V.S.: Intrinsic characterizations of distribution spaces on domains. Studia Math. **127**, 277–298 (1998)
320. Rychkov, V.S.: On restrictions and extensions of the Besov and Triebel–Lizorkin spaces with respect to Lipschitz domains. J. London Math. Soc. **60**, 237–257 (1999)
321. Safarov Yu., Vassiliev D.: The Asymptotic Distribution of Eigenvalues of Partial Differential Operators. Transl. of Math. Monographs, vol. 155. AMS, Providence, RI (1991)
322. Saks, S.: Theory of the Integral, 2nd ed. Monografie Matematyczne, vol. 7. Z Subwencji Funduszu Kultury Narodowej, Warsaw (1937)
323. Sarason, D.: Functions of vanishing mean oscillation. Trans. Amer. Math. Soc. **207**, 391–405 (1975)
324. Savaré, G.: Regularity and perturbation results for mixed second order elliptic problems. Commun. Partial Differ. Equations **22** (5–6), 869–899 (1997)
325. Savaré, G.: Regularity results for elliptic equations in Lipschitz domains. J. Funct. Anal. **152**, 176–201 (1998)
326. Schauder, J.: Über lineare elliptische Differentialgleichungen zweiter Ordnung. Math. Z. **38**, 257–282 (1934)
327. Schauder, J.: Numerische Abschätzungen in elliptischen linearen Differentialgleichungen. Studia Math. **5**, 34–42 (1934)
328. Schechter, M.: Coerciveness of linear partial differential operators for functions satisfying zero Dirichlet type boundary data. Commun. Pure Appl. Math. **11**, 153–174 (1958)
329. Schechter, M.: Integral inequalities for partial differential operators and functions satisfying general boundary conditions. Commun. Pure Appl. Math. **12**, 37–66 (1959)
330. Schechter, M.: General boundary value problems for elliptic partial differential equations. Commun. Pure Appl. Math. **12**, 457–486 (1959)
331. Schechter, M.: Remarks on elliptic boundary value problems. Commun. Pure Appl. Math. **12** (4), 561–578 (1959)
332. Schechter, M.: A generalization of the problem of transmission. Ann. Scuola Norm. Sup. Pisa **14**, 207–236 (1960)
333. Schechter, M.: Mixed boundary problems for general elliptic equations. Commun. Pure Appl. Math. **13** (2), 183–201 (1960)
334. Schechter, M.: Various types of boundary conditions for elliptic equations. Commun. Pure Appl. Math. **13** (3), 407–425 (1960)
335. Schwartz, L.: Théorie des Distributions. Hermann, Paris (1950)
336. Seeley, R.: Extensions of C^∞ functions defined in a half space. Proc. Amer. Math. Soc. **15**, 625–626 (1964)
337. Seeley, R.: Singular integrals and boundary value problems. Amer. J. Math. **88**, 781–809 (1966)

338. Seeley, R.T.: Norms and domains of the complex powers A_B^z. Amer. J. Math. **93** (2), 299–309 (1971)
339. Seeley, R.T.: Interpolation in L^p with boundary conditions. Studia Math. **44**, 47–60 (1972)
340. Shapiro, Z.Ya: The first boundary-value problem for an elliptic system of differential equations. Mat. Sb. **28** (70), 55–78 (1951)
341. Shapiro, Z.Ya.: On general boundary-value problems for elliptic equations. Izv. Akad. Nauk SSSR, Ser. Mat. **17**, 539–562 (1953)
342. Sheftel', Z.G.: General theory of boundary-value problems for elliptic systems with discontinuous coefficients. Ukr. Mat. Zh. **18** (3), 132–136 (1966)
343. Shen, Z.: Resolvent estimates in L_p for elliptic systems in Lipschitz domains. J. Funct. Anal. **133** (1), 224–251 (1995)
344. Shen, Z.: Necessary and sufficient conditions for the solvability of the L_p Dirichlet problem on Lipschitz domains. Math. Ann. **336**, 697–725 (2006)
345. Shen, Z.: The Neumann problem in L_p on Lipschitz and convex domains. J. Funct. Anal. **255** (7), 1817–1830 (2008)
346. Shen, Z.: The L_p regularity problem on Lipschitz domains. Trans. Amer. Math. Soc. **363** (3), 1241–1244 (2011)
347. Shneiberg, I.Ya.: Spectral properties of linear operators in interpolation families of Banach spaces. Mat. Issled. **9** (2), 214–227 (1974)
348. Shubin, M.A.: Pseudo-Differential Operators and Spectral Theory. Springer, Berlin (1987)
349. Shubin, M.A.: Lektsii ob uravneniyakh matematicheskoi fiziki (Lectures on Equations of Mathematical Physics). MTsNMO, Moscow (2001)
350. Slobodetskij, L.N.: Generalized Sobolev spaces and their application to boundary problems for partial differential equations. Amer. Math. Soc. Transl., Ser. 2 **57**, 207–275 (1966)
351. Slobodetskii, L.N.: Estimates in L_p of elliptic systems. Dokl. Akad. Nauk SSSR **123** (4), 618–619 (1958)
352. Smith, K.T.: Inequalities for formally positive integro-differential forms. Bull. Amer. Math. Soc. **67**, 368–370 (1961)
353. Sneddon, I.N.: Mixed Boundary Value Problems in Potential Theory. Elsevier, New York (1966)
354. Sobolev, S.L.: Applications of Functional Analysis in Mathematical Physics. Amer. Math. Soc., Providence, R.I. (1963)
355. Solonnikov, V.A.: On general boundary value problems for systems elliptic in the sense of A. Douglis and L. Nirenberg, I. Amer. Math. Soc. Transl., Ser. 2. **56**, 193–232 (1964)
356. Solonnikov, V.A.: On general boundary value problems for systems elliptic in the sense of A. Douglis– L. Nirenberg, II. Proc. Steklov Mat. Inst. **92** (1968), 3–32.
357. Solonnikov, V.A.: Estimates in L_p of solutions of elliptic and parabolic systems. Proc. Steklov Mat. Inst. **102**, 157–185 (1970)
358. Solonnikov, V.A., Ural'tseva, N.N.: Sobolev spaces. In: Izbrannye glavy analiza i vysshei algebry (Selected Chapters of Analysis and Higher Algebra), pp. 129–197. Izd. Leningrad. Univ., Leningrad (1981)
359. Stein, I.: Singular Integrals and Differentiability Properties of Functions. Princeton Univ. Press, Princeton, N.J. (1970)
360. Stein, E.M.: Topics in Harmonic Analysis Related to the Littlewood-Paley Theory. Annals of Mathematical Studies, vol. 63. Princeton Univ. Press, Princeton, N.J. (1970)
361. Stein, E.M., Weiss, G.: Introduction to Fourier Analysis on Euclidean Spaces. Princeton Univ. Press, Princeton, N.J (1971)
362. Steklov, V.A.: Obshchie metody resheniya osnovnykh zadach matematicheskoi fiziki (General Methods for Solving Basic Problems of Mathematical Physics). Khar'kovskoe Matematicheskoe Obshchestvo, Khar'kov (1901)
363. Stepanoff, W.: Sür les conditions d'existence de la differentielle totale. Mat. Sb. **32**, 511–526 (1925)
364. Stephan, E.: Boundary Integral Equations for Mixed Boundary Value Problems, Screen and Transmission Problems in $\mathbb{R}^3$. Habilitation Thesis, Technische Hochschule Darmstadt, Preprint 848, 1984.

365. Stephan, E.: A boundary integral equation method for three-dimensional crack problems in elasticity. Math. Methods Appl. Sci. **8**, 609–623 (1986)
366. Stephan, E.: Boundary integral equations for screen problems in $\mathbb{R}^3$. Integral Equations Oper. Theory **10**, 236–257 (1987)
367. Stephan, E.: Boundary integral equations for mixed boundary value problems in $\mathbb{R}^3$. Math. Nachr. **134**, 21–53 (1987)
368. Sternin, B.Yu.: Elliptic and parabolic problems on manifolds with a boundary consisting of components of various dimensions. Trans. Mosc. Math. Soc. **15**, 387–429 (1966)
369. Strichartz, R.T.: Multipliers on fractional Sobolev spaces. J. Math. Mech. **16** (9), 1031–1060 (1967)
370. Suslina, T.A.: Spectral asymptotics of variational problems with elliptic constraints in domains with piecewise smooth boundary. Russian J. Math. Phys. **6** (2), 214–234 (1999)
371. Taibleson, M.H.: On the theory of Lipschitz spaces of distributions on Euclidean n-space, I: Principal properties; II: Translation invariant operators, duality, and interpolation; III: Smoothness and integrability of Fourier transforms, smoothness of convolution kernels. J. Math. Mech. **13**, 407–479 (1964); **14**, 821–839 (1965); **15**, 973–981 (1966)
372. Tanabe, H.: Equations of Evolution. Pitman, London (1979)
373. Taylor, M.E.: Pseudodifferential Operators. Princeton Univ. Press, Princeton, NJ (1981).
374. Temam, R.: Navier–Stokes Equations. North-Holland, Amsterdam (1979)
375. Triebel, H.: Spaces of distributions of Besov type on Euclidean n-space: Duality, interpolation. Ark. Mat. **11**, 13–64 (1973)
376. Triebel, H.: Interpolation. Function Spaces. Differential Operators. North-Holland, Amsterdam (1978)
377. Triebel, H.: Theory of Function Spaces. Geest & Portig, Leipzig (1983)
378. Triebel, H.: Theory of Function Spaces II. Birkhäuser, Basel (1992)
379. Triebel, H.: Function spaces in Lipschitz domains and on Lipschitz manifolds. Characteristic functions as pointwise multipliers. Rev. Mat. Complut. **15** (2), 475–524 (2002)
380. Uhlmann, G.: Inverse boundary problems and applications. Astérisque **207**, 153–207 (1992)
381. Vainberg, B.R.: Asymptotic Methods in Equations of Mathematical Physics. Gordon & Breach, New York (1988).
382. Verchota, G.: Layer Potentials and Boundary Value Problems for Laplace's Equation in Lipschitz Domains. Ph.D. thesis, University of Minnesota (1982)
383. Verchota, G.: Layer potentials and regularity for the Dirichlet problem for Laplace's equation. J. Funct. Anal. **59**, 572–611 (1984)
384. Verchota, G.: The Dirichlet problem for polyharmonic equation in Lipschitz domains. Indiana Univ. Math. J. **39**, 671–702 (1990)
385. Verchota, G.C.: Potentials for the Dirichlet problem in Lipschitz domains. In: Král, J., Lukeš, J., Netuka, I., Veselý, J. (eds.) Potential Theory—ICPT 94: Proceedings of the International Conference on Potential Theory, held in Kouty, Czech Republic, August 13–20, 1994, pp. 167–187. Walter de Gruyter, Berlin, (1996)
386. Verchota, G.: The biharmonic Neumann problem in Lipschitz domains. Acta Math. **194**, 217–279 (2005)
387. Verchota, G.C., Vogel, A.L.: Nonsymmetric systems on nonsmooth planar domains. Trans. Amer. Math. Soc. **349** (11), 4501–4535 (1997)
388. Vishik, M.I.: On strongly elliptic systems of differential equations. Mat. Sb. **29** (71), 615–676 (1951)
389. Vishik, M.I.: On general boundary value problems for elliptic differential equations. Amer. Math. Soc. Transl., Ser. 2 **24**, 107–172 (1963)
390. Vishik, M.I., Grushin, V.V.: Degenerating elliptic differential and pseudo-differential operators. Russian Math. Surveys **25** (4), 21–50 (1970)
391. Vishik, M.I., Myshkis, A.D., Oleinik, O.A.: Partial differential equations. In: Matematika v SSSR za 40 let (40 Years of Mathematics in the USSR), vol. 1, pp. 563–636. Fizmatgiz, Moscow (1959)
392. Voititsky, V.I., Kopachevsky, N.D., Starkov, P.A.: Multicomponent conjugation problems and auxiliary boundary-value problems. J. Math. Sci. **170** (2), 131–172 (2010)

393. Volevich, L.R.: Solvability of boundary value problems for general elliptic systems. Amer. Math. Soc. Transl., Ser. 2 **67**, 182–225 (1968)

394. Vol'pert, A.I.: On the index and normal solvability of boundary-value problems for elliptic systems of differential equations on the plane. Tr. Mosk. Mat. Obshch. **10**, 41–87 (1961)

395. Weyl, G.: The method of orthogonal projection in potential theory. Duke Math. J. **7**, 411–444 (1940)

396. Whitney, H.: Analytic extensions of differentiable functions defined in closed sets. Trans. Amer. Math. Soc. **36**, 63–89 (1934)

397. Whitney, H.: Differentiable functions defined in closed sets. I. Trans. Amer. Math. Soc. **36**, 369–487 (1934)

398. Whitney, H.: Geometric Integration Theory. Princeton Univ. Press, Princeton, N.J. (1957)

399. Wolff, T.W.: A note on interpolation spaces. In: Lecture Notes in Mathematics, vol. 908, 199–204. Springer, Berlin (1982)

400. Wong, M.W.: An Introduction to Pseudo-Differential Operators. 2nd ed. Word Scientific, Singapore (1999)

401. Yagi, A.: Coincidence entre des espaces d'interpolation et des domaines de puissances fractionaires d'opérateurs. C. R. Acad. Sci. Paris, Sér. 1 **299** (6), 173–176 (1984)

402. Yakovlev, G.N.: Traces of functions in the space W_p^l on piecewise smooth surfaces. Math. USSR, Sb. **3** (4), 481–497 (1967)

403. Yosida, K.: Functional Analysis. Springer, Berlin (1965)

404. Zafran, M.: Spectral theory and interpolation of operators. J. Funct. Anal. **36**, 185–204 (1980)

405. Zanger, D.Z.: The inhomogeneous Neumann problem in Lipschitz domains. Commun. Partial Differ. Equations **25** (9–10), 1771–1808 (2000)

406. Zorich, V.A: Mathematical Analysis, Parts 1, 2. Springer, Berlin (2004)

407. Zygmund, A.: Trigonometric Series, 2nd. ed., vols. 1, 2. Cambridge Univ. Press, Cambridge (1959)

Index

A

a priori estimate for solutions
  of boundary value problems in Lipschitz
    domains   144, 150, 272
  of elliptic boundary value problems   80,
    81, 235
  of elliptic equations   68, 72
  of Fredholm equations   279
adjointness of operators
  in a Hilbert space   281
  in Banach spaces   280
approximation of a Lipschitz surface   117

B

Banach pair   195
Besov spaces   223, 231, 233, 304
Bessel potential spaces   1, 27, 57, 223

C

calculus of pseudodifferential operators   292
Calderón program   307
Calderón projections   180, 263, 299
closed manifold   24
coercivity   101, 108, 138, 140
  strong   101, 103, 108, 140
    on Lipschitz boundary   152, 181, 182
cokernel of an operator   276, 277
complemented subspace   208, 276
complete system of vectors   287
cone condition on the boundary   117
conormal derivative   137
  smooth   137
coretraction   208

D

Dirichlet boundary value problem   80, 108,
    137, 143, 237, 258, 307
Dirichlet-to-Neumann operator   94, 152
discrete norm   22, 125, 233
discrete representation of functions   22, 128
discrete spectrum   287
double-layer potential   174, 176
duality of spaces   10, 50, 61, 120, 210, 228

E

elliptic boundary value problem   78
ellipticity
  in Douglis–Nirenberg sense   76
  of a partial differential operator   66
    regular   78
  of a pseudodifferential operator   294
  strong   97, 107, 138
  with parameter   67, 76, 81
embedding theorems   3, 29, 59, 224, 229
estimate of intermediate norm   13
extension operator
  by zero   41
  Hestenes   35
  Rychkov universal   132, 227
  Seeley   296
extrapolation of invertibility   212

F

formally adjoint boundary value problems
  90
formally adjoint operators   73, 141
fractional powers of operators   220, 289
Fredholm operator   276

© Springer International Publishing Switzerland 2015   329
M.S. Agranovich, *Sobolev Spaces, Their Generalizations and Elliptic Problems
in Smooth and Lipschitz Domains*, Springer Monographs in Mathematics,
DOI 10.1007/978-3-319-14648-5

Printed in the United States
By Bookmasters